Stress in Yeasts and Filamentous Fungi

British Mycological Society Symposia Series

Series Editor:

Pieter van West
University of Aberdeen
Aberdeen Oomycete Group
Institute of Medical Sciences
Aberdeen, UK

The cover image represents a microcosm containing a *Pinus sylvestris* seedling with intact extraradical mycelium of the mycorrhizal fungus *Piloderma fallax* colonising patches of control (C) and limed (L) substrate taken from the fermentation horizon of a forest 15 years after the addition of the lime (photo courtesy of Dr. Andy Taylor, The Swedish University of Agricultural Sciences, Uppsala, Sweden).

STRESS IN YEASTS AND FILAMENTOUS FUNGI

Editors

Simon V. Avery
and
Malcolm Stratford
School of Biology
Institute of Genetics
University of Nottingham
University Park
Nottingham, UK

Pieter van West
University of Aberdeen
Aberdeen Oomycete Group
Institute of Medical Sciences
Aberdeen, UK

ELSEVIER

Amsterdam – Boston – Heidelberg – London – New York – Oxford
Paris – San Diego – San Francisco – Singapore – Sydney – Tokyo
Academic Press is an imprint of Elsevier

ACADEMIC PRESS

Academic Press is an imprint of Elsevier
84 Theobald's Road, London WC1X 8RR, UK
Radarweg 29, PO Box 211, 1000 AE Amsterdam, The Netherlands
30 Corporate Drive, Suite 400, Burlington, MA 01803, USA
525 B Street, Suite 1900, San Diego, CA 92101-4495, USA

First edition 2008
Reprinted 2008 (twice)

British Library Cataloguing in Publication Data
A catalogue record for this book is available from the British Library

Library of Congress Cataloging-in-Publication Data
A catalog record for this book is available from the Library of Congress

ISBN: 978-0-12-374184-4

Printed and bound in the United Kingdom

Transferred to Digital Print 2011

Contents

List of Contributors vii
Preface xi

1 Interactions between *Agaricus bisporus* and the pathogen *Verticillium fungicola*
Peter Mills, John Thomas, Martin Sergeant, Ana Costa, Patrick Collopy, Andy Bailey, Gary Foster and Mike Challen 1

2 Environmental fluxes and fungal interactions: Maintaining a competitive edge
Naresh Magan and David Aldred 19

3 Intracellular mycoparasites in action: Interactions between powdery mildew fungi and *Ampelomyces*
Levente Kiss 37

4 The population dynamical consequences of density dependence in fungal plant pathogens
Michael W. Shaw 53

5 Differences in stress responses between model and pathogenic fungi
Janet Quinn 67

6 The remarkably diverse pathogenicity effectors of the obligate oomycete *Hyaloperonospora parasitica*
Jim Beynon 87

7 Ace2 and fungal virulence
Helen Findon and Ken Haynes 97

8 Integrative analysis of yeast osmoregulation
Stefan Hohmann 109

9 Oxidative stress, fragmentation and cell death during bioreactor growth of filamentous fungi
Geoffrey D. Robson, Jeff Puci, William Throndset and Nigel Dunn-Coleman 129

10 Weak organic acid resistance of spoilage yeasts
Mehdi Mollapour and Peter W. Piper 143

11 Heavy metal pollution and genetic adaptations in ectomycorrhizal fungi
Jan V. Colpaert 157

12 Lichens and metals
O. William Purvis and Barbara Pawlik-Skowrońska 175

13 Responses of mycorrhizal fungi to stress
Roger D. Finlay, Björn D. Lindahl and Andy F.S. Taylor 201

14 Regulation of protein synthesis in yeast by oxidative stress
Daniel Shenton, Claire Mascarenhas and Chris M. Grant 221

15 Cell differentiation as a response to oxidative stress
Wilhelm Hansberg, Jesús Aguirre, Mauricio Ríos-Momberg, Pablo Rangel, Leonardo Peraza, Yésika Montes de Oca and Nallely Cano-Domínguez 235

16 Signalling and defences against oxidative stress in *Candida albicans*
Rebeca Alonso-Monge, Elvira Román, David M. Arana, Susanne Gola and Jesus Pla 259

17 Oxidant-specific protein folding during fungal oxidative stress: Activation and function of the Yap1p transcription factor in *Saccharomyces cerevisiae*
W. Scott Moye-Rowley 275

Colour Section 291

Subject Index 307

List of contributors

Jesús Aguirre, Departamento de Genética Molecular, Instituto de Fisiología Celular, Universidad Nacional Autónoma de México, Circuito exterior s/n, Ciudad Universitaria Copilco, Coyoacán, 04510 México

David Aldred, Applied Mycology Group, Cranfield Health, Cranfield University, Bedford MK45 4DT, UK.

Rebeca Alonso-Monge, Departmento de Microbiologia II, Faculted de Farmacia, Universidad Complutense de Madrid, Plaza Romón y Cajal S/N, 28040 Madrid, Spain

David M. Arana, Departmento de Microbiologia II, Faculated de Farmacia, Universidad Complutense de Madrid, Plaza Romón y Cajal S/N, 28040 Madrid, Spain

Andy Bailey, School of Biological Sciences, University of Bristol, Woodland Road, Bristol BS8 1UG, UK

Jim Beynon, Warwick HRI, Warwick University, Wellesbourne, Warwick, CV35 9EF, UK

Nallely Cano-Domínguez, Instituto de Fisiologia Celular, Universidad Nacional Autonoma de Mexico, Circuito Exterior S/N, Ciudad Universitaria Copilco, Coyoacan, 04510 Mexico

Mike Challen, Warwick HRI, University of Warwick, Warwick CV35 9EF, UK

Jan V. Colpaert, Hasselt University, Environmental Biology Group, Agoralaan, Gebouw D, B-3590, Diepenbeek, Belgium

Patrick Collopy, School of Biological Sciences, University of Bristol, Woodland Road, Bristol BS8 1UG, UK

Ana Costa, Warwick HRI, University of Warwick, Warwick CV35 9EF, UK

Nigel Dunn-Coleman, Genencor International, 925 Page Mill Road, Palo Alto, CA 94304, USA

Helen Findon, Department of Molecular Microbiology & Infection, Flowers Building, Imperial College, London SW7 2AZ, UK

Roger D. Finlay, Department of Forest Mycology & Pathology, SLU, Box 7026, Uppsala SE-750 07, Sweden

Gary Foster, School of Biological Sciences, University of Bristol, Woodland Road, Bristol BS8 1UG, UK

Susanne Gola, Centro Nacional De Biotecnología, Depto. Biotecnología Microbiana, C/Darwin 3, Campus de Cantoblanco, 28049 Madrid, Espana

Chris M. Grant, University of Manchester, Faculty of Life Sciences, The Michael Smith building, Oxford Road, Manchester M13 9PT, UK

Wilhelm Hansberg, Departamento de Bioquímica, Instituto de Fisiología Celular, Universidad Nacional Autónoma de México, Circuito exterior s/n, Ciudad Universitaria, Copilco, Coyoacán, 04510 México

Ken Haynes, Department of Molecular Microbiology & Infection, Flowers Building, Imperial College, London SW7 2AZ, UK

Stefan Hohmann, Department Cell and Molecular Biology, Göteborg University, Box 462, Göteborg 40530, Sweden

Levente Kiss, Plant Protection Institute of the Hungarian Academy of Sciences, Department of Plant Pathology, P.O. Box 102, H-1525 Budapest, Hungary

Björn D. Lindahl, Department of Forest Mycology & Pathology, SLU, Box 7026, Uppsala SE-750–07, Sweden

Naresh Magan, Applied Mycology Group, Cranfield Health, Cranfield University, Bedford MK45 4DT, UK

Claire Mascarenhas, University of Manchester, Faculty of Life Sciences, The Michael Smith Building, Oxford Road, Manchester M13 9PT, UK

Peter Mills, Warwick HRI, University of Warwick, Warwick CV35 9EF, UK

Mehdi Mollapour, Department of Molecular Biology and Biotechnology, The University of Sheffield, Fifth Court, Western Bank, Sheffield S10 2TN, UK

Yésika Montes de Oca, Departamento de Genética Molecular, Instituto de Fisiologia Celular, Universidad Nacional Autonoma de Mexico, Circuito Exterior S/N, Ciudad Universitaria Copilco, Coyoacan 04510, Mexico

W. Scott Moye-Rowley, 6-530 Bowen Science Building, Department of Molecular Physiology and Biophysics, University of Iowa, Iowa City, IA 52242, USA

Barbara Pawlik-Skowrońska, Centre for Ecological Research, PAS at Dziekanów Leśny, Experimental Station, Niecała 18, 20-080 Lublin, Poland

Leonardo Peraza, Departamento de Genética Molecular, Instituto de Fisiologia Celular, Universidad Nacional Autonoma de Mexico, Circuito Exterior S/N, Ciudad Universitaria Copilco, Coyoacan 04510, Mexico

Jesus Pla, Department of Microbiology, Faculty of Pharmacy, Universidad Complutense de Madrid, Plaza de Ramón y Cajal s/n, Madrid 28040, Spain

Peter W. Piper, Department of Molecular Biology and Biotechnology, The University of Sheffield, Fifth court, Western Bank, Sheffield S10 2TN, UK

Jeff Puci, Genencor International, 925 Page Mill Road, Palo Alto, CA 94304, USA

O. William Purvis, Department of Botany, Natural History Museum, Cromwell Road, London SW7 5BD, UK

Janet Quinn, Institute for Cell and Molecular Biosciences, Faculty of Medical Sciences, Newcastle University, Framlington Place, Newcastle upon Tyne NE2 4HH, UK

Pablo Rangel, Departamento de Genética Molecular, Instituto de Fisiologia Celular, Universidad Nacional Autonoma de Mexico, Circuito Exterior S/N, Ciudad Universitaria Copilco, Coyoacan 04510, Mexico

Mauricio Ríos-Momberg, Departamento de Bioquímica, Instituto de Fisiología Celular, Universidad Nacional Autónoma de México, UNAM, 04510 México, D. F., México

Geoffrey D. Robson, Faculty of Life Sciences, 1.800 Stopford Building, University of Manchester, Manchester M13 9PT, UK

Elvira Román, Departmento de Microbiologia II, Faculated de Farmacia, Universidad Complutense de Madrid, Plaza Romón y Cajal S/N, 28040 Madrid, Spain

Martin Sergeant, Warwick HRI, University of Warwick, Warwick CV35 9EF, UK

Michael W. Shaw, School of Biological Sciences, Plant Science Building, University of Reading, Lyle Tower, Whiteknights, Reading RG6 6AJ, UK

Daniel Shenton, University of Manchester, Faculty of Life Sciences, The Michael Smith Building, Oxford Road, Manchester M13 9PT, UK

Andy F.S. Taylor, Department of Forest Mycology & Pathology, SLU, Box 7026, Uppsala SE-750 07, Sweden

John Thomas, Warwick HRI, University of Warwick, Warwick CV35 9EF, UK

William Throndset, Genencor International, 925 Page Mill Road, Palo Alto, CA 94304, USA

Preface

Yeasts and filamentous fungi need to cope with stress, in one form or another, whether growing in the laboratory or in the natural environment, whether victims or offenders in interactions with other organisms. Fungal research gives invaluable insight to the environmental impact of stress, and to the molecular mechanisms of action and response. These considerations define the ethos of this volume, which considers stress in this broad sense, within the context of mycology.

The chapters herein are contributed by invited speakers at the 2006 Annual Scientific Meeting of the British Mycological Society, Birmingham, UK. The meeting shared the same title as this volume, Stress in Yeasts and Filamentous Fungi, and shared a similar blend of content dedicated to both environmental and cell and molecular aspects of stress effects and responses in fungi. This volume draws on the expertise of leaders in the field from both the yeast and filamentous-fungal communities.

The stresses associated with organism–organism interactions (e.g. predation, competition, etc.) are highlighted in contributions from Naresh Magan, Peter Mills, Levente Kiss, Michael Shaw and their co-authors, which consider these effects at the whole organism level in fungi in their natural environments. The stress interface between different organisms is probed further in the context of fungal pathogenicity by Jan Quinn and Ken Haynes with their respective work on the human pathogen *Candida* spp., and with an oomycete agent of downy mildew by Jim Beynon.

There is increasing awareness that stress may arise under the comparatively controlled conditions of fungal culture in the laboratory, aspects of which are tackled in the chapter by Geoff Robson and co-authors. Stress under controlled conditions may be exploited for control of fungal growth, and Mehdi Mollapour with Peter Piper examine how yeasts respond to stress provoked by weak acids that may be used for food preservation. Stefan Hohmann describes a systems biology approach to modelling such stress responses in yeast.

Another area that continues to attract research attention is anthropogenic stress towards fungi in the environment, for example via metal pollutants generated through industrial activities. The impacts that such stressors may have on different organisms and communities in the wild are explained by Jan Colpaert (ectomycorrhizal fungi), William Purvis (lichens) and Roger Finlay (mycorrhizal fungi), with their co-authors. Effects of anthropogenic stress on fungi in the environment are underpinned by effects originating at the cellular and molecular level. Many such stressors act as pro-oxidants, and different aspects of the responses to oxidative stress exhibited by fungal cells are explored in the contributions by Chris Grant (protein synthesis response), Scott Moye-Rowley

(Yap1-mediate transcriptional response), Jesus Pla (signalling response) and Wilhelm Hansberg (cell differentiation response) and their co-authors.

Finally, it is hoped that this volume will provide a valuable resource for those interested in the effects of stress on fungi, bringing together researchers who study these issues at different levels and who, consequently, may rarely gain exposure to each other's work. The content encompasses a breadth of scope from the bigger picture of stress effects on fungi in their natural habitats, down to the fascinating recent advances in our underlying molecular-level understanding. I hope that readers will benefit from seeing how their particular angle of interest fits within this broader view, while of course feeding their own particular interest and hopefully enjoying an all-round stimulating read!

Simon V. Avery
April 2007

CHAPTER 1

Interactions Between *Agaricus bisporus* and the Pathogen *Verticillium fungicola*

Peter Mills, John Thomas, Martin Sergeant, Ana Costa, Patrick Collopy, Andy Bailey, Gary Foster and **Mike Challen**

Contents

1. Introduction 2
2. Pathogen Variability 4
3. Extra-Cellular Enzyme Production 5
4. Screening for Resistance 8
5. Fungal Transformation 9
6. Molecular Interactions of *Verticillium* and *Agaricus* 10
7. Functional Characterisation of Host–Pathogen Response Genes 12
8. Future Work 14
Acknowledgements 15
References 15

Abstract

The interaction between *Agaricus bisporus* and *Verticillium fungicola* may be the most economically significant interaction between two fungi. Work described in this chapter covers diverse aspects of the biology and molecular interaction of the host and pathogen and has established a robust baseline for future study of a pathogenic system involving two members of the same kingdom.

Pathogen variability has been assessed, and specific genotypes important to major mushroom-producing countries have been characterised. This has enabled representative isolates to be identified for use in this study.

A wide range of cell-wall-degrading enzymes from *V. fungicola* have been identified, and principal component analysis showed a complex correlation between enzyme production and symptom expression. It is likely that some, or indeed many of these enzymes play a critical role in the pathogenicity of *Verticillium*.

Although it is generally accepted that *A. bisporus* has a relatively narrow genetic base, our work, and that of other groups, has shown that variability in host tolerance to infection with *V. fungicola* exists to a large extent to justify a detailed study to exploit it. We have developed

British Mycological Society Symposia Series
Published by Elsevier Ltd.

transformation technologies for the mushroom *A. bisporus* (Challen *et al.*, 2000; Leach *et al.*, 2004; Foster *et al.*, 2004b; Burns *et al.*, 2005, 2006), its pathogen *V. fungicola* (Amey *et al.*, 2002, 2003) and other pathogenic fungi (Rogers *et al.*, 2004; Gewiss-Mogensen *et al.*, 2006). Recent advances in gene suppression technologies for the host include deployment of anti-sense and RNAi hairpin constructs to down-regulate endogenous *A. bisporus* genes (Burns, 2004; Heneghan *et al.*, in press). Similarly gene knockout methodology has been established for the pathogen *V. fungicola* (Amey *et al.*, 2003; Foster *et al.*, 2004a).

An extensive range of genetic resources have been established, which include: *V. fungicola* pathogenicity simulated cDNA library (mushroom cell wall (MCW) agar), numerous *A. bisporus* fruiting cDNA libraries (macro-arrayed), host (*A. bisporus*)–pathogen (*V. fungicola*) lesions infection SSH libraries (forward and reverse subtractions for up- and down-regulated mushroom genes) and macro-arrayed cDNA infection library, genomic DNA libraries for *A. bisporus* and other homobasidiomycete mushrooms, EST fungal–fungal interaction database; transformation and expression vectors designed specifically for *Agaricus* and *Verticillium*, and vectors to expedite cloning of fragments for gene silencing experiments.

The resources now in place will enable us to determine, in a systematic manner, the impact of gene expression on the disease phenotype.

This work provides platform technologies for alternative methods of disease control, reduced pesticide use and sustainable crop production. Investigations into this model fungal–fungal interaction will also provide information on the research and development of fungal biological control agents of fungal plant pathogens.

1. INTRODUCTION

The homobasidiomycete fungus *Agaricus bisporus* (Lange) Imbach, the white button mushroom, is a high-value cultivated crop species with approximately 1.5 million tonnes p.a. cultivated for human consumption worldwide. Pathogen threats to *A. bisporus* include bacteria, viruses, mites, insects and fungi. The latter are the most detrimental and pose a significant threat to the industry.

Dry bubble, caused by *Verticillium fungicola* (Preuss) Hassebrauk, is the most common and serious fungal disease of *A. bisporus*. Symptoms vary (Figure 1) and may depend on a number of factors such as developmental stage at the time of infection and genetic variability of the host and include bubble, cap spotting and stipe blowout (Fletcher *et al.*, 1994). The least severe symptom is cap spotting (lesions) where *V. fungicola* appears to infect only the surface of the fruiting body causing brown lesions to appear (Dragt *et al.*, 1996). 'Stipe blow-out' is a more extreme symptom where the *A. bisporus* stipe swells and peels; it is thought this symptom is due to infection earlier in mushroom development than in the case of cap spotting (Ware, 1933). Stipe blowout occurs experimentally when high levels of *V. fungicola* are used to inoculate *A. bisporus*, at the primordial stage prior to major differentiation of the fruiting body (North and Wuest, 1993). In the most

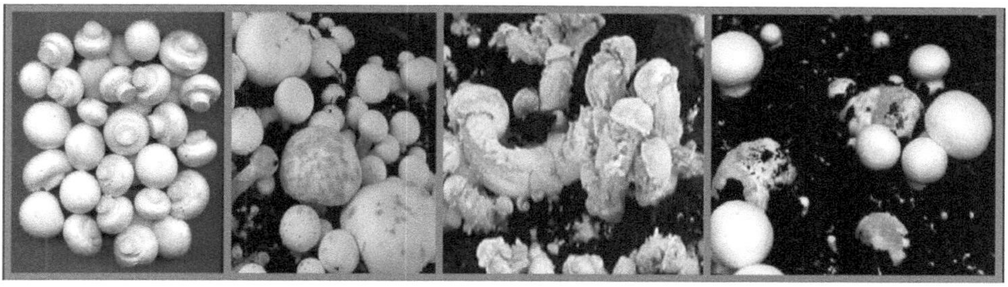

Figure 1 *Verticillium fungicola* infections of *Agaricus bisporus*. From left to right: healthy mushrooms, cap lesions, stipe blowout and dry bubble.

severe symptom, from which the disease derives its common name 'dry bubble', mycelia of *V. fungicola* and *A. bisporus* appear to grow together causing a 'mushroom mass' with little or no differentiation (Jeffries and Young, 1994). This only occurs when the early initials of *A. bisporus* (mushroom pins) are inoculated (North and Wuest, 1993). Hitherto this complex infection has been poorly characterised. Appressoria, penetration and intracellular growth of *Verticillium* hyphae in *A. bisporus* fruit bodies can be seen using light microscopy and TEM (Dragt *et al.*, 1996). EM evidence for *V. fungicola* action on host *A. bisporus* cell walls *in vitro*, and the production of hydrolytic enzymes in *V. fungicola* cultures grown on various carbon sources has also been demonstrated (Calonje *et al.*, 1997, 2000).

None of the commercially available *A. bisporus* spawns are resistant to this pathogen. *Verticillium* is currently controlled by the C14 demethylation inhibitor of egosterol biosynthesis, prochloraz-manganese, marketed as 'Sporgon'. That only a single fungicide control chemical is available is a very high-risk situation. If resistance to the fungicide were to become widespread (resistant strains have already been reported; Grogan and Gaze, 2000; Grogan *et al.*, 2000) or if fungicide control was to become unavailable for any reason, for example more stringent pesticide safety regulations, then the industry would be faced with an acute problem. In Holland, for example, there are increasing pressures against the use of 'Sporgon' because of perceived residue problems. Outbreaks of *Verticillium* can be devastating. The current 'boom and bust' situation exactly parallels the position in the 1970s where benomyl-type fungicides were used to control *Verticillium* and the pathogen soon developed resistance resulting in a major problem for the industry. The mechanisms of infection of *Agaricus* by fungal pathogens are poorly understood.

Advances in fungal–fungal interactions are significant in the context of mushroom (*A. bisporus*) pathology where *V. fungicola* is the single most geographically significant mycoparasite; it also serves as a relevant mycoparasitic model for the study of other mycoparasitic diseases (Williams *et al.*, 2003) and several biocontrol agents (e.g. *Coniothyrium*, *Trichoderma* and *Sporidesmium*) used against soil-borne fungal pathogens. However, unlike many of these systems where isolation of host or pathogen RNA may be limiting, the *Agaricus–Verticillium* interaction permits easy isolation of both host (basidiomycete) and pathogen (ascomycete) sequences from a live interaction, minimises problems with cross-hybridisations and facilitates likely provenance of isolated sequences.

The fungal kingdom encompasses a diverse range of organisms occupying numerous ecological niches with various lifestyles, from litter decomposers to human pathogens. Although there have been significant efforts to characterise interactions of the pathogenic fungi with plants and humans, studies of fungal pathogens of other fungi remain in their infancy.

In order to understand the underlying mechanisms involved in the interaction between *Verticillium* and *Agaricus* a number of experimental components were required. These included a measure of pathogen variability, an understanding of the possible mechanism of pathogenicity, pathogen and host transformation technologies and a measure of host resistance. Collectively these components were put in place at the Universities of Warwick and Bristol and have allowed a detailed study of the molecular interaction of the two component fungi.

2. PATHOGEN VARIABILITY

V. fungicola is classified into three varieties based on conidiophore morphology and maximum growth temperatures. *V. fungicola* var. *fungicola* is thought to be the primary cause of *Verticillium* disease in Europe, whereas *V. fungicola* var. *aleophilum* is thought to cause brown spotting and has been associated more commonly with *Agaricus bitorquis* and higher growing temperatures. A third variety *V. fungicola* var. *flavidum*, isolated from wild agarics appears to be weakly pathogenic to cultivated *Agaricus* species. To complicate the situation further *V. psalliotae* has been reported in the literature as a minor pathogen of mushrooms causing brown cap spotting. Further information on *Verticillium* varieties is available in Collopy *et al.* (2001). It is also interesting to note that *V. fungicola* var. *aleophilum* can be more aggressive to some strains of *A. bisporus* than var. *fungicola* (Largeteau *et al.*, 2005).

In an attempt to clarify the causal agent of disease in the UK isolates were collected from UK farms and also from other major mushroom-producing countries including France, Spain, Germany, Hungary, The Netherlands, New Zealand and the USA. Using a range of routine molecular tests including ribosomal DNA RFLP, the diversity of this group of isolates was assessed. Results indicated that all European isolates formed a single group and could be distinguished from American and New Zealand isolates. Mitochondrial DNA (mtDNA) RFLPs revealed further genetic variation within the three major groups. European isolates possessed four haplotypes and were quite distinct from the New Zealand and American isolates, which shared a common mtDNA RFLP pattern. Sequencing of the internal transcribed spacer (ITS) 1 region of the rDNA confirmed the genetic uniformity of the European isolates. Isolates from America and New Zealand possessed nearly identical ITS 1 sequence whilst having a 3.5% divergence from European isolates.

The outcome of the analyses of the variability of pathogenic *Verticillium* isolates from mushrooms has shown that there are four distinct taxa within

V. fungicola. The European isolates relate to the *ex*-type isolate of *V. fungicola* var. *fungicola*. The isolates from New Zealand are divided into two taxa of which one taxon conforms to the *ex*-type isolate of *V. fungicola* var. *aleophilum*; the fourth taxon comprised isolates from the USA. The taxon containing the European isolates is genotypically very different from the other three taxa.

The significance of these data is that there are considerable genetic differences in mushroom pathogenic *Verticillia* across the world but that geographical regions generally possess only one major genotype. This contradicts the previously accepted view that all varieties of *Verticillium* may be present in all regions.

It is also worth noting that the full sequence of the nitrate reductase gene has been obtained for *V. fungicola* (Amey *et al.*, unpublished data). The central portion of this gene was amplified and sequenced from a number of *V. fungicola* isolates and related fungi, and the resulting phylogenies were compared to those obtained from analysis of the rDNA ITS regions for these fungi. Both nitrate reductase and ITS analyses provided additional evidence that suggests the mushroom pathogenic *Verticillium* species are related more to other chitinolytic fungi such as the insect pathogens *Verticillium lecanii* and *Beauveria bassiana* than to the plant pathogenic *Verticillia*. These data suggest that *V. fungicola* could be reclassified separately from other *Verticillium* species. Although there were high similarities with the insect pathogenic *Beauveria* species it was clear that sufficient differences were evident at the genetic and biological levels for *V. fungicola* to be placed in a form genus of its own. A new genus called *Lecanicillium* has been proposed to include the majority of entomogenous and fungicolous taxa (Zare and Gams, 2001).

3. EXTRA-CELLULAR ENZYME PRODUCTION

Although *Verticillium* species cause a range of symptoms on *A. bisporus* the infection process is not fully understood because of the complexity, involving the presence of *Verticillium* spores and the subsequent interaction between mycelia of the pathogen and host (Calonje *et al.*, 1997).

Kalberer (1984) assayed *Verticillium* proteolytic enzymes, but was unable to prove their role in the infection process of mushrooms. Scanning electron microscopy (North and Wuest, 1993) and *in vitro* studies (Matthews, 1983) found interhyphal penetration by *V. fungicola* var. *fungicola* into mushroom fruiting body tissue, indicating that hydrolytic enzymes were involved in the mycoparasitism by *Verticillium*. Dragt *et al.* (1996) using light microscopy and TEM showed the formation of appressoria, penetration and intracellular growth of *Verticillium* hyphae in the *A. bisporus* fruiting body hyphae. The thin wall of mycoparasitised hyphal cells compared with the electron-dense wall of healthy mushroom cells again supported the view that wall-lytic enzymes played a role in the infection process.

In an attempt to correlate enzyme production by *V. fungicola* and symptom expression, a study was undertaken on 17 isolates (Table 1) selected on the basis of differences in pathology demonstrated in cropping trials or on the basis of

Table 1 *Verticillium* isolates studied

Warwick HRI No.	Designated species	Geographic origin
5-3	*V. fungicola* var. *Fungicola* (ICMP3343)	New Zealand
7-3	*V. Fungicola* var. *aleophilum*^{T} (CBS357.80)	Netherlands
7-4	*V. fungicola* var. *fungicola*^{T} (CBS440.34)	UK
14-4	*V. fungicola* var. *fungicola*	USA
19-4	*V. fungicola* var. *fungicola*	Taiwan
19-5	*V. fungicola* var. *fungicola*	South Africa
21-3	*V. fungicola* var. *fungicola*	Spain
26-10	*V. fungicola* var. *fungicola*	Netherlands
55-3	*V. fungicola* var. *fungicola*	USA
55-12	*V. fungicola* var. *fungicola*	USA
115-1	*V. fungicola* var. *fungicola*	UK
124-1	*V. fungicola* var. *fungicola*	Eire
150-1	*V. fungicola* var. *fungicola*	UK
157-1	*V. lecanii*	UK
170-3	*V. psalliotae* (IMI246435)	UK
171-1	*V. fungicola* var. *fungicola*	Brazil
174-1	*V. fungicola* var. *fungicola* (TMI65020)	Japan

Note: T, type strain.

molecular typing described above. Cultures were grown on minimal synthetic media supplemented with MCW material. Culture filtrates were assayed for enzymes likely to be involved in the breakdown of host cell walls.

A wide range of enzymes were produced. Lipase production from 17 *Verticillium* spp. was initially determined by a clearing zone assay on agar containing Tween 80. All isolates showed lipase activity. Esterase (C4 degradation) activities of *Verticillium* spp. were relatively uniform, with the notable exceptions of three *V. fungicola* var. *fungicola* isolates (two from the USA and one from South Africa) and one *V. psalliotae* isolate. There was generally an inverse relationship between *Verticillium* spp. enzymic activities and the carbon chain length of 4-*p*-nitrophenyl fatty acid substrates. Only 50% of *V. fungicola* var. *fungicola* isolates showed significant stearase (C18) activity.

Most *Verticillium* spp. produced high levels of 1,3-β-glucanase, though there was considerable variation between isolates with three European isolates *V. fungicola* var. *fungicola* showing much lower levels of activity.

Proteinase activities of *V. fungicola* var. *fungicola* 7-4^{T} was assayed using a synthetic peptide substrate. A pH profile of *Verticillium* serine proteinase activities (assayed using suc-ala-ala-pro-phe-*p*N substrate) showed it to be an alkaline proteinase, having activity optima between pH 7 and 10. This enzyme activity was inhibited by PMSF and partially inhibited by EDTA but not iodoacetamide or pepstatin, thus confirming it to be a serine proteinase. Three aminopeptidase activities were found using glu-*p*-NA, arg-*p*-NA and leu-*p*-NA substrates; an isolate of *V. psalliotae* produced the highest activities of these enzymes.

The *Verticillium* spp. studied were all able to produce three principal types of chitiolytic enzyme: (i) 1–4, β-*N*-acetlyglucosaminidase (substrate GlcNAc), which splits diacetylchitobiose etc. into *N*-acetylglucosamine monomers; (ii) an exochitinase (substrate $[GlcNAc]_2$), which catalyses progressive release of diacetylchitobiose in a stepwise fashion; and (iii),endochitinase(s) (substrate $[GlcNAc]_{3–4}$), which randomly cleave sites over the entire length of chitin micofibrils to produce soluble low-molecular-mass multimers of acetylglucosamine.

Initial chitiolytic activities were performed using *p*-nitrophenol substrates; large variations were found between *Verticillium* isolates. Subsequent chitinase assays were performed using a more sensitive fluorescent substrate.

Principal component analysis of data from two sources, mushroom crops of *A. bisporus* A12 inoculated with individual *Verticillium* spp. and, enzyme profiles of 17 *Verticillium* spp. grown on *A. bisporus* A12 cell walls, was used to establish a tentative relationship between percentage disease and *Verticillium* spp. enzyme profiles. Hydrolytic enzymes play a crucial part in the mycoparasitism of *Agaricus* by *Verticillium*. Cap spotting was associated with *Verticillium* isolates that produced high levels of chitinases and stearase, but had low esterase activity towards MCW material; whereas *Verticillium* isolates that caused 'dry bubble' symptoms were notable for higher esterase and endochitinase activities allied to lower stearase, serine proteinase and other chitinase activities.

The chitiolytic system of *Verticillium* spp. may be a key element in pathogenicity of the fungus towards *Agaricus* fruiting bodies. Similar chitinolytic enzyme systems have been demonstrated as crucial in the success of entomopathogenic fungi (St. Leger *et al.*, 1997) and pathogenic and ectomycorrhizal basidiomycetes (Hodge *et al.*, 1995). The chitinolytic system of the biological control agent *Trichoderma harzianum* is made up of two β-1,4 -*N*-acetylglucosaminidases and four endochitinases (Haran *et al.*, 1996); their expression during mycoparasitism was regulated in a very specific and finely tuned manner that was affected by the host (Zeilinger *et al.*, 1999).

Induction of enzymes in liquid cultures supplemented with MCW material is a good indicator of the enzyme's role in causing disease symptoms in mushroom crops. Principal component analysis of our data has revealed a link between disease expression and *Verticillium* enzyme production thus providing evidence towards a mode of action for the mycoparasite. Dragt *et al.* (1996) unequivocally demonstrated that the interaction between pileal hyphae of *A. bisporus* and of the pathogen *V. fungicola* var. *fungicola* took place through a mucilaginous extracellular matrix (ECM); such a matrix would facilitate *Verticillium* extracellular degradative enzyme activities.

For entry, phytopathogenic fungi usually secrete a cocktail of hydrolytic enzymes. These enzymes are also required for the saprophytic lifestyle; therefore, they are unlikely to represent the tools specifically developed by fungi to implement pathogenesis, and each individual hydrolytic enzyme may not be absolutely necessary for penetration (Knogge, 1996). This does not, however, preclude the adaptation of enzyme structure or biosynthetic regulation to the specific needs of a pathogen, say *Verticillium* on mushrooms. Alternatively, or in combination with hydrolytic enzymes, some fungi have developed specialised penetration organs (appressoria) to penetrate the cuticle of host plants and these have now been observed associated with the intracellular infection of *Agaricus* by *V. fungicola* var. *fungicola* (Dragt *et al.*, 1996).

It would be informative to determine which pathogen enzymes are active and how they are induced during mycoparasitism. Genes encoding polysaccharide- and protein-degrading enzymes have been isolated from a wide range of saprophytic fungi, including the closely related insect pathogen *Beauveria*. To determine the contribution made by specific cell-wall-degrading enzymes in colonisation of mushroom host tissue and symptom formation, further work would be required. Similarly, assessment of the biochemical response by *Agaricus* to infection is being addressed by the purification of major novel proteins induced by challenge inoculations between *Agaricus* spp. and cell-free supernatants from *Verticillium* liquid cultures.

4. SCREENING FOR RESISTANCE

One hundred and seventy 'wild' accessions of *Agaricus* were obtained from the *Agaricus* Resource Program (ARP; Kerrigan, 1991). From this collection, 46 ARP and commercial strains were fruited in two randomised trials. Pot cultures were inoculated with *V. fungicola* var. *fungicola* strain 150-1, and both cap spotting and dry bubble symptoms were expressed. Six strains displaying either high or low levels of susceptibility to *V. fungicola* 150-1 were selected for further study. No ARP strain showed absolute resistance to *Verticillium* with respect to cap spotting. However, some pathogen–host permutations did not produce 'bubble' symptoms. Similarly, tolerance to *V. fungicola* has been observed in other wild strains of *A. bisporus* (Dragt *et al.*, 1995; Largeteau *et al.*, 2005). Wild germplasm offers prospects both for the production of disease-tolerant varieties and for the identification of genes involved in the hosts defence pathways.

It is interesting to speculate about the nature of partial 'resistance' of *Agaricus* to *Verticillium*; a number of possible 'resistance' strategies can be postulated.

The cell wall of a fungus is a complex structure composed of glucans (α and β), glycans, chitins and lesser quantities of proteins and lipids (Bartnicki-Garcia, 1968). The work of a Spanish group (Calonje *et al.*, 1997), and the results reported here have identified that numerous endo- and exopolysaccharidases, proteinases and lipolytic enzymes are produced by *Verticillium* spp. when grown on a minimal synthetic liquid medium containing *A. bisporus* fruiting body, or stipe, hyphal cell walls. The gross chemical composition of individual strains of *A. bisporus* mycelial and hyphal cell walls is similar (Mol and Wessels, 1990), though significant differences in gross hyphal wall composition were detected in four commercial *A. bisporus* strains (Calonje *et al.*, 1995). Studies using light and electron microscopy have shown stipe hyphae to be more vacuolated than those of fruiting bodies (Umar and Van Griensven, 1997). It is expected that the overall architecture and chemical composition (including polysaccharide structure) of hyphae aggregated into stipes and fruiting bodies may influence susceptibility to mycoparasitism. Thus, current studies compare *Verticillium* hydrolytic enzyme production on cell wall preparations from ARP strains more, or less, tolerant to *Verticillium* disease in cropping trials.

Dragt *et al.* (1995) in a study of 100 ARP accessions found several that were partially resistant to *Verticillium* disease. Histological studies of the infection process in one of the partially resistant strains (characterised by a brown roughened cap with a white stipe) gave some evidence on the nature of resistance. Necrosis of the outer part of the mushroom cap tissue and deposition of brown pigments, possibly fungitoxic melanins, in the underlying layers were demonstrated. The number of damaged or collapsed *A. bisporus* cells was reduced. Melanins represent a class of dark pigments that result from oxidative polymerisation of various types on phenols; they are insoluble in water and extremely resistant to enzymatic degradation (Hegnauer *et al.*, 1985).

5. FUNGAL TRANSFORMATION

Agrobacterium tumefaciens transformation of fungi is now well established and has been successfully used on a range of filamentous fungi (De Groot *et al.*, 1998; Gouka *et al.*, 1999; Covert *et al.*, 2001; Mullins *et al.*, 2001). This method has dramatically improved the efficacy of fungal gene manipulation (Mullins and Kang, 2001).

For *Verticillium*, *Agrobacterium*-mediated transformation was shown to be substantially more efficient in terms of time and effort than PEG-mediated transformation (Amey *et al.*, 2002). It was possible to consistently obtain more than 4,000 transformants per 10^5 conidia with *Agrobacterium*-mediated transformations compared to PEG-mediated protoplast transformation where transforming efficiency was unpredictable. Observations also suggested that *Agrobacterium*-mediated transformation mainly results in single-copy integration events and indicates that this method is ideal for targeted mutagenesis of *Verticillium*.

Before the observation that *A. tumefaciens* could transfer its T-DNA to a wide range of fungi (De Groot *et al.*, 1998) it was very difficult to transform the mushroom. Early protoplast-based transformations proved unreliable and inefficient (Challen *et al.*, 2000). Agrobacterium-mediated transformation of *A. bisporus* with a hygromycin resistance marker has proved a relatively efficient and remarkably robust method (Leach *et al.*, 2004). A wide range of mushroom tissue types can be used for transformations, such as germinating basidiospores (De Groot *et al.*, 1998) and mycelia (Mikosch *et al.*, 2001), but the most efficient transformations are obtained using gill tissue (Chen *et al.*, 2000). The GFP reporter gene can also be expressed in *A. bisporus* provided it is coupled with a 5′ intron sequence (Foster *et al.*, 2004b; Burns *et al.*, 2005). A molecular toolkit comprising interchangeable promoter, marker gene and terminator sequences has also been developed to expedite expression studies in mushrooms (Burns *et al.*, 2006). These advances in transformation of *Verticillium* and *Agaricus* have enabled us to apply functional approaches to our studies on genes involved in the fungal–fungal interaction. Using β-1,6-glucanase and other pathogen genes, we have established gene knockout methodology for *V. fungicola* (Amey *et al.*, 2003; Foster *et al.*, 2004a).

Recently we have also developed gene-silencing methodologies for mushroom species (Eastwood *et al.*, 2006; Heneghan *et al.*, in press) and we are using these to modify expression of disease response genes in *A. bisporus.*

6. MOLECULAR INTERACTIONS OF *VERTICILLIUM* AND *AGARICUS*

Recent advances in *Agaricus* and *Verticillium* molecular genetic technologies provide an excellent opportunity to further dissect the infection processes and to gain an understanding of the genetic, biochemical and physiological processes that regulate this fungal–fungal interaction. Whilst there are numerous examples of studies focussed on plant–fungal interactions, few studies have investigated fungal–fungal interactions, particularly with relevance to an important crop species.

Hitherto knowledge on the cellular stress and immune responses of eukaryotes has been largely confined to organisms in the plant and animal kingdoms. These have included such diverse organisms as mammals, insects, fish, plants and even lower invertebrates such as nematode worms. However, very little information is available on the cellular responses within the fungal kingdom to stress and pathogen attack.

A series of experiments were designed to elucidate the interactions occurring during infection. Lesions were induced on *A. bisporus* by inoculation of an experimental crop with a spore suspension of *V. fungicola*; the lesions were then excised and stored prior to extraction of mRNA. Total RNA from both lesion and healthy mushroom material was obtained and suppressive subtractive hybridization (SSH) was used to identify candidate genes involved in the interaction.

To enrich for host genes up-regulated during infection, our SSH was principally performed as a forward subtraction using the lesion mRNA as 'tester' against healthy mushroom 'driver' mRNA. Additionally we performed SSH cloning in reverse to help identify host genes down-regulated during infection. Products with an insert size of 91–764 bp were cloned and more than 400 recombinants sequenced. A summary of putative biological functions is illustrated in Figure 2.

Bioinformatic analysis of these sequences revealed approximately 80 *Agaricus* uni-genes (transcripts) differentially expressed upon infection within the following classifications: metabolism (14%), energy (6%), protein synthesis (4%), protein destination (4%), transport (4%), cell rescue, defence, death, ageing (8%), cellular communication, signal transduction (5%), cellular biogenesis (5%), cell growth and division (1%), cellular organisation (8%), transposon and insertion sequence proteins (1%), hypothetical or no ascribed function (40%). Amongst the host genes identified some were known to be involved in other host–pathogen systems, for example stress/defence (superoxide dismutase), signal transduction (MAP kinase), cell wall structure (chitin synthase, and B 1,3-glucan binding protein), primary metabolism (7-phosphate DAHP synthase and glyceraldehyde phosphate dehydrogenase and protein synthesis and processing (60S ribosomal proteins).

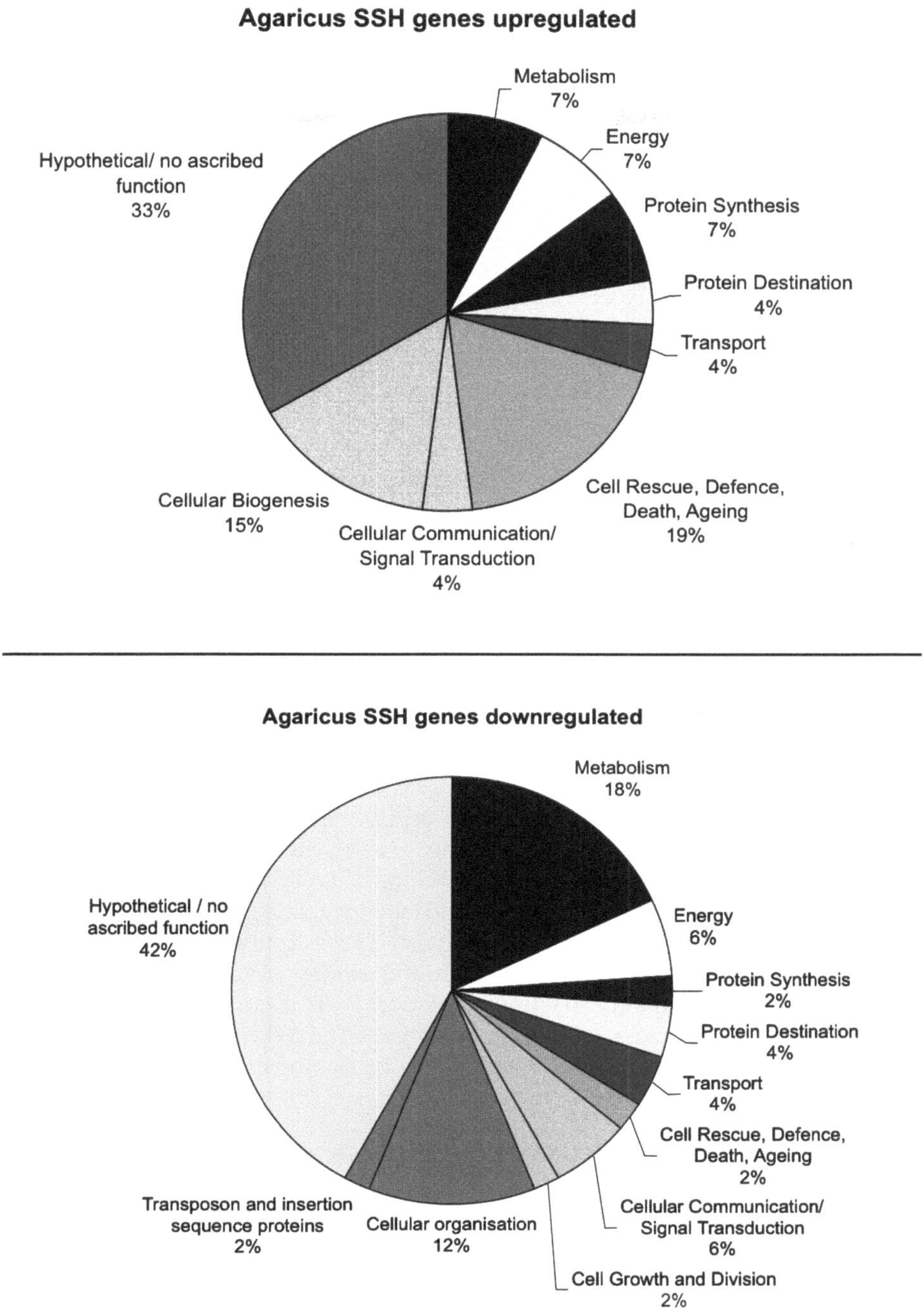

Figure 2 Putative biological function of *A. bisporus* response genes. SSH cloning was used to recover genes up- and down-regulated during infection with *V. fungicola*.

The SSH also revealed more than 50 up-regulated *Verticillium* genes from the following classes: metabolism (6%), protein synthesis (24%), protein destination (2%), transport (4%), cell rescue, defence, death, ageing (6%), cellular communication/signal transduction (4%), cell wall biogenesis (6%), cellular organisation (4%), transposon and insertion sequence proteins (4%), hypothetical or no ascribed function (41%). Amongst the pathogen genes identified were those associated with the cytoskeleton (β-tubulin), signal transduction (ADP ribosylation), membrane transporters, primary metabolism (gluconate kinase and β-glucosidase, chromatin DNA and RNA metabolism (histone H3 protein and the chromatin assembly complex) and protein synthesis (60S proteins).

Importantly, approximately 40% of the SSH derived transcripts had no significant homolog in public databases and therefore could not be ascribed a putative biological function. Functional genomic approaches (gene silencing or suppression) will therefore be needed to fully characterise genes involved in the *Agaricus–Verticillium* interaction.

To help expedite out functional analysis, an infection cDNA library was prepared,comprising more than 50,000 clones, stored as ordered micro-titre plates and macro-arrayed to facilitate easy screening. Validation of our cDNA infection library has been demonstrated by recovery of SSH sequences (92% efficacy) and is representative for both high- and low-abundance genes. Full or extended length versions of our SSH sequences have been easily recoverable from the cDNA infection library. The integrity of differentially expressed transcripts and efficiency of the SSH subtractions have been confirmed using both virtual northerns and also quantitative PCR (Q-PCR) approaches (Figure 3). These experiments have also shown that cross-hybridisation between the ascomycete pathogen and basidiomycete host is minimal.

Complementary to our SSH analyses additional 'pathogenicity simulation' *Verticillium* cDNAs were recovered using growth on a MCW medium. This approach has yielded more than 600 pathogenicity ESTs spanning an equally diverse range of putative functions, including: protein synthesis (12%), protein destination (3%), energy (6%), transcription (3%), transposon and insertion sequence (1%), cell rescue, defence, death and ageing (6%), development (1%), transport facilitation (2%), intracellular transport (<1%), cell growth and division (<1%), cell wall biogenesis (5%), cellular organisation (3%), metabolism (9%), cell communication/signalling (1%), hypothetical or no ascribed function (41%).

7. FUNCTIONAL CHARACTERISATION OF HOST–PATHOGEN RESPONSE GENES

Transgene-mediated insertional mutagenesis has proved a reasonably effective tool for identification and functional characterisation of genes in pathogenic fungi (Rogers *et al.*, 2004; Gewiss-Mogensen *et al.*, 2006). In *Verticillium* we are able to use targeted gene disruption through homologous recombination to inactivate pathogenicity genes involved in the interaction with *Agaricus* (Amey *et al.*, 2003; Foster *et al.*, 2004a) and to demonstrate that β-1-6-glucanase knockout

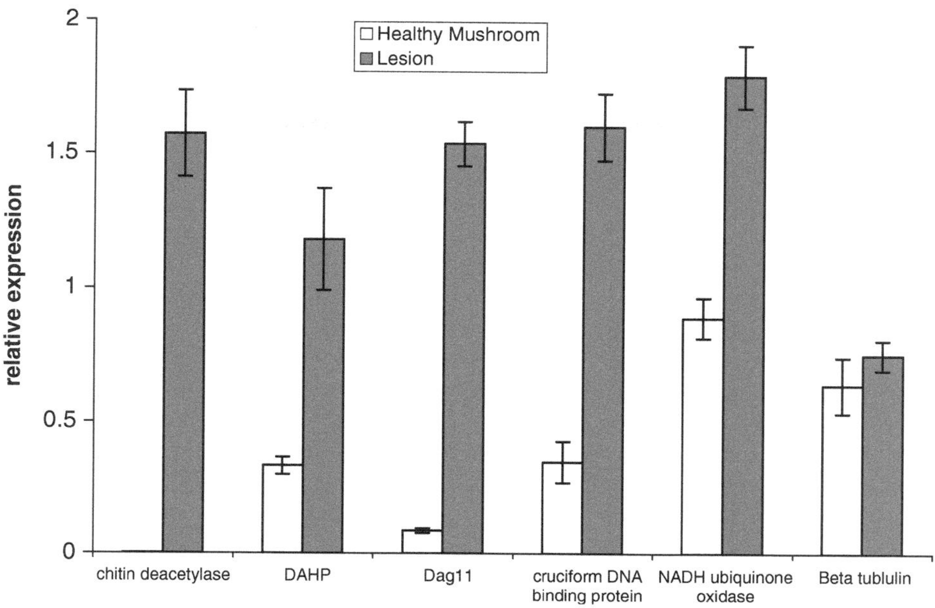

Figure 3 Relative expression levels for specific *Agaricus bisporus* genes Identified in healthy mushroom and infected lesion tissues. Expression was measured by Q-PCR and normalised against 18S. The standard employed for each gene consisted of dilutions of pooled cDNA from lesion samples. Control gene is β-tubulin, where there is no significant difference between lesion and healthy tissues. Five genes identified by SSH are all up-regulated on infection. Relative comparisons between genes are not appropriate.

transformants exhibit reduced virulence. Similarly we recently used gene disruption to investigate the role of a *Verticillium MAPK1* gene (Collopy *et al.*, unpublished).

Insertional mutagenesis or gene knockout approaches are likely to prove less productive in *A. bisporus*, which has multinucleate, heterokaryotic cells. We have developed alternative approaches for gene suppression in homobasidiomycete mushrooms (Heneghan *et al.*, in press) and have observed that both RNAi hairpin constructs (Eastwood *et al.*, 2006) and conventional anti-sense constructs (Burns, 2004) can down-regulate genes in *Agaricus*.

To characterise *Agaricus* genes involved in response to pathogen infection we have focussed our efforts on RNAi hairpin-mediated gene-suppression vectors introduced via *Agrobacterium* transformation. Specifically, using constructs based on the pRNAi-DE001 vector assembled at Warwick (Eastwood *et al.*, unpublished) we have introduced hairpin dsRNA of *A. bisporus* genes to initiate RNAi-mediated gene suppression. Target gene PCR amplicons (200–500 bp) are cloned in sense and anti-sense orientations separated by a non-coding loop, derived from the *E. coli gusA* gene or using an *A. bisporus* intron sequence. Transcription of self-complementary hairpins is routinely regulated by *A. bisporus gpd2* (5′-utr) promoter and *A. nidulans trpC* terminator (3′-utr) sequences. Several genes were targeted but most effort has been focussed on the *A. bisporus* chitin deacetylase and 3-deoxy-7-phosphoheptulonate synthetase (DAHP synthetase) genes. Chitin deacetylase was chosen because it was massively up-regulated during infection. We have observed that hairpin-mediated gene suppression is easier to detect on highly expressed genes and is more likely to be detected as a mutant phenotype. Silencing of DAHP synthetase has had a profound effect on the interactions of other plant–pathogen systems (Jones *et al.*, 1995).

Down-regulation of specific transformants was initially assessed in mycelia using Q-PCR approaches. Numerous transformants that exhibited a variety of expression levels were recovered for both gene targets. Several down-regulated chitin-deacetylase (CHID) transformants were recovered, and in one of those, chitin-deacetylase expression was down-regulated to about 1% of the non-transformed control. Interestingly, some transformants exhibited increased chitin-deacetylase expression, approximately fivefold increase compared to controls. The DAHP synthetase hairpin transformants displayed similar patterns of gene expression. Transformants exhibiting 30–40% of wild-type DAHP synthetase expression were identified. Again some DAHP transformants showed increased gene expression, up to 10-fold increase compared to controls.

Five transformants each for CHID and DAHP were taken into infection trails in the Warwick HRI mushroom unit and infected with wild-type *V. fungicola*. Lesion samples and control mushrooms were used to prepare RNA and analyse gene expression in response to infection. There were no gross changes in the phenotype of the host–pathogen interaction in terms of size and/or morphology of lesions. However, more detailed experiments using various levels of inoculum would be required to determine whether we had substantially altered the host–pathogen interaction.

The most down-regulated CHID transformant identified in mycelia experiments proved unstable in the fruiting experiments, and will need retesting under more controlled conditions to maintain the transgenes. Another CHID transformant demonstrated approximately 50% reduced expression in the lesion tissue and there was some qualitative evidence that lesions in this transformant differed from the control host strain. In a third CHID transformant, expression of chitin deacetylase was more than double that of the control, and was consistent with the observations made with the mycelial Q-PCR.

With the DAHP transformants, four of the five tested showed reduced target gene expression compared to the control. In two of these transformants, there was no difference in the expression of the target gene in either non-infected or lesion tissues. In the fifth transformant, which was enhanced in mycelium Q-PCR, there was no significant difference with the control.

8. FUTURE WORK

The interaction between *A. bisporus* and *V. fungicola* is arguably the most economically significant interaction between two fungi. Our work has addressed diverse aspects of the biology and molecular interactions between the host and pathogen and provides a robust platform for future studies into the pathogenic system between two members of the same kingdom.

We have established advanced tools and genetic resources to permit molecular dissection and functional characterisation of this intriguing fungal–fungal interaction. The ability to suppress or silence specific response genes in the host or pathogen, coupled with microarray methodologies, presents exciting opportunities for the characterisation of specific genes and infection response mechanisms.

ACKNOWLEDGEMENTS

The authors would like to acknowledge the Department for Environment, Food and Rural Affairs for funding; Richard Amey for his contribution to the work on transformation of *Verticillium* and fungal systematics; Meena Muthumeenakshi for her contribution to global variation of *Verticillium* isolates and Terry Fermor and Steve Lincoln for his work on *Verticillium* enzyme production.

REFERENCES

Amey, R.C., Athey-Pollard, A., Burns, C., Mills, P.R., Bailey, A. and Foster, G.D. (2002). PEG-mediated and *Agrobacterium*-mediated transformation in the mycopathogen *Verticillium fungicola. Mycological Research,* **106**, 4–11.

Amey, R.C., Mills, P.R., Bailey, A. and Foster, G.D. (2003). Investigating the role of a *Verticillium fungicola* beta-1,6-glucanase during infection of *Agaricus bisporus* using targeted gene disruption. *Fungal Genetics and Biology,* **39**, 264–275.

Bartnicki-Garcia, S. (1968). Cell wall chemistry, morphogenesis and taxonomy of fungi. *Annual Review of Bacteriology,* **22**, 87–108.

Burns, A.C. (2004). Transformation and transgene expression in the homobasidiomycete fungi *Agaricus bisporus* and *Coprinus cinereus.* PhD Thesis, Biological Sciences, Bristol, 297 pp.

Burns, C., Gregory, K.E., Kirby, M., Cheung, M.K., Riquelme, M., Elliott, T.J., Challen, M.P., Bailey, A. and Foster, G.D. (2005). Efficient GFP expression in the mushroom *Agaricus bisporus* and *Coprinus cinereus* requires introns. *Fungal Genetic and Biology,* **42**, 191–199.

Burns, C., Leach, K.M., Elliott, T.J., Challen, M.P., Foster, G.D. and Bailey, A. (2006). Evaluation of *Agrobacterium*-mediated transformation of *Agaricus bisporus* using a range of promoters linked to hygromycin resistance. *Molecular Biotechnology,* **32**, 129–138.

Calonje, M., Gargia Mendoza, C., Galan, B. and Novaies-Ledieu, M. (1997). Enzymic activity of the mycoparasite *Verticillium fungicola* on *Agaricus bisporus. Mycological Research,* **101**, 1341–1347.

Calonje, M., Garcia Mendoza, C., Perez Cabo, A., Bernardo, D. and Novaes-Ledieu, M. (2000). Interaction between the mycoparasite *Verticillium fungicola* and the vegetative mycelial phase of *Agaricus bisporus. Mycological Research,* **104**, 988–992.

Calonje, M., Mendosa, C.G., Cabo, A.P. and Novaes-Ledieu, M. (1995). Some significant differences in wall chemistry among four commercial *Agaricus bisporus* strains. *Current Microbiology,* **30**, 111–115.

Challen, M., Gregory, K.E., Sreenivasaprasad, S., Rogers, C.C., Cutler, S.B., Diaper, D.C., Elliott, T.J. and Foster, G.D. (2000). Transformation technologies for mushrooms. *Mushroom Science,* **15**, 165–172.

Chen, X., Stone, M., Schlagnhaufer, C. and Romaine, C.P. (2000). A fruiting body tissue method for efficient Agrobacterium-mediated transformation of *Agaricus bisporus. Applied and Environmental Microbiology,* **66**, 4510–4513.

Collopy, P.D., Largeteau-Mamoun, M.L., Romaine, C.P. and Royse, D.J. (2001). Molecular phylogenetic analyses of *Verticillium fungicola* and related species causing dry bubble disease of the cultivated button mushroom, *Agaricus bisporus. Phytopathology,* **91**, 905–912.

Covert, S.F., Kapoor, P., Lee, M.-H., Briley, A. and Nairn, C.J. (2001). *Agrobacterium tumefaciens*-mediated transformation of *Fusarium circinatum*. *Mycological Research*, **105**, 259–264.
De Groot, M.J.A., Bundock, P., Hooykaas, P.J.J. and Beijersbergen, A.G.M. (1998). *Agrobacterium* tumefaciens-mediated transformation of filamentous fungi. *Nature Biotechnology*, **16**, 839–842.
Dragt, J.W., Geels, F.P., de Bruijin, W.C. and van Griensven, L.J.L.D. (1996). Intracellular infection of the cultivated mushroom *Agaricus bisporus* by the mycoparisite *Verticillium fungicola* var. *fungicola*. *Mycological Research*, **100**, 1082–1086.
Dragt, J.W., Geels, F.P., Rutjens, A.J. and Van Griensven, L.J.L.D. (1995). Resistance in wild types of *Agaricus bisporus* to the mycoparasite *Verticillium fungicola* var. *fungicola*. *Mushroom Science*, **14**, 670–683.
Eastwood, D.C., Jenkins, H., Challen, M.P. and Burton, K.S. (2006). Reducing argininosuccinate lyase transcript levels by RNAi in *Agaricus bisporus*. In: *8th European Congress on Fungal Genetics*, Vienna, Austria, April 8–11, 2006. Abstracts, IIIp-33.
Fletcher, J.T., White, P.F. and Gaze, R.H. (1994). *Mushrooms: Pest and Disease Control*, 2nd edn. Intecept, Andover, UK.
Foster, G.D., Amey, R.C., Bailey, A. and Mills, P.R. (2004a). Investigation of the interaction between *Verticillium fungicola* and *Agaricus bisporus*. *Mushroom Science*, **16**, 479–487.
Foster, G.D., Burns, A.C., Bailey, A.M., Challen, M., Elliott, T. and Burton, K. (2004b). Tools for *Agaricus* and *Coprinus* transformation and analysis of gene expression. *Mushroom Science*, **16**, 59–65.
Gewiss-Mogensen, E., Challen, M.P. and Strange, R.N. (2006). Reduction in solanapyrone phytotoxin production by *Ascochyta rabiei* transformed with *Agrobacterium tumefaciens*. *FEMS Microbiology Letters*, **255**, 255–261.
Gouka, R.J., Gerk, C., Hooykaas, P.J.J., Bundock, P., Musters, W., Verrips, C.T. and De Groot, M.J.A. (1999). Transformation of *Aspergillus awamori* by *Agrobacterium tumefaciens*-mediated homologous recombination. *Nature Biotechnology*, **17**, 598–601.
Grogan, H.M. and Gaze, R.H. (2000). Fungicide resistance among *Cladobotryium* spp.-causal agent of cobweb disease of the edible mushroom *Agaricus bisporus*. *Mycological Research*, **104**, 357–364.
Grogan, H.M., Keeling, C. and Jukes, A.A. (2000). In vivo response of the mushroom pathogen *Verticillium fungicola* (dry bubble) to prochloraz-manganese. In: *Proceedings of the Brighton Crop Protection Conference: Pests and Diseases, BCPC*, Farnham, Surrey, pp. 273–278.
Haran, S., Schickler, H., Oppenheim, A. and Chet, I. (1996). Differential expression of *Trichoderma harzianum* chitinases during mycoparasitism. *Phytopathology*, **86**, 980–985.
Hegnauer, H., Nyhlen, L.E. and Rast, D.M. (1985). Utrastructure of native and synthetic *Agaricus bisporus* melanins — implications as to the compartmentation of melanogensis in fungi. *Experimental Mycology*, **9**, 221–229.
Heneghan, M.N., Costa, A.M.S.B., Challen, M.P., Mills, P.R., Bailey, A. and Foster, G.D. (In press). A comparison of methods for successful triggering of gene silencing in *Coprinus cinereus*. *Molecular Biotechnology*.
Hodge, A., Alexander, I.J. and Gooday, G.W. (1995). Chitinolytic enzymes of pathogenic and ectomycorrhizal fungi. *Mycological Research*, **99**, 935–941.
Jeffries, P. and Young, T.W.K. (1994). Interfungal parasitic relationships. *CAB International*, **8**, 172–175.
Jones, J.D., Henstrand, J.M., Handa, A.K., Herrmann, K.M. and Weller, S.C. (1995). Impaired wound induction of 3-deoxy-D-arabino-heptulosonate-7-phosphate (DAHP) synthase and altered stem development in transgenic potato plants expressing a DAHP synthase antisense construct. *Plant Physiology*, **108**, 1413–1421.
Kalberer, P. (1984). Some properties of extra-cellular proteolytic enzyme of *Verticillium fungicola*, a pathogen of the cultivated mushroom *Agaricus bisporus*. *Phytopathology Z*, **110**, 213–220.
Kerrigan, R.W. (1991). What on earth is the *Agaricus* recovery program? *The Mycologist*, **5**, 22.
Knogge, W. (1996). Fungal infection of plants. *The Plant Cell*, **8**, 1711–1722.
Largeteau, M.L., Baars, J.P.P, Regnault-Roger C. and Savoie J.-M. (2005). Wild strains of *Agaricus bisporus*: a source of tolerance to dry bubble disease. Abstracts — *Congress Genetics and Cellular Biology of Basidiomycetes VI*, Pamplona, Spain, 3–6 June 2005.

Leach, K. Odon, V., Zhang, C., Kim, H.K., Henderson, J., Warner, P. Challen, M. and Elliott, T. (2004). Progress in *Agaricus bisporus* transformation: *Agrobacterium* methodologies and development of novel marker genes. *Mushroom Science*, **16**, 93–102.

Matthews, C.T. (1983). The pathology and ecology of *Verticillium fungicola* (Preuss) Hassebrauk. PhD Thesis, University of Bath, UK.

Mikosch, T.S.P., Lavrijssen, B., Sonnenberg, A.S.M. and van Griensven, L. (2001). Transformation of the cultivated mushroom *Agaricus bisporus* (Lange) using T-DNA from *Agrobacterium tumefaciens*. *Current Genetics*, **39**, 35–39.

Mol, P.C. and Wessels, J.G.H. (1990). Differences in wall structure between substrate hyphae and hyphae of fruit-body stipes in *Agaricus bisporus*. *Mycological Research*, **94**, 472–479.

Mullins, E.D., Chen, X., Romaine, P., Raina, R., Geiser, D.M. and Kang, S. (2001). *Agrobacterium*-mediated transformation of *Fusarium oxysporum*: an efficient tool for insertional mutagenesis and gene transfer. *Phytopathology*, **91**, 173–180.

Mullins, E.D. and Kang, S. (2001). Transformation: a tool for studying fungal pathogens of plants. *Cellular Molecular Life Science*, **58**, 2043–2052.

North, L.H. and Wuest, P.J. (1993). The infection process and symptom expression of *Verticillium* disease of *Agaricus bisporus*. *Canadian Journal of Pathlogy*, **15**, 74–80.

Rogers, C.W., Challen, M.P., Green, J.R. and Whipps, J.M. (2004). Use of REMI and *Agrobacterium* mediated transformation to identify pathogenicity mutants of the biocontrol fungus, *Coniothyrium minitans*. *FEMS Microbiology Letters*, **241**, 207–214.

St. Leger, R.J., Joshi, L. and Roberts, D.W. (1997). Adaption of proteases and carbohydrates of saprophytic, phytopathogenic and entomopathogenic fungi to the requirements of their ecological niches. *Microbiology*, **143**, 1983–1992.

Umar, M.H. and Van Griensven, L.J.L.D. (1997). Morphological studies on the life span, developmental stages, senescence and death of fruitbodies of *Agaricus bisporus*. *Mycological Research*, **101**, 1409–1422.

Ware, W.M. (1933). A disease of cultivated mushrooms caused by *Verticillium malhousei* sp. nov. *Annals of Botany*, **47**, 763–785.

Williams, J., Clarkson, J.M., Mills, P.R. and Cooper, R.M. (2003). Saprotrophic and mycoparasitic components of aggressiveness of *Trichoderma harzianum* groups toward the commercial mushroom *Agaricus bisporus*. *Applied and Environmental Microbiology*, **69**, 4192–4199.

Zare, R. and Gams, W. (2001). A revision of *Verticillium* section Prostrata. VI. The genera *Lecanicillium* and *Simplicillium* gen. nov. *Nova Hedwigia*, **73**, 1–50.

Zeilinger, S., Galhaup, C., Payer, K., Woo, S.L., Mach, R.L., Fekete, C., Lorito, M. and Kubicek, C.P. (1999). Chitinase gene expression during mycoparasitic interactions of *Trichoderma hazianum* with its hosts. *Fungal Genetics and Biology*, **26**, 131–140.

CHAPTER 2

Environmental Fluxes and Fungal Interactions: Maintaining a Competitive Edge

Naresh Magan and **David Aldred**

Contents

1. Introduction 19
2. Environmental Factors and Competitiveness based on Mycelial Interactions 21
3. Niche Overlap, Environmental Factors and Competitiveness 24
4. Role of Secondary Metabolites in Maintaining a Competitive Edge 25
5. Conclusions 33

References 33

Abstract

In all ecosystems, fungi inevitably come into contact with each other as they germinate and grow to become established in specific niches. They use different ecological strategies and have different physiological attributes, which enable them to compete against other species in different nutritionally diverse niches. The outcome of such interactions and thus community structure is dependent on factors such as carbon utilisation patterns and fluctuating environmental factors. The ability to grow over a wider water-availability range than bacteria or actinomycetes also gives fungi an advantage in dominating certain ecological domains. The ability to produce secondary metabolites, volatiles or a battery of enzymes provides a competitive edge to a specific fungal species to become established and maintain a competitive edge. This chapter examines the way by which fungi are able to maximise their abilities to tolerate environmental stress to enable them to compete effectively and exclude other species during colonisation of different ecosystems.

1. INTRODUCTION

Fungi are ubiquitous in nature and have evolved over time to colonise a wide range of ecosystems. The ability of fungal species to colonise and occupy specific

British Mycological Society Symposia Series
Published by Elsevier Ltd.

ecological niches is determined by their capacity to effectively compete against other microorganisms that make up the community. They compete directly with each other for resources by releasing extracellular enzymes into their immediate environment. Their activity is further impacted by prevailing abiotic factors such as temperature, water availability, gas balance and pH (Magan, 1997). Stress factors may be long lived or transient resulting in a dynamic fungal community structure, which is often in a state of flux. To overcome such changes fungi use different primary strategies to survive and prosper in different ecological niches. They can use combative (C-selected) strategies, e.g., *Trichoderma* species, Xylariales and basidiomycetes, which maximise occupation and exploitation of food matrices in relatively unstressed and undisturbed conditions and where recalcitrant compounds such as lignin need to be degraded; stress (S-selected) strategies which allow survival and endurance of continuous environmental stress, e.g., *Saccharomyces rouxi, Eurotium, Penicillium* and *Aspergillus* species; and ruderal (R-selected) strategies, e.g., *Rhizopus, Mucor* species, characterised by species with a high reproductive potential and short life span which facilitates successful exploitation in severely disturbed but nutrient-rich conditions. These three can merge to result in secondary strategies (C-R, S-R, C-S, C-S-R), which form part of a continuum with some transition between them. The main attributes of these three primary groupings are summarised in Figure 1. If one considers a specific ecological niche, e.g., the maize ecosystem, Marin *et al.* (2004) suggested that different strategies are used by spoilage fungi depending on the moisture content (m.c.) of the maize. For example, some species, which have adapted to xerophilic conditions, e.g., *Eurotium* species, use specific strategies to colonise partially

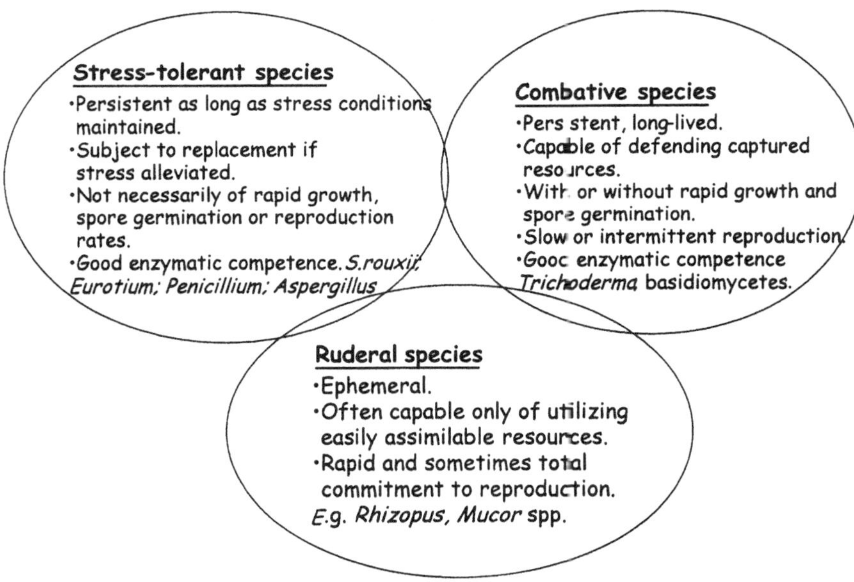

Figure 1 The main types of ecological characteristics and competencies based on ecological Strategies.

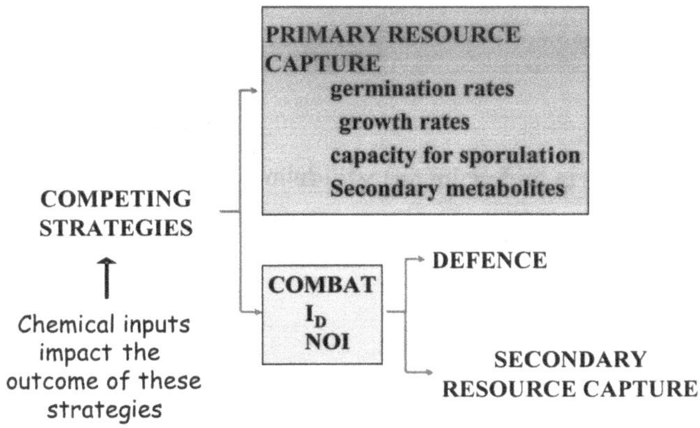

Figure 2 Schematic of the types of interactions which can occur in relation to primary and secondary resource capture.

dried maize (13–14% m.c.). In contrast, wet drying maize (>18% m.c.) is colonised by a range of *Aspergillus, Penicillium* species and some mycotoxigenic species. They are involved in primary resource capture, combat and defence of the niche resulting in secondary resource capture. Primary resource capture involves utilization of readily available carbon sources, especially water-soluble sugars. This promotes access and utilization via secondary resource capture of less readily available resources such as cellulose, hemicelluloses and occupation of territory. Combat involves defence of the colonised territory by an already established occupant. The combative species are able to keep captured resources by storage of lipid bodies, glycogen or compatible solutes with or without fast growth and spore germination, slow or periodic reproduction, enzymatic competence and the production of mycotoxins (Cooke and Whipps, 1993; Magan, 1997). A possible key component of fungi that uses S- and C-selected strategies to occupy ecological niches is the production of secondary metabolites. Of course, the impact of fluctuating environmental factors, xenobiotics and other abiotic and biotic stresses will all have an impact on, and influence the outcome of such interactions (Figure 2).

2. ENVIRONMENTAL FACTORS AND COMPETITIVENESS BASED ON MYCELIAL INTERACTIONS

Most vegetation is colonized by a mixed consortium of bacteria, yeasts and filamentous fungi. It is thus inevitable that interspecific and intraspecific interactions will occur depending on the microbial diversity and the nutritional status of the resource. Furthermore, environmental factors may exert a selective pressure influencing community structure and dominance of individual species. It is important to understand the type of interactions that occur between fungi under different environmental regimes to enable better prediction of not just dominance by a fungal species, but also the potential role of secondary metabolites, volatiles

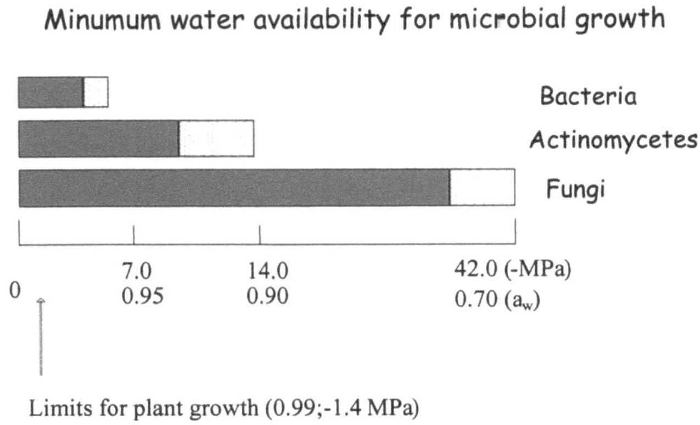

Figure 3 Comparison of the relative ranges and limits of water availabilities over which different microorganisms can grow.

and other extralites. One key factor is the prevailing water availability measured as water activity (a_w) or water potential (Ψ). These are accurate measures of actual water available for microbial growth. There are different ranges of water availability for bacteria, actinomycetes and fungi. This is important as it dictates, to a large extent, the type of interactions that can occur in microbial communities (Figure 3). It is important to note that fungal activity occurs over much wider water availability range than that for other microorganisms. This also means that fungal interactions can occur far below the wilting point of plants (−1.4 MPa Ψ). The ability to tolerate such stresses, especially when interacting with temperature and other abiotic factors can result in some fungal species being at a competitive advantage.

Many studies have used *in vitro* dual culture assays to examine interactions between fungi, especially when screening for potential biocontrol agents of fungal and bacterial plant diseases and in relation to decay of plant material, e.g., yeasts, *Trichoderma* species (control of *Botrytis*, range of foliar and soil-borne pathogens), *Ulocladium atra* (*Botrytis* in a range of crops), *Epicoccum nigrum* (foliar pathogens) and basidiomycetes such as *Phlebiopsis gigantea* (*Heterobasidion* control), and interactions between wood decay fungi (Boddy, 1993; Rayner, 1996). Wicklow (1980) used *in vitro* interactions between hyphae of different fungi based on: (a) intermingling of hyphae; (b) inhibition on contact; (c) inhibition at a distance; (d) dominance by one species over another on contact; and (e) inhibition by one species at a distance with the dominant species continuing to grow. They used these categories to develop an Index of Antagonism by giving numerical scores to each interaction type. This enabled antagonistic interactions between *Aspergillus flavus* and a range of species including *Trichoderma viride*, *Penicillium funiculosum* and *P. oxalicum* to be identified. It was found that other species such as *A. niger*, *Alternaria alternata*, *Cladosporium* species and *Candida guilliermondi* were not very antagonsitic to *A. flavus*. However, these studies did not examine the impact or dynamics of interacting environmental factors and influence on dominance of a species.

Subsequent studies by Magan and Lacey (1984, 1985) modified this scoring system to give a higher numerical score to fungi able to dominate *in vitro* in dual culture than those dominating by antagonism and developed an Index of Dominance (I_D) to assist with interpreting patterns of colonisation and dominance in the stored grain ecosystem (Magan *et al.*, 2003). The I_D was found to significantly change with water availability (water activity, a_w), temperature and also nutritional status. Of 15 species, the most competitive species in wheat grain in the UK were found to be *P. brevicompactum, P. hordei, P. roqueforti, A. fumigatus* and *A. nidulans*. Decreasing the a_w, conditions increased competitiveness of *P. brevicompactum*. Only *Fusarium culmorum* could compete with storage moulds, at >0.93–0.95 a_w. One important finding was that actual growth rate was not directly related to dominance. Previously, Ayerst (1969) had suggested that speed of germination and growth were key determinants of colonisation of nutrient rich matrices such as grain. The I_D approach has been adapted over the years for many studies of food-based ecosystems (Marin *et al.*, 1998; Lee and Magan, 1999; Magan *et al.*, 2003). The impact of interacting conditions of $a_w \times$ temperature on interactions and total I_D of *F. culmorum* against different species is shown in Table 1. This shows that the type of interspecific interactions, which can occur

Table 1 Interactions and index of dominance (I_D) scores for *Fusarium culmorum* in dual culture with other grain fungi on 2% wheat agar at three a_w levels and two temperatures. Key to fungi: F, *Fusarium*; A, *Alternaria*; C, *Cladosporium*; M, *Microdochium*; P, *Penicillium*

Temperature		15°C		I_D
Water activity (a_w)	0.995	0.98	0.95	
a_w+Species				
F. Graminearum	0/4	0/4	2/2	2/10
F. Poae	2/2	1/1	2/2	5/5
A. Tenuissima	0/4	0/4	2/2	2/10
C. Herbarum	2/2	4/0	2/2	8/4
M. nivale	0/4	1/1	3/3	4/8
M. majus	0/4	0/4	2/2	2/10
P. verrucosum	2/2	4/0	2/2	8/4
Total I_D	6/12	10/16	15/15	31/53
		25°C		
F. Graminearum	1/1	0/4	2/2	3/7
F. Poae	2/2	2/2	2/2	6/6
A. Tenuissima	4/0	4/0	4/0	12/0
C. Herbarum	4/0	4/0	4/0	16/0
M. Nivale	2/2	4/0	4/0	12/2
M. Majus	4/0	4/0	4/0	16/0
P. Verrucosum	4/0	4/0	2/2	10/2
Total I_D	21/5	22/6	22/6	65/17

Source: Hope (2004).

varies with water availability and temperature. In some cases when either is changed dominance can become modified to mutual antagonism or mutual intermingling. Furthermore, the nutritional makeup of the substrate may further impact the competitive ability of a species.

Recent studies on environmental impacts of competition between the wood-decay pathogen *Heterobasidion annosum* and the biocontrol species *Plebiopsis gigantea* have shown that under wet conditions the biocontrol agent is very effective at outcompeting the pathogen. However, when Ψ and temperature were decreased (drier conditions), then the pathogen was able to regain an ecological advantage by either being mutually antagonistic to the biocontrol agent or dominating the *Phlebiopsis* species (Swanwick and Magan, unpublished data).

Recently, studies on the competing abilities of fungal plant pathogens have been considered *in vitro* and *in situ* in fruit (*P. digitatum*, *P. expansum* and *Geotrichum candidum*; Plaza *et al.*, 2004). Growth rates of a single pathogen were influenced, usually inhibited, by changing temperature in the range 10–30°C and 0.995 a_w. However, when a_w was reduced to 0.95–0.90 a_w (=95% equilibrium relative humidity (ERH)), there was less impact of interactions on the relative growth rates of the pathogens. In citrus fruit lesions, an increase in increment was larger when *P. digitatum* was inoculated in separate wounds from *P. italicum*. When co-inoculated in the same wound, interactions and competition resulted in a significant effect on the rate of lesion expansion. It appears that under drier conditions, those fungi that can tolerate or survive desiccation effectively may be more competitive and thus able to displace less-adapted species (Atlas and Bartha, 1993). In agar-based studies, overall *P. italicum* was more competitive than the others examined. However, *in situ* in fruit, *P. digitatum* expanded much faster than *P. italicum* over the range of temperatures tested, resulting in an advantage in rapid establishment and primary access to the resources, in this case citrus fruit tissue. Of course other factors may also influence the dominance of specific species such as volatile production patterns, ability to produce enzymes and tolerance of essential oils (Eckert and Ratanayake, 1994). Sporulation effects could also be important as could secondary metabolites or transient antibiotic compounds.

3. NICHE OVERLAP, ENVIRONMENTAL FACTORS AND COMPETITIVENESS

Alternative approaches have been examined to try and understand the relative competitiveness of different species within fungal communities colonising different resources. It was suggested by Wilson and Lindow (1994a, 1994b) that the co-existence of microorganisms, particularly on plant surfaces, may be mediated by nutritional-resource partitioning. They suggested that *in vitro* carbon utilisation patterns could be used to determine niche overlap indices (NOI) and thus the level of ecological similarity. Based on the range of similar C-sources utilised and those unique to an individual isolate/species, they suggested that NOI values of >0.9 were indicative of co-existence between species in an ecological niche, while scores of <0.9 represented occupation of separate niches. This approach was modified by Marin *et al.* (1998), Lee and Magan (1999) and Magan *et al.* (2003) to include a multifactorial approach by including water availability, temperature and range of C-sources into the system. This approach demonstrated that, based on utilization of maize C-sources, the NOIs for mycotoxigenic moulds such as *F. verticillioides* and *F. proliferatum* were >0.90 at >0.96 a_w at 25 and 30°C, indicative of co-existence with *Penicillium* spp., *A. flavus* and *A. ochraceus*. However, for some species, pairing with *F. verticillioides* resulted in NOI values <0.80 indicating occupation of different niches. Interestingly, in these studies no direct correlation could be found between I_D and NOI methods (Marin *et al.*, 2004). The results suggested that niche overlap was in a state of flux and significantly influenced by both temperature and water availability. The nutrient status is very important. Lee and Magan (1999) demonstrated that comparison of C-sources in a standard BIOLOG test plate [95 C-sources] with those only relevant to maize grain [18 C-sources] gave very different results in terms of niche size and NOI under different environmental conditions. This approach confirms that interactions and dominance are dynamic, not static, and emphasizes the importance of taking account of such fluxes in any integrated approach to understanding the activity of fungi in an ecosystem.

Recent studies have tried to also combine the efficacy of environmental factors with the presence of chemical control agents, e.g., fungistatic or fungicidal compounds (Arroyo, 2003; Cairns *et al.*, 2003). It is known that food grade preservatives based on aliphatic acids (e.g., calcium propionate, potassium sorbate) are fungistats and that sub-optimal concentrations can lead to a stimulation of growth and also sometimes secondary metabolites, in this case the mycotoxin ochratoxin (Arroyo *et al.*, 2005). Figure 4 shows that when pH or temperature is modified, the C-source utilisaton patterns changes for a range of food spoilage moulds. Figure 5 shows that for one specific mycotoxigenic mould, *P. verrucosum*, the effect of adding the preservatives calcium propionate and potassium sorbate at 300 ppm further affects the C-source utilization patterns based on those present in bread. An example of the effect of environmental factors on niche size and NOI between *P. verrsucosum* and other spoilage moulds is shown in Table 2. This shows that using BIOLOG C-sources the effects on total C-source utilization changes with a_w and temperature. Furthermore, as conditions change niche exclusion can occur resulting in species occupying different niches in the same matrix. This again exemplifies the fact that interactions change as environmental and other chemical inputs into an ecosystem occur and that this will impact fungal community structure and diversity.

4. ROLE OF SECONDARY METABOLITES IN MAINTAINING A COMPETITIVE EDGE

Interactions between fungi, occupation and dominance in a specific niche could all be influenced by whether the necessary battery of enzymes and secondary metabolites can be produced under the prevailing abiotic and biotic regime. The role of secondary metabolites can be important in some niches and must be

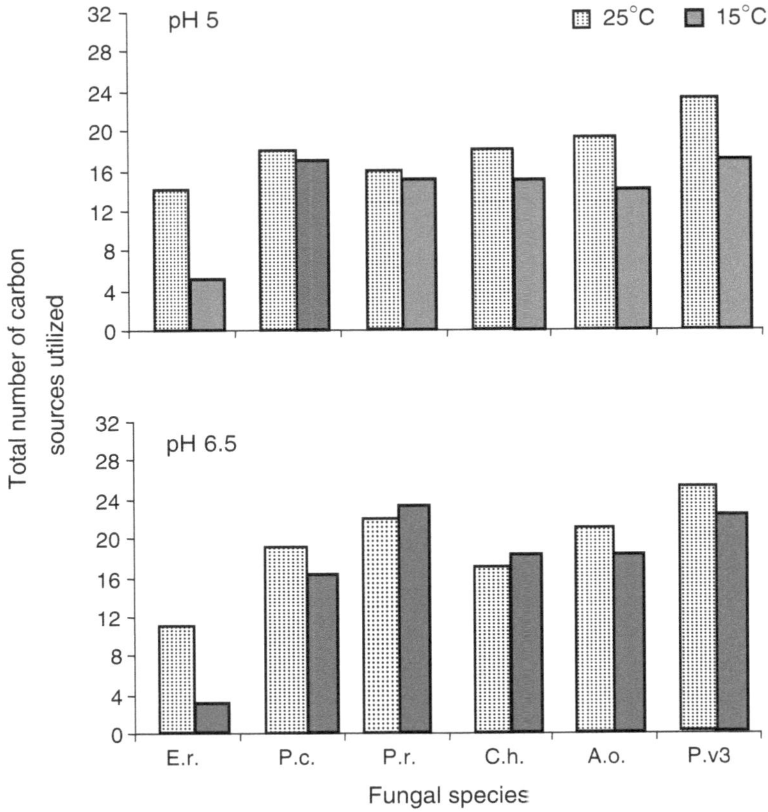

Figure 4 Example on the utilisation patterns of nutritional C-sources present in bread in relation to temperature and pH. Key to Fungal Species: E.r., *Eurotium rubrum*; P.c., *Penicillium chrysogenum*; P.r., *P. roqueforti*; C.h., *Cladosporium herbarum*; A.o., *Aspergillus ochraceus*; P. v3, *P. verrucosum*.

considered in a wider ecological context. Many studies on production of mycotoxins by spoilage fungi, especially in culture or in naturally contaminated food products have involved quantification at a single point in time. This often gives no idea of the temporal production rates and the conditions under which they were produced on often nutritionally heterogeneous food matrices. In nature, for fungi to compete effectively in a mixed microbial community they often have to be very competitive to become established. Thus, rapid early production of secondary metabolites, e.g., mycotoxins, could be one way in which they are able to spatially retain a competitive edge in a range of fluctuating conditions. Studies that have examined guttation or exudate production in penicillia suggest that these can contain significant amounts of metabolites and the significance of this in natural ecosystems needs to be determined because these metabolites could have ecological significance (Prof. M. Gareis, personal communication). Recent

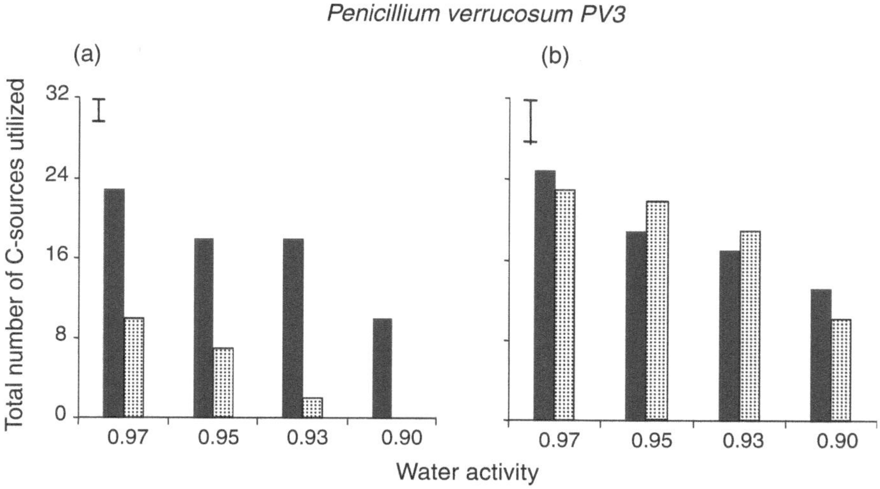

Figure 5 Effect of presence of (a) calcium propionate and (b) potassium sorbate (300 ppm) on total number of carbon sources utilised at different water activity (a_w) conditions by *Penicillium verrucosum.* bars indicate LSD. Solid bars, 15°C, speckled bars, 25°C.

Table 2 Example of the impact of environmental factors on niche size and niche overlap index (NOI) between *Penicillium verrucosum* and other spoilage fungi. Key to fungi: P, *Penicillium*; A, *Aspergillus*; F, *Fusarium*

Water activity (a_w)	0.995		0.930	
	Niche size	NOI_{Pv}/NOI_{Sp}	Niche size	NOI_{Pv}/NOI_{Sp}
15 °C				
P. verrucosum	92		74	
A. ochracheus	71	0.71/0.92	52	0.50/0.74
F. culmorum	90	0.88/0.92	63	0.58/0.74
P. aurantiogriseum	90	0.78/0.92	75	0.69/0.74
25 °C				
P. verrucosum	85		69	
A. ochraceus	88	0.84/0.85	63	0.55/0.69
F. culmorum	89	0.82/0.85	51	0.45/0.69
P. aurantiogriseum	74	0.84/0.85	64	0.59/0.60

Source: Magan *et al.* (2003).
Note: Niche size, total number of C-sources utilised by a species based on BIOLOG GN plate of 95 C-sources; NOI_{Pv}, total number of C-sources utilised in common divided by the total number utilised by *P. verucosum* only, under each set of conditions; NOI_{Sp}, total number of C-sources utilised in common, divided by the total utilised by the competitor, under each set of conditions.

studies by Puel *et al.* (2005) show that mycophenolic acid kinetics by *Byssochlamys nivea* was correlated with biomass with rapid early production in the first 10 days, increasing over 30 days before decreasing, probably due to nutrient exhaustion. These studies were conducted in defined liquid culture only. This is supported by studies on the polyketide synthase (*pks*) gene expression, which is related to ochratoxin A (OTA) production in *P. nordicum* (*Pn*) and *P. verrucosum* (*Pv*). The OTA polyketide synthase gene for *P. verrucosum* and *P. nordicum* (OTA*pksPv* and OTA*pksPn*) have been shown to be rapidly produced in the first few days with optimum production after about 9 days (Geisen, 2004). This does not support the view that secondary metabolites are produced solely in late exponential and stationary phase. They must be produced earlier for an ecological reason as part of an overall strategy to colonise, occupy and defend territory in ecological niches.

Since mycotoxigenic moulds do not occur in ecological niches alone, they are often part of a diverse community of microorganisms in which they need to compete effectively, and sometimes exclude other species from colonising territory by using the range of strategies detailed previously. One such strategy must include the rapid production of the necessary hydrolytic enzymes to gain access to the nutrients and then produce or activate synthesis of appropriate secondary metabolites or antibiotics to maintain a hold on territory. Of course, the ability to use some C-sources, which other species cannot use over a range of conditions may provide an advantage in producing secondary metabolites as part of maintaining a competitive edge over other species. Mycotoxigenic species such as *P. verrucosum* (ochratoxin, citrinin) and *Fusarium* section Liseola species (fumonisins) are able to compete effectively against other spoilage species such as *Alternaria* and *Cladosporium* species, partially by their ability to utilise carbon sources over a range of environmental conditions, and perhaps also by the production of secondary metabolites.

For example, *Fusarium* section Liseola species is considered to have endophytic phases of growth in maize tissue. Fumonisin FB_1 production may be an important component for becoming established and also retaining its niche. Growth of a range of species including *A. alternata*, *P. expansum*, *Botrytis cinerea* and *F. graminearum* can be inhibited by FB_1 at different concentrations, although generally fumonisin producers are not affected (Keyser *et al.*, 1999). However, field studies with co-inoculation of producers and non-producers of the fumonisin producing species, *F. verticillioides* and *F. graminearum*, in field infection of maize showed that interactions between these two species did not directly affect mycotoxin production or accumulation. Such studies suggest that FB_1 production may not be directly associated with a competition strategy.

Other studies however suggest that interactions between mycotoxigenic species and other mycoflora in an ecological niche is further affected by environmental factors resulting in a fluid situation where interactions can vary and dominant species may change. For example, fumonisin production by *F. verticillioides* and *F. proliferatum* was inhibited by the presence of some species (*A. flavus*, *A. ochraceus*, *P. implicatum*) regardless of water availability or temperature. However, interactions with *A. niger* resulted in a significant stimulation in FB_1 production on maize grain (Marin *et al.*, 1998). Furthermore, co-culture (intraspecific interactions) of both *Fusarium* species resulted in a stimulation of FB_1 production than when they were grown individually. This could be due to the production of synthesis precursors or to the fact that competition may encourage these spoilage fungi to produce mycotoxins at an earlier stage. It can be speculated that perhaps some mycotoxigenic species may produce quorum sensing (QS) like compounds that could influence the activity of other species competing for the same ecological niche.

Studies of different mycotoxin producing species have shown that when *Fusarium* section Liseola species were competing with *F. graminearum*, FB1 accumulation decreased at 15°C regardless of moisture content. However, at 25°C *F. verticillioides* produced higher amounts of FB1 in the presence of *F. graminearum* than when cultured alone. In contrast, *F. proliferatum* produced less than in the absence of the competitor. There was however, no effect on production of zearalenone by *F. graminearum*, although there was some stimulation under specific environmental conditions (15°C, 0.98 a_w). However, the production of the trichothecene deoxynivalenol (DON) was significantly stimulated at slightly reduced water availability (0.98 a_w; Velluti *et al.*, 2000a, 2000b). Table 2 summarises some studies where interaction with species has resulted in a stimulation or inhibition of secondary metabolites (mycotoxin) production by other competing species on grain matrices. Some of these are summarised below.

Studies have examined interactions between *A. ochraceus* and other spoilage fungi and found that the effect of interactions on growth and OTA production was more pronounced *in vitro* than *in situ* on maize grain (Lee and Magan 1999, 2000a, 2000b). Interaction of *Fusarium* section Lisoela species and *A. parasiticus* showed that there was no effect on FB1 production by the former species. However, the *Fusaria* were competitive and inhibited aflatoxin production over a range of environmental conditions (Marin *et al.*, 2001). Field studies of co-inoculation with these same species (*F. verticillioides* and *A. flavus*) also resulted in lower aflatoxin accumulation than when the aflatoxigenic species was used alone (Zummo and Scott, 1992).

Table 3 Effect of interactions between *F. culmorum* and other species on deoxynivalenol (DON) and nivalenol (NIV) (ng/g grain) production on irradiated wheat grain at two water activity (a_w) levels at 25°C. Key to fungi: F, *Fusarium*; A, *Alternaria*; M, *Microdochium*; *M. majus*, *M. nivale* var. *majus*; P, *Penicillium*

	Mycotoxin			
	DON		NIV	
a_w	0.995	0.955	0.995	0.955
F. culmorum	7669	447	289	298
F. culmorum + *C. herbarum*	634	0	316	412
F. culmorum + *M. nivale*	451	444	0	288
F. culmorum + *M. majus*	0	440	292	0
F. culmorum + *P. verrucosum*	3264	450	0	0

Source: Magan *et al.* (2003).
Note: LSD ($p = 0.05$: DON = 180; NIV = 123).

Recent studies have examined the interaction between *F. culmorum* and a range of other fungi, which colonise ripening ears and grain during harvesting, to examine their impact on DON and NIV (nivalenol) production. Table 3 shows that on wheat grain the presence of fungi such as *Cladosporium herbarum*, *Alternaria tenuissima*, *Microdochium nivale* and *M. majus* and the mycotoxigenic species *P. verrucosum* showed that in most cases the presence of other species inhibited DON production but in some cases increased NIV production. This suggests that wherever a range of mycotoxins is produced, there may be a switch from one synthetic pathway to another as part of the strategy to maintain a competitive edge (Magan *et al.*, 2002). Although DON is produced in higher concentrations than NIV, the latter is in fact more toxic. Similar results were obtained when some essential oils were examined to control growth and DON/NIV production by *F. culmorum* (Hope *et al.*, 2003). They found that some essential oils could inhibit growth and DON production under some environmental factors. However, there was a switch to NIV and sometimes other trichothecenes. It was thus suggested that in the stored grain ecosystem complex interactions occurred between germination, establishment and mycotoxin production and environmental factors which were influenced by the presence and activity of other species (Magan *et al.*, 2004). Table 4 summarises a range of studies, which have shown that interactions between species can result in decreases as well as stimulation of secondary metabolite production.

Recent innovative studies have shown that *Penicillium* species can produce QS inhibitors, which can affect the functioning of bacteria such as *Pseudomonas aeruginosa*. Microorganisms exist in mixed consortia and this elegant study showed that penicillic acid and patulin were two metabolites that were biologically active and significantly affected the QS-controlled gene expression of homoserine lactones by *P. aeruginosa* (Rasmussen *et al.*, 2005). This suggests that quite complex interactions might occur between microorganisms and that fungal secondary metabolites may have a much wider influence in mixed communities than thought of previously. Many gram-negative bacteria produce N-acetyl homoserine lactones (AHLs) which co-ordinate expression of virulence in response to surrounding bacterial populations as QS compounds. The potential interaction of fungal secondary metabolites with such QS AHLs needs to be investigated in more detail under different environmental stress conditions as Rasmussen *et al.* (2005) showed that 33 of 50 extracts from *Penicillium* species produced QS-inhibitor compounds.

Studies at Cranfield have examined the partitioning of OTA between spores, biomass and medium for *A. ochraceus* and *A. carbonarius* species from Europe. These studies showed that for *A. ochraceus*, 60% of the OTA was in the biomass, and about 20% in the spores and medium. In contrast, for many *A. carbonarius* isolates, under optimum temperature and water availability conditions (20°C, 0.98–0.95 a_w), at least 60% of OTA was in the spores, 30% in the mycelial biomass and only 20% in the medium (Mitchell, Aldred and Magan, unpublished data). This suggests that during germination, establishment could be aided by the presence of toxic secondary metabolites, which may give a competitive advantage in primary and secondary resource capture. It is possible

Table 4 Interactions between mycotoxigenic fungi and other species and outcomes on mycotoxin production

Species interactions	Effects on mycotoxins	References
F. verticillioides+*Trichoderma viride*	Reduced fumonisin B1 by 85%	Yates *et al.* (1999).
F. verticillioides+*A. niger, A. flavus, A. ochraceus, P. implicatum*	Inhibited fumonisin FB1, except at 0.98 a_w	Marin *et al.* (1998)
F. verticillioides+*F. proliferatum* (co-culture on maize)	Enhanced fumonisin B1	Marin *et al.* (1998)
F. verticillioides+*F. graminearum*	<Fumonisin at 15°C; >at 25°C >Deoxynivalenol (DON) at 0.98 a_w, no effect on zearalenone	Velluti *et al.* (2002a) Velluti *et al.* (2002b)
A. flavus+different fungi (maize grain)	Inhibition of aflatoxin depending on species	Cuero *et al.* (1988)
A. ochraceus+other species	Stimulation in some cases *in vitro;* inhibition of ochratoxin in maize	Lee and Magan (1999, 2000b)
A. flavus+*Hyphopichia burtonii*+*Bacillus amyloliquefaciens* (maize grain)	Stimulation of aflatoxin	Cuero *et al.* (1987)
Fusarium culmorum+grain fungi (wheat grain)	Predominantly inhibition of DON	Magan *et al.* (2003)
Penicillium verrucousm+*A. alternata, F. culmorum, E. repens, P. aurantiogriseum, A. ochraceus*	Some species inhibited while others stimulated ochratoxin A (OTA) on wheat grain	Cairns *et al.* (2003)

that abiotic or biotic stress impacts physiologically in a similar way and results in expression of genes or gene products for the rapid biosynthesis of such mycotoxins.

The impact of environmental conditions on induction of the OTA biosynthesis genes in *P. nordicum* was examined (Geisen, 2004). Real time PCR specific for OTA*pksPn*) for *P. nordicum* demonstrated the induction of the transcription factor to be correlated with biosynthesis of OTA. Thus the induction of the molecular signal can be used to support the production kinetics of mycotoxins in the environment. The kinetics using this approach supports the early synthesis of OTA, within 3–5 days. The effect of temperature, pH, and ionic solute concentration all showed a parallel expression of OTA*pksPn* gene and OTA production. Interestingly, for pH and ionic solute stress treatments used there was less relationship between expression and OTA production and temperature (15–30°C). It was speculated that temperature has less effect on OTA*pksPn* gene expression although it does have an impact on OTA production. The maximum amount of ionic stress imposed was 50 g/l, which only equates to about 0.975 a_w. *Penicillium* species, especially *P. verrucosum* grows and optimally produces OTA at about 0.95 a_w (Cairns-Fuller *et al.*, 2005). Thus an opportunity does exist to use this approach to confirm the role of mycotoxins in ecological competitiveness during colonisation of food matrices and their role in gaining territorial advantage.

It is interesting to note that sometimes fungi are able to break down mycotoxins if necessary and synthesise more when required. For example, *A. carbonarius* produces OTA over 5–15 days and then there is a decrease followed by another cycle of production. Abrunhosa *et al.* (2002) found that a range of fungi isolated from grapes could degrade OTA produced by *A. carbonarius*. Varga *et al.* (2002) showed that OTA can be degraded by *Aspergillus* species. Recently, Varga *et al.* (2005) have also demonstrated the capacity of *Rhizopus* species to degrade a range of mycotoxins. For example, they could degrade 95% of OTA in 16 days, even on moist wheat. This suggests that depending on the ecological conditions and the strategy for competitive exclusion of competitors, enzyme systems may be activated to enable such degradation pathways to be instituted.

Metabolomic production profiles for mycotoxins and other secondary metabolites may indeed be a method for such fungi to respond in a fluid way to changes in competition from other species alone or interactions with fluxes of changing abiotic factors. Recently, Aldred *et al.* (2005) showed that ecological niches from which fungi are isolated significantly influence metabolomic profiles and titres of individual metabolites. Production was very different for species from damp rainforest ecosystems and from the phyllosphere. Thus the production of mycotoxins and secondary metabolites in general must be seen in an ecological context.

Opportunities do now exist to use the modern genomic and metabolomic-based microarrays to examine in more detail the interactions zones between species and to compare this with species involved in primary resource capture to understand the relative up and down regulation of genes during interactions. The role of metabolites can be directly examined by using specific arrays that contain the key secondary metabolite gene clusters and determine the expression of specific metabolites during intra- and interspecific fungal interactions. The role of volatile production in the absence and presence of competitors may be another avenue that could provide more useful evidence to help understand the various physiological pathways involved during competition.

5. CONCLUSIONS

- Fungi use different ecological strategies for primary resource capture and subsequently secondary resource capture.
- Abiotic factors play a major role in influencing the competitive ability for dominating in an ecological niche.
- Transient changes in water availability, temperature, pH and xenobiotic compounds all impact on fungal interactions and determine dominance within fungal communities.
- Niche size and NOI approach has shown that interactions are in a state of flux and will vary with changing abiotic and biotic factors.
- Evidence suggests that mycotoxins produced by fungi (and other microorganisms) confer a competitive advantage to the producer. This role may not always be clear (especially in laboratory situations), and probably operates only within natural ecosystems.
- Under stress conditions the need for competitive advantage is increased and this is when mycotoxins probably become important.
- In natural systems the ecological milieu determines the community structure, and the predominant species may be partially dependent on stress tolerance and the ability to produce a battery of secondary metabolites. This array of metabolites could have different functions against other microorganisms, mites and nematodes.

REFERENCES

Abrunhosa, L., Serra, R. and Venancio, A. (2002). Biodegradation of ochratoxin A by fungi isolated from grapes. *Journal of Agriculture and Food Chemistry*, **50**, 7492–7496.

Aldred, D., Penn, J. and Magan, N. (2005). Water availability and metabolomic profiles of *Epicoccum nigrum* and *Sarophorum palmicola* grown in solid substrate fermentation systems. *The Mycologist*, **19**, 18–23.

Atlas, R.M. and Bartha, R. (1993). *Microbial Ecology. Fundamentals and Applications*. The Benjamin/Cummings Publishing Co., San Francisco, CA 94111, USA.

Arroyo, M. (2003). *Natural antifungal systems for prevention of mould spoilage in bakery products*. PhD Thesis, Bedford, UK.

Arroyo, M., Aldred, D. and Magan, N. (2005). Environmental factors and weak organic acid interactions have differential effects on control of growth and ochratoxin A production by *Penicillium verrucosum* isolates in bread. *International Journal of Food Microbiology*, **98**, 223–231.

Ayerst, G. (1969). The effects of moisture and temperature on growth and spore germination in some fungi. *Journal of Stored Products Research*, **5**, 127–141.

Boddy, L. (1993). Saprotrophic cord-forming fungi: Warfare strategies and other ecological aspects. *Mycological Research*, **94**, 641–655.
Cairns, V., Hope, R. and Magan, N. (2003). Environmental factors and competing mycoflora affect growth and ochratoxin production by *Penicillium verrucosum* on wheat grain. *Aspects of Applied Biology*, **68**, 81–90.
Cairns-Fuller, V., Aldred, D. and Magan, N. (2005). Water, temperature and gas composition interactions affect growth and ochratoxin A production by isolates of *Penicillium verrucosum* on wheat grain. *Journal of Applied Microbiology*, **99**, 1215–1221.
Cooke, R.C. and Whipps, J.M. (1993). Ecophysiology of Fungi. Blackwell, Oxford, U.K.
Cuero, R., Lacey, J. and Smith, J.E. (1988). Mycotoxin production by *Aspergillus flavus* and *Fusarium graminearum* in irradiated maize grains in the presence of other fungi. *Journal of Food Protection*, **51**, 452–456.
Eckert, J.W. and Ratanayake, M. (1994). Role of volatile compounds from wounded oranges in induction of germination of *Penicillium digitatum* conidia. *Phytopathology*, **84**, 746–750.
Geisen, R. (2004). Molecular monitoring of environmental conditions influencing the induction of ochratoxin A biosynthesis genes in *Penicillium nordicum*. *Molecular and Nutritional Food Research*, **48**, 532–540.
Hope, R. (2004). Ecology and control of *Fusarium* on cereals, PhD Thesis, Cranfield University, Bedford, U.K.
Hope, R., Jestoi, M. and Magan, N. (2003). Multi-target environmental approach for control of growth and toxin production by *Fusarium culmorum* using essential oils and antioxidants. In: *Advances in Stored Product Protection* (P. Credland, D.M. Armitage, C.H. Bell, P.M. Cogan, eds.), pp. 486–492. Cabi International, Wallingford, Oxon, U.K.
Keyser, Z., Vismer, H.F., Klaasen, J.A., Snijman, P.W. and Marasas, W.F.O. (1999). The antifungal effect of fuminisin FB1 on Fusarium and other fungal species. *South African Journal of Science*, **95**, 455–458.
Lee, H.B. and Magan, N. (1999). Environmental factors and nutritional utlization patterns affect niche overlap indices between *Aspergillus ochraceus* and other spoilage fungi. *Letters in Applied Microbiology*, **28**, 300–304.
Lee, H.B. and Magan, N. (2000a). Impact of environment and interspecific interactions between spoilage fungi and *Aspergillus ochraceus* and ochratoxin production on maize. *International Journal of Food Microbiology*, **61**, 11–16.
Lee, H.B. and Magan, N. (2000b). Environmental influences on *in vitro* interspecific interactions between *A. ochraceus* and other maize spoilage fungi on growth and ochratoxin production. *Mycopathologia*, **146**, 43–47.
Magan, N. (1997). Fungal in extreme environments. In: *MYCOTA Vol. IV: Environmental and Microbial Relationships* (D.T. Wicklow and B. Soderstrom, eds.), pp. 271–280. Springer Verlag, Berlin.
Magan, N. and Lacey, J. (1984). The effect of water activity, temperature and substrate on interactions between field and storage fungi. *Transactions of the British Mycological Society*, **82**, 83–93.
Magan, N., Hope, R., Cairns, V. & Aldred, D. (2003). Post-harvest fungal ecology: impact of fungal growth and mycotoxin accumulation in stored grain. *European Journal of Plant Pathology*, **109**, 723–730.
Magan, N., Hope, R., Colleate, A. and Baxter, E.S. (2002). Relationship between growth and mycotoxin production by *Fusarium* species, biocides and environment. *European Journal of Plant Pathology*, **108**, 685–690.
Magan, N. and Lacey, J. (1985) Interactions between field and storage fungi on wheat grain. *Transactions of the British Mycological Society*, **85**, 29–37.
Magan, N., Sanchis, V. and Aldred, D. (2004). Role of spoilage fungi in seed deterioration, Chapter 28. In: *Fungal Biotechnology in Agricultural, Food and Environmetal Applications* (D.K. Aurora, ed.), pp. 311–323. Marcell Dekker, New York.
Marin, S., Albareda, X., Ramos, A.J. and Sanchis, V. (2001). Impact of environment and interactions of *Fusarium verticillioides* and *F. proliferatum* with *Aspergillus parasiticus* on fumonisin B1 and aflatoxins on maize grain. *Journal of Science of Food and Agriculture*, **81**, 1060–1068.

Marin, S., Magan, N., Ramos, A.J. and Sanchis, V. (2004). Fumonisin-producing strains of *Fusarium*: A review of their ecophysiology. *Journal of Food Protection*, **67**, 1792–1805.

Marin, S., Sanchis, V., Ramos, A.G. and Magan, N. (1998). Environmental factors, interspecific interactions, and niche overlap between *Fusarium moniliforme* and *F. proliferatum* and *Fusarium graminearum, Aspergillus* and *Penicillium* spp. isolated from maize. *Mycological Research*, **102**, 831–837.

Plaza, P., Usall, J., Teixido, N. and Vinas, I. (2004). Effect of water activity and temperature on competing abilities of common postharvest citrus fungi. *International Journal of Food Microbiology*, **90**, 75–82.

Puel, O., Tadrist, S., Galtiere, P., Oswald, I.P. and Delaforge, M. (2005). Bysochlamys nivea as a source of mycophenolic acid. *Applied and Environmental Microbiology*, **71**, 550–553.

Rasmussen, T.B., Skindersoe, M.E., Bjarnsholt, T., Phipps, R.K., Christensen, K.B, Jensen, P.O., Andersen, J.B., Koch, B., Larsen, T.O., Hentzer, M., Eberl, L., Hoiby, N. and Givskov, M. (2005). Identity and effects of quorum sensing inhibitors produced by *Penicillium* species. *Microbiology*, **151**, 1325–1340.

Rayner, A.D.M. (1996). Has chaos theory a place in environmental mycology? In: *Fungi and Environmental Change* (J.C. Frankland, N. Magan and G.M. Gadd, eds.), pp. 317–341. Cambridge University Press, Cambridge.

Varga, J., Peteri, Z., Tabori, K., Teren, J. and Vagvolgyi, C. (2005). Degradation of ochratoxin A and other mycotoxins by *Rhizopus* species. *International Journal of Food Microbiology*, **99**, 321–328.

Varga, J., Rigo, K. and Teren, J. (2002). Degradation of ochratoxin A by Aspergillus species. *International Journal of Food Microbiology*, **59**, 1–7.

Velluti, A.S., Marin, S., Betucci, A., Ramos, A.J. and Sanchis, V. (2000a). The effect of fungal competition on colonisation of maize grain by *Fusarium moniliforme, F. proliferatum* and *F.graminearum* and on fumonisin B1 and zearalenone formation. *International Journal of Food Microbiology*, **59**, 59–66.

Velluti, A.S. Marin, S., Gonzalez, R., Ramos, A.J. and Sanchis, V. (2000b). Fumonisin B1, zearalenone and deoxynivalenol production by *Fusarium moniliforme, F.proliferatum* and *F.graminearum* in mixed cultures on irradiated maize kernels. *Journal of Science of Food and Agriculture*, **81**, 88–94.

Wicklow, D.T. (1980). Patterns of fungal associations within maize kernels in North Corolina. *Plant Disease*, **72**, 113–115.

Wilson, M. and Lindow, S. (1994a). Coexistence among epiphytic bacterial populations mediated through nutritional resource partitioning. *Applied and Environmental Microbiology*, **60**, 4468–4477.

Wilson, M. and Lindow, S. (1994b). Ecological similarity and co-existence of epiphytic ice-nucleating (Ice+) *Pseudomonas syringae* strains and a non-ice nucleating (Ice−) biological control agent. *Applied and Environmental Microbiology*, **60**, 3128–3137.

Yates, I.E., Meredith, F., Smart, W., Bacon, C.W. and Jaworski, A.J. (1999). *Trichoderma Viride* suppresses fumonisin production by *Fusarium moniliforme. Journal of Food Protection*, **68**(7), 1326–1332.

Zummo, N. and Scott, G.E. (1992). Interaction of *Fusarium moniliforme* and *Aspergillus flavus* on kernel infection and aflatoxin contamination in maize ears. *Plant Disease*, **76**, 771–773.

CHAPTER 3

Intracellular Mycoparasites in Action: Interactions Between Powdery Mildew Fungi and *Ampelomyces*

Levente Kiss

Contents

1. Introduction 38
2. Details of the Mycoparasitic Interaction 40
   2.1 Life Cycle 40
3. Mycoparasitism as a Biotic Stress for the Parasitized Fungi 45
   3.1 Direct Effects of Mycoparasitism: Impact on Growth and Sporulation of the Mycohosts 45
   3.2 Indirect Effects in a Tri-Trophic Relationship: Impact of Mycoparasitism on the Mildew-Infected Plants 47
4. Other Fungal Antagonists of Powdery Mildew Fungi 47
   4.1 Confusion of Ampelomyces with Other Pycnidial Mycoparasites of Powdery Mildews 48
5. Conclusions 48
Acknowledgement 49
References 49

Abstract

Pycnidial fungi of the genus *Ampelomyces* are widespread intracellular mycoparasites of powdery mildew fungi worldwide. Their pycnidia are produced in hyphae, conidiophores and immature ascomata of their mycohosts. Thus, they suppress both the asexual and the sexual reproduction of the invaded powdery mildew mycelia, and then destroy them completely. Conidia of *Ampelomyces* are released from the intracellular pycnidia by the rupture of the pycnidial wall; conidia then germinate on the host plant surfaces, penetrate the intact hyphae of powdery mildew mycelia found in their vicinity and invade them internally growing from cell to cell through the septal pores of the mycohost. The early stage of mycoparasitism is apparently biotrophic, but the invaded cytoplasm then begins to die and a necrotrophic interaction results. Toxin production has not been detected in *Ampelomyces*, so it might act directly by invasion and destruction of the

British Mycological Society Symposia Series
Published by Elsevier Ltd.

host cytoplasm. Experimental data showed that parasitized powdery mildew colonies can continue their growth, but their sporulation is stopped soon after *Ampelomyces* penetrated their mycelia. It is concluded that these mycoparasites represent a stress factor in the life cycle of their mycohosts but their role in the natural control of powdery mildew infections requires further investigations.

1. INTRODUCTION

Parasites, by definition, have a negative effect on host fitness because their activities represent a biotic stress for the host organisms (Jarosz and Davelos, 1995). Mycoparasites, that is, fungi that parasitize other fungi, are a diverse group of parasites which include fungi that absorb nutrients from their mycohosts through haustoria, or other special interfaces between their cell walls and membranes, or invade the mycelia of the mycohosts internally, growing from cell to cell in the hyphae, spores, fruiting bodies and other structures of their mycohosts, and absorbing nutrients from there (Jeffries and Young, 1994). The structural and ultra-structural aspects of the diverse types of mycoparasitism are relatively well known (Jeffries and Young, 1994; Jeffries, 1997); however, mycoparasites are still a relatively less studied group of parasites from an ecological point of view (Hirsch and Braun, 1992; Cooke and Whipps, 1993; Jeffries, 1995, 1997; Kiss, 2001). Our knowledge on the role of mycoparasites in natural inter-fungal relationships is still very limited (Hirsch and Braun, 1992; Jeffries, 1995; Kiss, 1998). In a broader ecological context, in multi-trophic relationships, little is known about the interactions among parasitized fungi, their mycoparasites and a variety of organisms other than fungi (Hirsch and Braun, 1992; Kiss, 2001). In short, little is known about both the amount and the significance of the biotic stress caused by mycoparasites in their mycohosts under natural conditions. This is interesting to note because the use of a number of mycoparasites in the biological control of various plant pathogenic fungi is largely based on their supposed importance in the natural control phenomena (Cooke and Whipps, 1993).

The interactions between powdery mildew fungi, obligate biotrophic parasites of many plants, and their *Ampelomyces* mycoparasites, are one of the most evident cases of mycoparasitic relationships in nature, because this relationship is common worldwide and takes place exclusively on aerial plant surfaces, thus facilitating its direct observation (Kiss, 1998, 2004). Sometimes the natural occurrence of *Ampelomyces* mycoparasites in powdery mildew colonies is visible in the field even to the naked eye (Figure 1). Historically, pycnidial fungi, belonging to the genus *Ampelomyces*, were among the first mycoparasites to be studied in detail (De Bary, 1870) and also among the first fungi used as potential biocontrol agents of economically important plant pathogens (Yarwood, 1932). This paper reviews their mode of action in the mycohost mycelia and also the experimental data on the damage (the amount of stress) caused by them in the powdery mildew colonies.

Figure 1 Powdery mildew-infected *Lycium halimifolium*, a common solanaceous plant in eastern Europe and elsewhere. The brownish patches in the white powdery mildew colonies are masses of intracellular pycnidia of *Ampelomyces*. This is a common and an evident case of a tri-trophic relationship in the phyllosphere. **(See Colour Section)**

2. DETAILS OF THE MYCOPARASITIC INTERACTION

2.1 Life Cycle

Pycnidia of *Ampelomyces* are commonly found in the cells of the hyphae, conidiophores (Figures 2–3B) and immature ascomata (Figure 3C) of powdery mildew fungi worldwide (Falk *et al.*, 1995a, 1995b; Kiss *et al.*, 2004). Conidia are produced in these intracellular pycnidia and are unicellular, hyaline, mostly guttulate and are embedded in a mucilaginous matrix inside the pycnidia. In the presence of water, these matrices swell and conidia are released from intracellular pycnidia as a cirrhus by the rupture of the pycnidial wall (Figure 3C). In approximately 10–20 h, under conditions of high humidity, conidia germinates and the emerging hyphae of the mycoparasites can then penetrate the hyphae and conidia (Figures 3D–4B) of powdery mildews in their vicinity. The concentration of *Ampelomyces* conidia on the leaves is an important factor as Gu and Ko (1997) showed experimentally that germination rapidly decreases above a concentration of 10^6 cfu ml^{-1} due to the production of a self inhibitor. After penetration, the hyphae of the mycoparasite continue their growth internally (Figure 4A–D) and produce their intracellular pycnidia after 5–8 days in the mycelia of their host fungi (Figures 2–3B and 4D(II)). A high relative humidity of the environment enhances the internal growth and sporulation of *Ampelomyces* (Jarvis and Slingsby, 1977; Philipp and Crüger, 1979).

Conidia can be dispersed within the plant canopy by rain-splash or water runoff from plant surfaces. *Ampelomyces* can also spread to long distances as

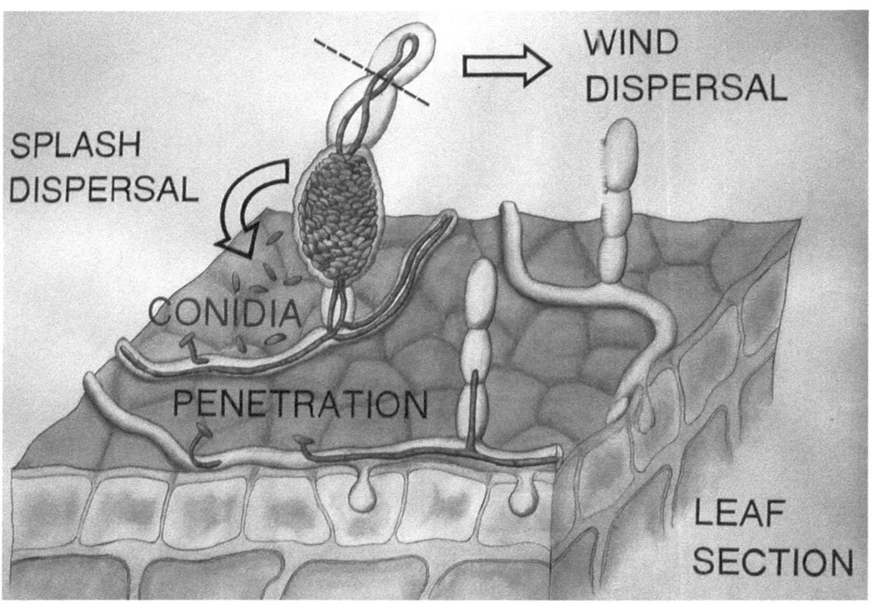

Figure 2 The life cycle of *Ampelomyces* mycoparasites in an asexually reproducing powdery mildew colony. The parasitism of immature ascomata is not represented in the figure. **(See Colour Section)**

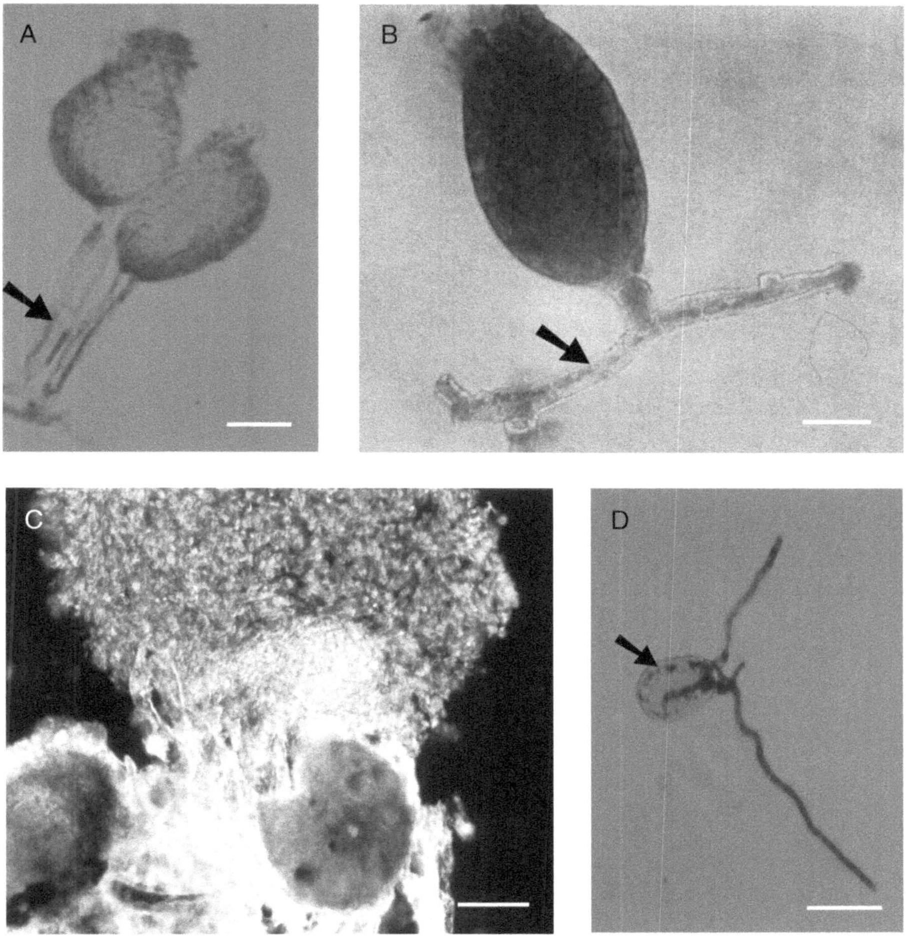

Figure 3 Aspects of the interactions between *Ampelomyces* mycoparasites and powdery mildew fungi. A: intracellular hyphae (arrow) and two pycnidia of *Ampelomyces*, slightly stained with cotton blue, and produced in the conidiophores of *Golovinomyces depressa* infecting *Arctium minus*. Bar equal 30 μm. B: a pycnidium of *ampelomyces* in the basal cell of a conidiophore of *G. orontii* infecting *Cucurbita pepo*. The arrow indicates an intracellular hypha. The *Ampelomyces* structures were stained with cotton blue. Bar equal 25 μm. C: conidia of *Ampelomyces* released from a parasitized ascoma of the apple powdery mildew fungus, *Podosphaera leucotricha*, after over-wintering in a bud on an apple tree. Bar equal 50 μm. D: intracellular hyphae of *Ampelomyces* (arrow), stained with cotton blue, emerging from a conidium of *E. syringae-japonicae* infecting *Ligustrum vulgare*, after placing powdery mildew conidia collected from the field on water agar for 24 h. Bar equal 20 μm. **(See Colour Section)**

hyphal fragments in parasitized and detached powdery mildew conidia (Figures 2 and 3D) (Jarvis and Slingsby, 1977; Speer, 1978; Sundheim, 1982). When these parasitized air-borne conidia land close to any powdery mildew colony under humid conditions, the outgrowing hyphae of *Ampelomyces*

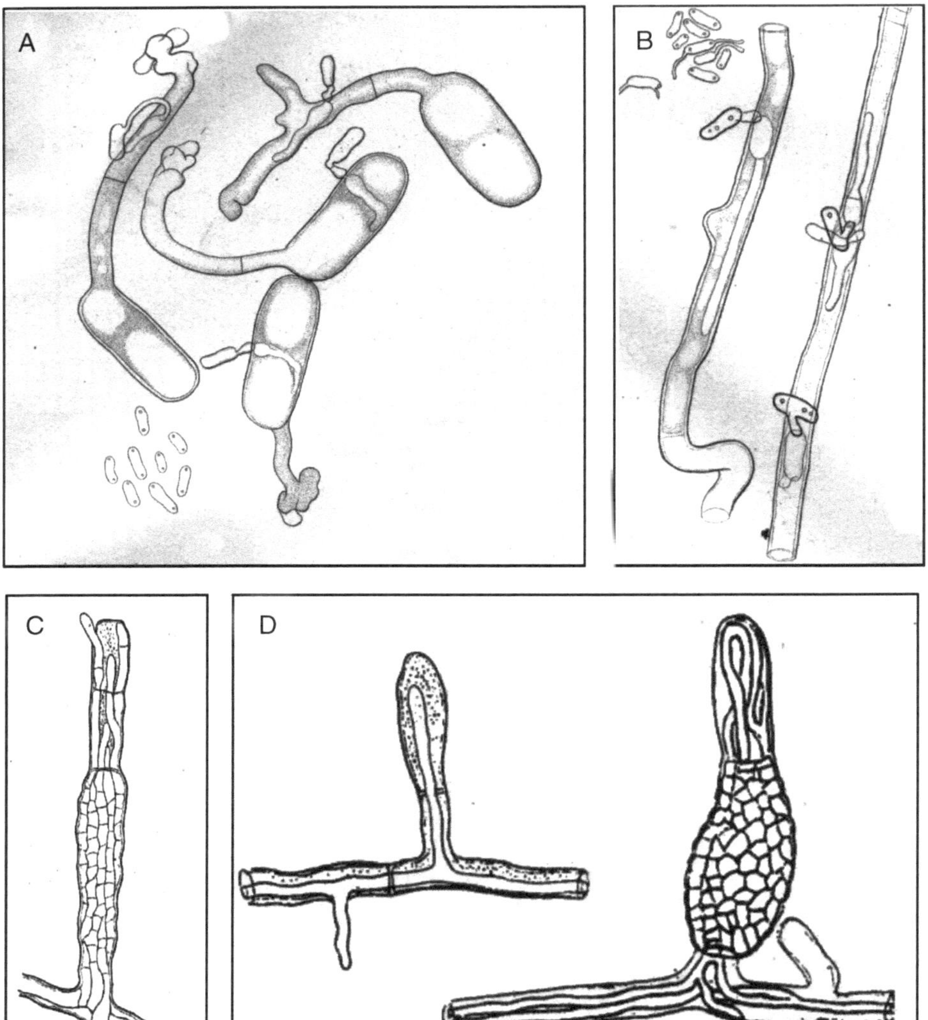

Figure 4 Drawings from De Bary (1870) showing the details of the interactions between *Ampelomyces* mycoparasites and powdery mildew fungi. A: Penetration of the germ tubes of *Ampelomyces* conidia into germinating conidia of *E. heraclei* infecting *Anthriscus silvestris*. B: Penetration of the germ tubes of *Ampelomyces* into the hyphae of *Neoerysiphe galeopsidis* infecting *Galeopsis tetrahit*. C: A young, immature pycnidium of *Ampelomyces* produced in the basal cell of a conidiophore of *N. galeopsidis* infecting *G. tetrahit*. The intracellular hyphae of *Ampelomyces* emerged from the cells of the conidiophore after keeping the parasitized powdery mildew mycelium in water for a few hours. D(I): An outgrowing *Ampelomyces* hypha from the same powdery mildew species after keeping the parasitized powdery mildew in water for a few hours. D(II): An *Ampelomyces* pycnidium in the basal cell of a conidiophore of the same powdery mildew species.

(Figure 3D) can penetrate their mycelia. Thus, *Ampelomyces* can be transported far from its original fungal host and can parasitize other powdery mildew species. After penetration, the hyphae of *Ampelomyces* invade the host mycelium internally (Figures 2 and 4A–D), and produce their pycnidia mostly in the conidiophores (Figures 2–3B and 4D(II)) and young, immature ascomata (Figure 3C) of powdery mildews. Occasionally, they also produce pycnidia in the invaded hyphal cells. The life cycle starts again when pycnidia are mature. The sexual stage is unknown in *Ampelomyces*.

Until recently, little was known about the over-wintering of *Ampelomyces*. De Bary (1870), Emmons (1930) and Yarwood (1939) found that the fungus can produce pycnidia saprophytically in the senescent or dead-plant tissues at the end of the season, and suggested that these structures served as over-wintering structures for *Ampelomyces* in the field. However, Falk *et al.* (1995a, 1995b) considered that saprophytic pycnidia produced in leaf debris were not significant for the over-winter survival of *Ampelomyces* in North American vineyards. They showed that pycnidia of *Ampelomyces* survived until the next season mainly in the parasitized ascomata produced on the bark of grapevine stocks. Similarly, Marboutie *et al.* (1995) found that pycnidia of *Ampelomyces* over-wintered in the parasitized ascomata of *Podosphaera pannosa* on the bark of peach trees. Conidia of *Ampelomyces* are probably rain dispersed from the parasitized ascomata in spring, and then germinate on the young leaves and, if powdery mildew colonies are present, penetrate their hyphae and invade their mycelia (Falk *et al.*, 1995a, 1995b).

A detailed study of the over-wintering of *Ampelomyces* on apple trees and 13 other woody and herbaceous host-plant species revealed that these mycoparasites can survive the winter in many different ways on the host-plant surfaces. Experiments carried out using detached mildew-infected leaves maintained *in vitro* demonstrated that the over-wintered pycnidia of *Ampelomyces* collected from the host plants in the spring, and produced in both the conidiophores and the immature ascomata (Figure 3C) of powdery mildews during the previous season, can initiate the life cycle of these mycoparasites (Szentiványi and Kiss, 2003). Conidia found in some of the over-wintered pycnidia and, in addition, cells of the pycnidial walls of empty pycnidia germinated in the spring and gave rise to new intracellular pycnidia of *Ampelomyces* when powdery mildew colonies were inoculated with them *in vitro*. Similar experiments showed that the thick-walled, brownish resting hyphae of *Ampelomyces*, described for the first time by De Bary (1870), could also serve as sources of primary inocula of *Ampelomyces* in the spring (Szentiványi and Kiss, 2003). So, this study demonstrated that the conidia, the pycnidial cells and the cells of the resting hyphae of *Ampelomyces* produced in the mycelia of powdery mildews during the previous season can initiate the life cycle of these mycoparasites in the spring.

2.1.1 De Bary's Pioneering Studies on *Ampelomyces*

In the 19th century, soon after *Ampelomyces quisqualis* was described succinctly as a pycnidial fungus associated with powdery mildew colonies covering grapevine leaves (Cesati, 1852), a number of mycologists (e.g., Mohl, 1852; Tulasne, 1856) considered that its pycnidia are special-fruiting bodies of the powdery mildew fungi with which they were found (Figure 5). The first thorough study of *A. quisqualis* was carried out by De Bary (1870) who clearly recognized that this fungus is an intracellular parasite of the Erysiphales. He named it as *Cicinnobolus cesatii* and showed that its hyphae grow internally in the mycelia of powdery mildews from cell to cell through the septal pores (Figure 4A–D(I)), and produce their pycnidia in one or two cells of the hyphae, conidiophores (Figure 4D(II)), conidium initials and ascomata of their fungal hosts. Much later, Speer (1978) criticized De Bary's drawings as these do not show that the intrahyphal hyphae of *Ampelomyces* become narrower when they penetrate the septal pores and then thicken again in the next cell. However, this is clearly shown in some of De Bary's drawings (e.g., Figure 4A) that illustrate the penetration of the germinating conidia of *Ampelomyces* into germinating conidia of a powdery mildew fungus.

De Bary (1870) proved experimentally that the intracellular hyphae could grow out from the parasitized cells when placed in water for a few hours (Figure 4C and D(I)). In addition, De Bary (1870) carried out cross-inoculation experiments demonstrating that an *Ampelomyces* mycoparasite collected from a given powdery mildew species could also produce intracellular pycnidia in mycelia of other powdery mildew species. This was subsequently supported by many experiments (e.g., Philipp and Crüger, 1979; Sztejnberg *et al.*, 1989; Szentiványi and Kiss, 2003; Liang *et al.*, 2007).

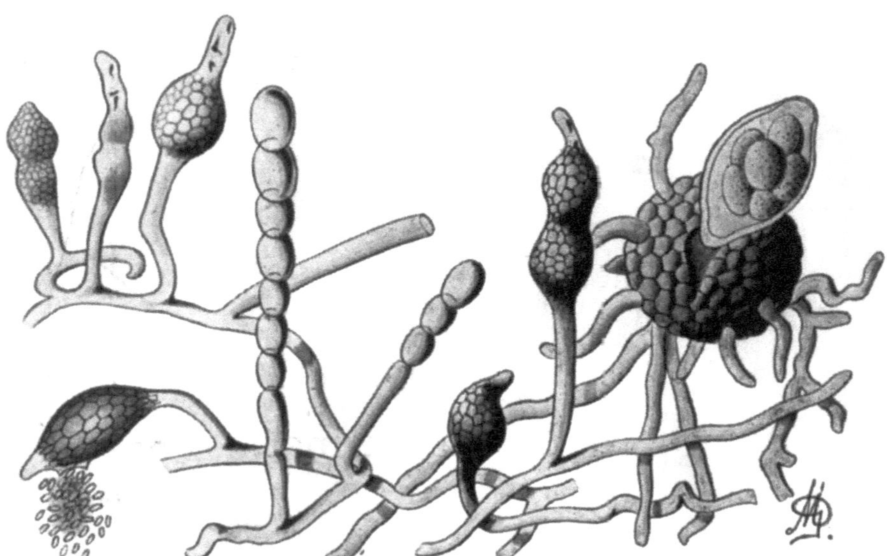

Figure 5 In the 19th century, some mycologists like Mohl (1852) and Tulasne (1856) thought that *Ampelomyces* pycnidia and conidia belong to the life cycle of the erysiphales. Thus, these were considered as special fruiting bodies of the powdery mildew fungi with which they were found. We show here one of Tulasne's drawings re-drawn by Moesz (1912).

De Bary's pioneering work was the first detailed study of an inter-fungal parasitic relationship. Emmons (1930) subsequently published an extensive cytological study describing in detail the penetration, growth and sporulation of *Ampelomyces* in the ascomata of powdery mildews. The potential for biocontrol was realized around the same time and Yarwood (1932) reported the treatment of powdery mildew infected plants with a conidial suspension of *Ampelomyces*, thus carrying out the first biocontrol experiment against a plant pathogenic fungus.

2.1.2 Mode of Action of *Ampelomyces*

Some biocontrol fungi have more than one mechanism to antagonize their hosts (Jeffries and Young, 1994). *Trichoderma* spp., for example, produce antifungal compounds, act as mycoparasites, induce plant defence mechanisms and can stimulate plant growth (Howell, 2003; Harman *et al.*, 2004). In contrast, *Ampelomyces* acts directly by invasion and destruction of host cytoplasm. *Ampelomyces* kills the parasitized powdery mildew hyphae by causing a rapid degeneration of the cell contents (Hashioka and Nakai, 1980). Toxin production has not been detected in *Ampelomyces* (Beuther *et al.*, 1981) in contrast to other pycnidial mycoparasites, such as *Coniothyrium minitans* (Machida *et al.*, 2001; McQuilken *et al.*, 2002, 2003).

The presence of the mycohosts is recognized by *Ampelomyces*, and a water-soluble substance from powdery mildew conidia was shown to stimulate the germination of *Ampelomyces* conidia *in vitro* (Gu and Ko, 1997). Directed growth of germ-tubes of *Ampelomyces* towards powdery mildew hyphae has also been observed (Sundheim and Krekling, 1982). As with phytopathogenic fungi, penetration of the host cell wall is likely to involve both enzymatic and mechanical processes. Appressorium-like structures were reported at the point of penetration (Figure 4A) (De Bary, 1870; Sundheim and Krekling, 1982). Extracellular lytic enzymes have been identified in liquid cultures of *Ampelomyces*, which may play a role in the degradation of the powdery mildew hyphal walls during penetration (Philipp, 1985). Rotem *et al.* (1999) demonstrated that an exo-beta-1,3-glucanase is excreted both in culture and during mycoparasitism, and showed that culture filtrates of an *Ampelomyces* isolate could cause degradation of hyphal walls of the cucumber powdery mildew fungus in the absence of active mycelium.

3. MYCOPARASITISM AS A BIOTIC STRESS FOR THE PARASITIZED FUNGI

3.1 Direct Effects of Mycoparasitism: Impact on Growth and Sporulation of the Mycohosts

Ampelomyces mycoparasites suppress both the asexual and the sexual sporulation of the attacked powdery mildew mycelia by colonizing and destroying the conidiophores, and the immature ascomata, respectively. The early stage of mycoparasitism is apparently biotrophic, but the invaded cytoplasm then begins to die and a necrotrophic interaction results (Hashioka and Nakai, 1980; Sundheim and Krekling, 1982). Philipp *et al.* (1984) showed that the parasitized powdery mildew colonies can continue their radial growth, but their sporulation stopped soon after *Ampelomyces* penetrated their mycelia. Similarly, Shishkoff and McGrath (2002) demonstrated that *Ampelomyces* could not stop the spread of powdery mildew colonies on detached leaves maintained *in vitro* but did reduce the amount of inoculum produced by each colony.

Microbial antagonists, which act through antibiosis against powdery mildews, such as *Pseudozyma* spp. or *Tilletiopsis* spp., can kill powdery mildew colonies rapidly, causing complete plasmolysis of their cells in 24–48 h (Bélanger and Labbé, 2002). As *Ampelomyces* acts against powdery mildews through mycoparasitism, without producing antifungal compounds, it destroys the invaded powdery mildew colonies only slowly, in 5–10 days, depending on the ambient temperature, relative humidity and other abiotic factors. During that period of time, some of the conidiophores of the invaded mycelium still produce fresh conidia, although these might contain intracellular hyphae of *Ampelomyces* (Figures 2, 3D, and 4C), and, thus, might contribute to the long-distance dispersal of the mycoparasites.

The sporulation rate of powdery mildew colonies depends on the inoculum density, physiological patterns of the host plant, abiotic factors and so on (Yarwood, 1957; Rouse *et al.*, 1984; Bushnell, 2002) When the sporulation and spread of these pathogens on the host-plant surfaces is intense, *Ampelomyces* mycoparasites can only slowly follow the spread of powdery mildew colonies. In these cases, they lag behind the spread of the disease, reducing its severity and limiting its negative effect on the infected plants to a certain extent. However, according to a mathematical model of the interactions between fungal parasites and their mycoparasites (Shaw, 1994; Shaw and Peters, 1994), mycoparasites might cause apparently random fluctuations in the abundance of their mycohosts from year to year, even in an absolutely constant environment. Thus, it is not obvious how environmental factors could be distinguished from intrinsic population instability in the field data on quantitative relationships between mycohosts and their mycoparasites.

In the field, *Ampelomyces* mycoparasites may not be observed on many plants until late in the growing season, long after powdery mildews have infected the aerial parts of their host plants (Gadoury and Pearson, 1988; Rankovic, 1997; Kiss, 1998). The situation is different for powdery mildew species that over-winter in buds, such as the apple powdery mildew fungus, *Podosphaera leucotricha*, and start their life cycles early in the season, because the mycoparasites might over-winter in the same buds, or on other parts of the host plant, in close vicinity of the primary powdery mildew inocula (Szentiványi and Kiss, 2003; Szentiványi *et al.*, 2005). However, even in these cases, the spread of the mycohosts in the spring is much more rapid after bud break than that of the over-wintered mycoparasites (Szentiványi and Kiss, 2003). Thus, powdery mildew epidemics usually reach damaging levels before their growth and sporulation are arrested by *Ampelomyces* in the field (Gadoury and Pearson, 1988; Falk *et al.*, 1995a, 1995b).

The phyllosphere is a much more dynamic environment than the rhizosphere as the leaves have a limited lifetime compared to roots (Fokkema and Schippers, 1986; Andrews 1990, 1992). Thus, both powdery mildews and *Ampelomyces* have a shorter window of infection compared to the soil-borne plant pathogenic fungi and their mycoparasites (Kiss, 2001). The only powdery mildew structures that are exposed for a longer period of time to mycoparasitic attack are the over-wintering ascomata on the bark of woody crops such as grapevine. A part of the ascomata is destroyed by *Ampelomyces* every year in many powdery mildew species in the field (Emmons, 1930; Speer, 1978; Falk *et al.*, 1995a; Rankovic, 1997; Kiss, 1998; Füzi, 2003). However, biocontrol of grapevine mildew ascomata through mycoparasitism by *Ampelomyces* was not effective (Falk *et al.*, 1995a). From an ecological perspective, it looks like there are a number of biotic and abiotic factors that do not seem favourable for the activity of *Ampelomyces* against powdery mildew fungi in the field, yet the widespread, natural occurrence of these fungi would suggest otherwise.

3.2 Indirect Effects in a Tri-Trophic Relationship: Impact of Mycoparasitism on the Mildew-Infected Plants

The presence of *Ampelomyces* in the powdery mildew mycelium results in a reduction of the negative effects of the pathogenic fungus on the infected plant. Abo-Foul *et al.* (1996) showed that infected cucumber plants regained vigour after being treated with *Ampelomyces*, which killed the pathogen. Eight days after treatment with a conidial suspension of *Ampelomyces*, chlorophyll content and also CO_2-fixation in the infected cucumber leaves were almost the same as in uninfected controls. These results were supported by Romero *et al.* (2003), who found that a treatment with *Ampelomyces* significantly increased the chlorophyll content of detached and mildew-infected melon leaves maintained *in vitro*. We have also found that the chlorophyll content of powdery mildew-infected *Lycium* leaves collected from the field, and heavily parasitized by *Ampelomyces* (Figure 1), is significantly higher than that of the mildew-infected leaves on which the mycoparasites were not present (Kiss and Szentiványi, unpublished data). The relationship between host plants, powdery mildew fungi and *Ampelomyces* mycoparasites could be further studied as a model system for tri-trophic interactions in fungal and plant ecology (Kiss, 2001; Kiss *et al.*, 2004).

4. OTHER FUNGAL ANTAGONISTS OF POWDERY MILDEW FUNGI

A review of all known fungal antagonists of powdery mildews, including those found in the field and those tested as potential biocontrol agents of powdery mildew infections without any record of a natural antagonistic relationship, revealed that more than 40 fungal taxa could suppress the growth and sporulation of these plant pathogens (Kiss, 2003). This suggests that a high number of fungi, including mycoparasites and other antagonists, are associated with powdery mildew colonies in the field. These fungi use powdery mildew mycelia as nutrient sources and might exploit powdery mildew fungi for other purposes, as well. Very little is known about these interactions under natural conditions (Raghavendra Rao and Pavgi, 1978; Mathur and Mukerji, 1981;

Hijwegen and Buchenauer, 1984; Hirsch and Braun, 1992; Szentiványi *et al.*, 2006). Most antagonists of powdery mildews, other than *Ampelomyces*, were evaluated as potential biocontrol agents only (Bélanger and Labbé, 2002; Kiss, 2003; Kiss *et al.*, 2004;) without paying attention to their possible role in the natural control of powdery mildew infections.

4.1 Confusion of Ampelomyces with Other Pycnidial Mycoparasites of Powdery Mildews

Recent results suggested that a number of pycnidial fungi, other than *Ampelomyces*, could also parasitize powdery mildew colonies in the field. For example, Sullivan and White (2000) suggested the possibility that some pycnidial fungi isolated from powdery mildew colonies collected from the field, and characterized by a 3–4 mm radial growth/day in culture at room temperature, were confused with *Phoma glomerata* isolates. Sullivan and White (2000) found that such isolates, identified as 'fast-growing *Ampelomyces* isolates' in earlier works (e.g., Mhaskar and Rao, 1974; Rudakov, 1979; Kiss, 1997; Kiss and Nakasone, 1998), typically came from sessile pycnidia found on mildew-infected leaves, whilst the so-called 'slow-growing *Ampelomyces* isolates', with a radial growth rate of 0.1–1 mm day^{-1} in culture, came from intracellular pycnidia typical of the 'true' *Ampelomyces*. ITS1 sequence analysis showed that the 'fast-growing isolates' clustered in a clade typified by *Phoma* species, while the 'slow-growing isolates' were distinct (Sullivan and White, 2000). This study supported the earlier results obtained by Kiss and Nakasone (1998). It seems likely that the 'fast-growing *Ampelomyces* isolates' are, in fact, *Phoma* species, whilst the true *Ampelomyces* isolates are always slow growing in culture and always produce intracellular pycnidia in powdery mildew mycelia (Sullivan and White, 2000; Szentiványi *et al.*, 2005; Liang *et al.*, 2007). This suggests that a number of pycnidial fungi, confused with *Ampelomyces* in earlier works, are also natural mycoparasites of powdery mildews, although their mycoparasitic activity has not been characterized in detail yet.

5. CONCLUSIONS

The high number of known fungal antagonists of powdery mildews (for review see Kiss, 2003) suggests that these plant pathogens are often attacked and damaged by mycoparasites and other antagonists in the field. Among these, pycnidial fungi belonging to the genus *Ampelomyces* are the most widespread and the oldest known natural enemies of the Erysiphales (Kiss *et al.*, 2004). Many aspects of the interactions between powdery mildews and *Ampelomyces* are well known, similar to the interactions between some species of the Erysiphales and a few other powdery mildew antagonists, especially *Pseudozyma* spp. (Bélanger and Deacon, 1996; Avis *et al.*, 2000; Avis and Bélanger, 2001), *Tilletiopsis* spp. (e.g., Hoch and Provvidenti, 1979; Urquhart and Punja, 2002) and *Lecanicillium lecanii* (syn. *Verticillium lecanii*) (e.g., Askary *et al.*, 1997), which destroy the attacked parts of powdery mildew mycelia much faster, through the production of hydrolytic enzymes and antifungal compounds. However, nothing is known about defence reactions of the powdery mildew mycelium against the attack of any of these antagonists. In some mycoparasitic interactions, papillae or other structural barriers are produced to stop the invaders (e.g., Vajna, 1985a, 1985b; Kiss, 2001); however, such structures have not been reported in powdery mildews so far. This would suggest that powdery mildews simply tolerate the biotic stress caused by mycoparasites and other antagonists; however, our knowledge on this aspect of inter-fungal parasitic relationships is, in general, too scarce to be able to formulate any general conclusions in this matter.

ACKNOWLEDGEMENT

The author acknowledges the support of a János Bolyai Research Fellowship of the Hungarian Academy of Sciences.

REFERENCES

Abo-Foul, S., Raskin, V.I., Sztejnberg, A. and Marder, J.B. (1996). Disruption of chlorophyll organization and function in powdery mildew-diseased cucumber leaves and its control by the hyperparasite *Ampelomyces quisqualis*. *Phytopathology*, **86**, 195–199.

Andrews, J.H. (1990). Biological control in the phyllosphere: Realistic goal or false hope? *Canadian Journal of Plant Pathology*, **12**, 300–307.

Andrews, J.H. (1992). Biological control in the phyllosphere. *Annual Review of Phytopathology*, **30**, 603–635.

Askary, H., Benhamou, N. and Brodeur, J. (1997). Ultrastructural and cytochemical investigations of the antagonistic effect of *Verticillium lecanii* on cucumber powdery mildew. *Phytopathology*, **87**, 359–368.

Avis, T.J. and Bélanger, R.R. (2001). Specificity and mode of action of the antifungal cis-9-heptadecenoic acid produced by *Pseudozyma flocculosa*. *Applied and Environmental Microbiology*, **67**, 956–960.

Avis, T.J., Boulanger, R. and Bélanger, R.R. (2000). Synthesis and biological characterization of (Z)-9-heptadecenoic and (Z)-6-methyl-9-heptadecenoic acids, fatty acids with antibiotic activity produced by *Pseudozyma flocculosa*. *Journal of Chemical Ecology*, **26**, 987–1000.

Bélanger, R.R. and Deacon, J.W. (1996). Interaction specificity of the biocontrol agent *Sporothrix flocculosa*: A video microscopy study. *Phytopathology*, **86**, 1317–1323.

Bélanger, R.R. and Labbé, C. (2002). Control of powdery mildews without chemicals: Prophylactic and biological alternatives for horticultural crops. In: *The Powdery Mildews: A Comprehensive Treatise* (R.R. Bélanger, W.R. Bushnell, A.J. Dik and T.L.W. Carver, eds.), pp. 256–267. APS Press, St. Paul, MN, USA.

Beuther, E., Philipp, W.D. and Grossmann, F. (1981). Untersuchungen zum Hyperparasitismus von *Ampelomyces quisqualis* auf Gurkenmehltau (*Sphaerotheca fuliginea*). *Phytopathologische Zeitschrift*, **101**, 265–270.

Bushnell, W.R. (2002). The role of powdery mildew research in understanding host-parasite interaction: past, present, and future. In: *The Powdery Mildews: A Comprehensive Treatise* (R.R. Bélanger, W.R. Bushnell, A.J. Dik and T.L.W. Carver, eds.), pp. 1–12. APS Press, St. Paul, MN, USA.

Cesati, V. (1852). Ampelomyces quisqualis Ces. Botanische Zeitung, **10**, 301–302.

Cooke, R.C. and Whipps, J.M. (1993). *Ecophysiology of Fungi*. Blackwell Scientific Publications, Oxford.

De Bary, A. (1870). *Eurotium, Erysiphe, Cicinnobolus*, nebst Bemerkungen über die Geschlechtsorgane der Ascomyceten. In: *Beiträge zur Morphologie und Physiologie der Pilze* (A. De Bary and M. Woronin, eds.), pp. 1–95. Verlag von C. Winter, Frankfurt a. M., Germany.

Emmons, C.W. (1930). *Cicinnobolus cesatii*, a study in host-parasite relationships. *Bulletin of the Torrey Botanical Club*, **57**, 421–439.
Falk, S.P., Gadoury, D.M., Cortesi, P., Pearson, R.C. and Seem, R.C. (1995a). Parasitism of *Uncinula necator* ascomata by the mycoparasite *Ampelomyces quisqualis*. *Phytopathology*, **85**, 794–800.
Falk, S.P., Gadoury, D.M., Pearson, R.C. and Seem, R.C. (1995b). Partial control of grape powdery mildew by the mycoparasite *Ampelomyces quisqualis*. *Plant Disease*, **79**, 483–490.
Fokkema, N.J. and Schippers, B. (1986). Phyllosphere versus rhizosphere as environments for saprophytic colonization. In: *Microbiology of the Phyllosphere* (N.J. Fokkema and J. Van den Heuvel, eds.), pp. 137–159. Cambridge University Press, London, UK.
Füzi, I. (2003) Natural parasitism of *Uncinula necator* cleistothecia by *Ampelomyces* hyperparasites in the south-western vineyards of Hungary. *Acta Phytopathologica et Entomologica Hungarica*, **38**, 53–60.
Gadoury, D.M. and Pearson, R.C. (1988). Initiation, development, dispersal, and survival of ascomata of *Uncinula necator* in New York vineyards. *Phytopathology*, **73**, 1413–1421.
Gu, Y.H. and Ko, W.H. (1997). Water agarose medium for studying factors affecting germination of conidia of *Ampelomyces quisqualis*. *Mycological Research*, **101**, 422–424.
Harman, G.E., Howell, C.R., Viterbo, A., Chet, I. and Lorito, M (2004). *Trichoderma* species—opportunistic, avirulent plant symbionts. *Nature Reviews, Microbiology*, **2**, 43–56.
Hashioka, Y. and Nakai, Y. (1980). Ultrastructure of pycnidial development and mycoparasitism of *Ampelomyces quisqualis* parasitic on Erysiphales. *Transactions of the Mycological Society of Japan*, **21**, 329–338.
Hijwegen, T. and Buchenauer, H. (1984). Isolation and identification of hyperparasitic fungi associated with Erysiphaceae. *Netherlands Journal of Plant Pathology*, **90**, 79–84.
Hirsch, G. and Braun, U. (1992). Communities of parasitic microfungi. In: *Fungi in Vegetation Science* (W. Winterhoff, ed.), pp. 225–250. Kluwer Academic, Netherlands.
Hoch, H.C. and Provvidenti, R. (1979). Mycoparasitic relationships: Cytology of the *Sphaerotheca fuliginea-Tilletiopsis* sp. interaction. *Phytopathology*, **69**, 359–362.
Howell, C.R. (2003). Mechanisms employed by *Trichoderma* species in the biological control of plant diseases: The history and evolution of current concepts. *Plant Disease*, **87**, 4–10.
Jarosz, A.M. and Davelos, A.L. (1995). Effects of disease in wild plant populations and the evolution of pathogen aggressiveness. *New Phytologist*, **129**, 371–387.
Jarvis, W.R. and Slingsby, K. (1977). The control of powdery mildew of greenhouse cucumber by water spray and *Ampelomyces quisqualis*. *Plant Disease Reporter*, **61**, 728–730.
Jeffries, P. (1995). Biology and ecology of mycoparasitism. *Canadian Journal of Botany*, **73**(Suppl. 1), S1284–S1290.
Jeffries, P. (1997). Mycoparasitism. In: *The Mycota. A Comprehensive Treatise on Fungi as Experimental Systems for Basic and Applied Research* (D.T. Wicklow and B.E. Söderström, eds.), pp. 149–164. Springer-Verlag, Berlin, Germany.
Jeffries, P. and Young, T.W.K. (1994). *Interfungal Parasitic Relationships*. CAB International, Wallingford, UK.
Kiss, L. (1997). Genetic diversity in *Ampelomyces* isolates, hyperparasites of powdery mildew fungi, inferred from RFLP analysis of the rDNA ITS region. *Mycological Research*, **101**, 1073–1080.
Kiss, L. (1998). Natural occurrence of *Ampelomyces* intracellular mycoparasites in mycelia of powdery mildew fungi. *New Phytologist*, **140**, 709–714.
Kiss, L. (2001). The role of hyperparasites in host plant–parasitic fungi relationships. In: *Biotic Interactions in Plant-Pathogen Associations* (M.J. Jeger and N.J. Spence, eds.), pp. 227–236. CAB International, Wallingford, UK.
Kiss, L. (2003). A review of fungal antagonists of powdery mildews and their potential as biocontrol agents. *Pest Management Science*, **59**, 475–483.
Kiss, L. (2004). How dangerous is the use of fungal biocontrol agents to nontarget organisms? *New Phytologist*, **163**, 453–455.
Kiss, L. and Nakasone, K.K. (1998). Ribosomal DNA internal transcribed spacer sequences do not support the species status of *Ampelomyces quisqualis*, a hyperparasite of powdery mildew fungi. *Current Genetics*, **33**, 362–367.

Kiss, L., Russell, J., Szentiványi, O., Xu, X. and Jeffries, P. (2004). Biology and biocontrol potential of *Ampelomyces* mycoparasites, natural antagonists of powdery mildew fungi. *Biocontrol Science and Technology*, **14**, 635–651.

Liang, C., Yang, J., Kovács, G.M., Szentiványi, O., Li, B., Xu, X.M. and Kiss, L. (2007). Genetic diversity of *Ampelomyces* mycoparasites isolated from different powdery mildew fungi in China inferred from analyses of rDNA ITS sequences. *Fungal Diversity*, **24**, 225–240.

Machida, K., Trifonov, L.S., Ayer, W.A., Lu, Z.-X., Laroche, A., Huang, H.C., Cheng, K.J. and Zantige, J.L. (2001). 3(2 H)-Benzofuranones and chromanes from liquid cultures of the mycoparasitic fungus *Coniothyrium minitans*. *Phytochemistry*, **58**, 173–177.

Marboutie, G., Combe, F., Boyer, E. and Berne, A. (1995). Travaux sur l'*Ampelomyces quisqualis* (1993–1994). *IOBC/WPRS Bulletin*, **18**, 75–78.

Mathur, M. and Mukerji, K.G. (1981). Antagonistic behaviour of *Cladosporium spongiosum* against *Phyllactinia dalbergiae* on *Dalbergia sissoo*. *Angew Bot*, **55**, 75–77.

McQuilken, M.P., Gemmell, J., Hill, R.A. and Whipps, J.M. (2003). Production of macrosphelide A by the mycoparasite *Coniothyrium minitans*. *FEMS Microbiology Letters*, **219**, 27–31.

McQuilken, M.P., Gemmell, J. and Whipps, J.M. (2002). Some nutritional factors affecting production of biomass and antifungal metabolites of *Coniothyrium minitans*. *Biocontrol Science and Technology*, **12**, 443–454.

Mhaskar, D.N. and Rao, V.G. (1974). The mycoparasite *Ampelomyces quisqualis* Ces. in artificial culture II. Effect of environmental factors. *Phytopathologia Mediterranea*, **13**, 147–154.

Moesz, G. (1912). Powdery mildew. *Urania*, **13**, 1–15. [In Hungarian.]

von Mohl, H. (1852). Die Traubenkrankheit. *Botanische Zeitung*, **10**(9–16), 31–33.

Philipp, W.D. (1985). Extracellular enzimes and nutritional physiology of *Ampelomyces quisqualis* Ces., hyperparasite of powdery mildew *in vitro*. *Journal of Phytopathology*, **114**, 274–283.

Philipp, W.D. and Crüger, G. (1979). Parasitismus von *Ampelomyces quisqualis* auf Echten Mehltaupilzen an Gurken und anderen Gemüsearten. *Zeitschrift für Pflanzenkrankheiten und Pflanzenschutz*, **86**, 129–142.

Philipp, W.D., Grauer, U. and Grossmann, F. (1984). Ergänzende Untersuchungen zur biologischen und integrierten Bekämpfung von Gurkenmehltau unter Glass durch *Ampelomyces quisqualis*. *Zeitschrift für Pflkanzenkrankheiten und Pflanzenschutz*, **91**, 438–443.

Raghavendra Rao, N.N. and Pavgi, M.S. (1978). Two mycoparasites on powdery mildews. *Sydowia*, **30**, 145–147.

Rankovic, B. (1997). Hyperparasites of the genus *Ampelomyces* on powdery mildew fungi in Serbia. *Mycopathologia*, **139**, 157–164.

Romero, D., Rivera, M.E., Cazorla, F.M., De Vincente, A. and Pérez-Garcia, A. (2003). Effect of mycoparasitic fungi on the development of *Sphaerotheca fusca* in melon leaves. *Mycological Research*, **107**, 64–71.

Rotem, Y., Yarden, O. and Sztejnberg, A. (1999). The mycoparasite *Ampelomyces quisqualis* expresses exgA encoding an exo-b-1,3-glucanase in culture and during mycoparasitism. *Phytopathology*, **89**, 631–638.

Rouse, D.I., MacKenzie, D.R. and Nelson, R.R. (1984). Density dependent sporulation of *Erysiphe graminis* f. sp. *tritici*. *Phytopathology*, **74**, 1176–1180.

Rudakov, O.L. (1979). Fungi of the genus *Ampelomyces* Ces. Ex Schlecht. *Mikologiya i Fitopatologiya*, **13**, 104–110. [In Russian.]

Shaw, M.W. (1994). Seasonally induced chaotic dynamics and their implications in models of plant disease. *Plant Pathology*, **43**, 790–801.

Shaw, M.W. and Peters, J.C. (1994). The biological environment and pathogen population dynamics: uncertainty, coexistence and competition. In: *Ecology of Plant Pathogens* (J.P. Blakeman and B. Williamson, eds.), pp. 17–37. CAB International, Wallingford, UK.

Shishkoff, N. and McGrath, M.T. (2002). AQ10 biofungicide combined with chemical fungicides or AddQ spray adjuvant for control of cucurbit powdery mildew in detached leaf culture. *Plant Disease*, **86**, 915–918.

Speer, E.O. (1978). Beitrag zur Morphologie von *Ampelomyces quisqualis* Ces. *Sydowia*, **31**, 242–246.

Sullivan, R.F. and White, J.F. Jr. (2000). *Phoma glomerata* as a mycoparasite of powdery mildew. *Applied and Environmental Microbiology*, **66**, 425–427.
Sundheim, L. (1982). Control of cucumber powdery mildew by the hyperparasite *Ampelomyces quisqualis* and fungicides. *Plant Pathology*, **31**, 209–214.
Sundheim, L. and Krekling, T. (1982). Host-parasite relationships of the hyperparasite *Ampelomyces quisqualis* and its powdery mildew host *Sphaerotheca fuliginea*. *Journal of Phytopathology*, **104**, 202–210.
Szentiványi, O. and Kiss, L. (2003). Overwintering of *Ampelomyces* mycoparasites on apple trees and other plants infected with powdery mildews. *Plant Pathology*, **52**, 737–746.
Szentiványi, O., Kiss, L., Russell, J.C., Kovács, G.M., Varga, K., Jankovics, T., Lesemann, S., Xu, X.M. and Jeffries, P. (2005). *Ampelomyces* mycoparasites from apple powdery mildew identified as a distinct group based on single-stranded conformation polymorphism analysis of the rDNA ITS region. *Mycological Research*, **109**, 429–438.
Szentiványi, O., Varga, K., Wyand, R., Slatter, H. and Kiss, L. (2006). *Paecilomyces farinosus* destroys powdery mildew colonies in detached leaf cultures but not on whole plants. *European Journal of Plant Pathology* , **115**, 351–356.
Sztejnberg, A., Galper, S., Mazar, S. and Lisker, N. (1989). *Ampelomyces quisqualis* for biological and integrated control of powdery mildew in Israel. *Journal of Phytopathology*, **124**, 285–295.
Tulasne, L.R. (1856). Nouvelles observations sur les Erysiphe. *Annales des Sciences Naturelles Botanique*, **6**, 299–324.
Urquhart, E.J. and Punja, Z.K. (2002). Hydrolytic enzymes and antifungal compounds produced by *Tilletiopsis* species, phyllosphere yeasts that are antagonists of powdery mildew fungi. *Canadian Journal of Microbiology*, **48**, 219–229.
Vajna, L. (1985a). Phytopathogenic *Fusarium oxysporum* Schlecht as a necrotrophic mycoparasite. *Journal of Phytopathology*, **114**, 338–347.
Vajna, L. (1985b). Mutual parasitism between *Trichoderma hamatum* and *Trichoderma pseudokoningii*. *Journal of Phytopathology*, **113**, 300–303.
Yarwood, C.E. (1932). *Ampelomyces quisqualis* on clover mildew. *Phytopathology*, **22**, 31. (Abstract)
Yarwood, C.E. (1939). An overwintering pycnidial stage of *Cicinnobolus*. *Mycologia*, **31**, 420–422.
Yarwood, C.E. (1957). Powdery mildews. *Botanical Reviews*, **23**, 235–301.

CHAPTER 4

The Population Dynamical Consequences of Density Dependence in Fungal Plant Pathogens

Michael W. Shaw

Contents

1. Biotic Stress 54
2. Competition 54
3. Synergism 58
4. Multi-trophic Interactions 59
5. Population Stress 62
References 63

Abstract

Almost all stages of a plant pathogen life cycle are potentially density dependent. At small scales and short time spans appropriate to a single-pathogen individual, density dependence can be extremely strong, mediated both by simple resource use, changes in the host due to defence reactions and signals between fungal individuals. In most cases, the consequences are a rise in reproductive rate as the pathogen becomes rarer, and consequently stabilisation of the population dynamics; however, at very low density reproduction may become inefficient, either because it is co-operative or because heterothallic fungi do not form sexual spores. The consequence will be historically determined distributions. On a medium scale, appropriate for example to several generations of a host plant, the factors already mentioned remain important but specialist natural enemies may also start to affect the dynamics detectably. This could in theory lead to complex (e.g. chaotic) dynamics, but in practice heterogeneity of habitat and host is likely to smooth the extreme relationships and make for more stable, though still very variable, dynamics. On longer temporal and longer spatial scales evolutionary responses by both host and pathogen are likely to become important, producing patterns which ultimately depend on the strength of interactions at smaller scales.

British Mycological Society Symposia Series
Published by Elsevier Ltd.

1. BIOTIC STRESS

At the population and higher levels, stress can only really be recognised by its effects on the population growth rate. While this must be determined by its effects on individuals a severe stress on a population may result from stresses to individuals which require little physiological adaptation. As such, one of the most critical sources of stress is the biotic environment, including conspecifics, hosts and natural enemies. In this review, I shall try to look at some of the population consequences of these biotic stresses; I have attempted to discuss ideas rather than produce a comprehensive review.

2. COMPETITION

Conspecific individuals within a certain range compete for local resources. A shortage of nutrients represents a stress signal, and we would expect directed ecological responses to this signal—for example, it might be appropriate to invade a host by a different route, to excrete antimicrobials (causing stress responses in others and possibly in conspecifics) or to sequester nutrients likely to be critical. During free-living phases of the life cycle—on a passive substrate—the distance over which competition can occur depends on how fast resources are removed by uptake relative to replacement by diffusion or bulk transport. These scales are likely to be of the order of millimetre or less, corresponding to densities of 100 individuals or more per square centimetre. These densities are common for phylloplane microflora (Fokkema, 1981; Magan and Lacey, 1986; Southwell *et al.*, 1999) or in experiments with inoculated spores. For example, Kema *et al.* (1996) studied infection by *Mycosphaerella graminicola* conidia at least 10^5 spores per square centimetre. Kessel *et al.* (2005) modelled the competition between *Ulocladium atrum* and *Botrytis cinerea* based on detailed measurements and showed how differing abilities determined the outcome of competition.

It is not uncommon for the early infection stages of fungal pathogens to rapidly exceed the volume of the propagule (e.g. an *M. graminicola* conidium 50×2 μm has the same volume as just 200 μm of hypha; similar considerations would apply to any small-spored fungus). In this case growth must involve uptake of exogenous nutrients, and competition will be important at relatively low spore densities. This is less likely to be the case for spores which are larger relative to hyphae, such as rusts. However, simple nutrient competition should always be more or less inversely proportional to the density of individuals, although the steepness of the proportionality will depend on the colony form. Starting at a density where colonies are just interacting, then with a 10-fold increase in density the area and resource available per individual will be reduced to one-tenth. In terms of physiological change, one would expect a rapid response to this situation, with declining response subsequently. This is what is observed for plant–plant resource competition (e.g. Silvertown and Lovett-Doust, 1993, p. 52).

This picture fits well for some pathogens, such as *Puccinia graminis* rust (Fig. 1), but for others the relationship is quite different. In *M. graminicola* infecting wheat

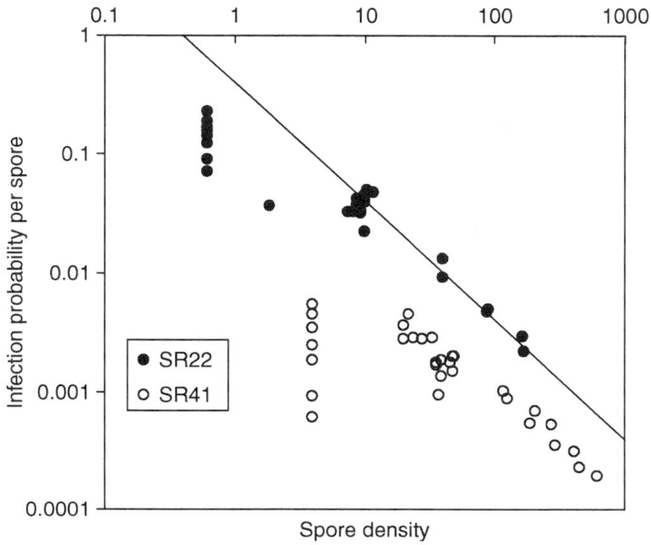

Figure 1 Relationship between spore density of two clones of *Puccinia graminis* f.sp. *tritici* on a leaf after inoculation and the number of established infections.
Data recalculated from Newton *et al.* (1997).

leaves, for example, infectivity is inversely proportional to the 0.7 power of the number of spores in a drop over four orders of magnitude, down to roughly 3 spores per square centimetre (Fig. 2). In *Phaeosphaeria nodorum*, Eyal and Scharen (1977) showed a similar slope (Fig. 2). What is happening here is that the infection process by *M. graminicola* involves germination, growth over the surface of the leaf, stomatal penetration and a non-symptomatic growth phase followed by a transition to an exponential increase in biomass and the initiation of pycnidia, after which there is little further expansion of a lesion (Kema *et al.*, 1996). The rapid expansion of biomass is correlated with a transition to a biochemistry suitable for a nutrient-rich environment and activation of enzymes to deal with active oxygen species (Keon *et al.*, 2005). The process will only be successfully concluded if plant defences are not activated during the early stages of attack and, more speculatively, if sufficient area can be invaded during the non-symptomatic phase to allow nutrient availability to compensate for attack by the plant. It seems plausible that the probability of triggering plant defences might be non-linearly related to the number of hyphae attempting to grow in the intercellular spaces. In this case the actual stress experienced by each spore is due to the plant, but is greater as more attacks are taking place simultaneously. Whatever the precise mechanism, the point here is that density dependence at sites of infection is extremely strong, or put another way, individuals experience extreme stress from the presence of other individuals at densities which would be considered inconveniently low in experimental settings (e.g. when looking for infection points on a leaf surface by electron or even light microscopy of fixed and sectioned material, it would involve heroic labour to examine a square centimeter of leaf surface).

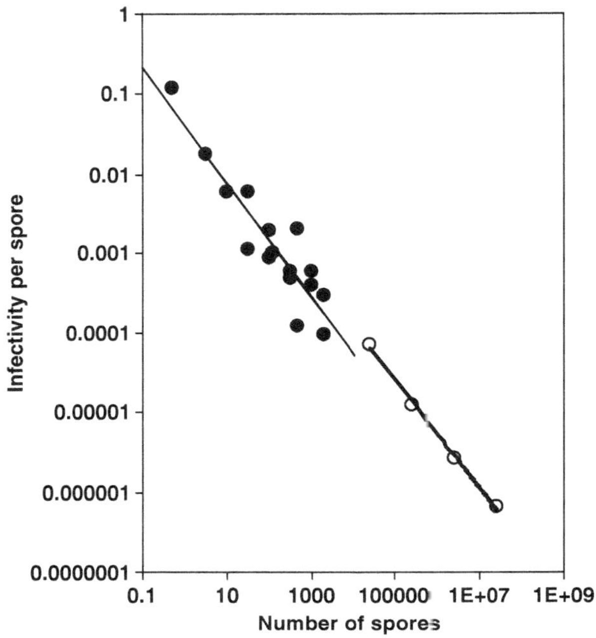

Figure 2 Relationship between spore density of non-biotrophic ascomycete pathogens and resulting infection. Solid symbols: infectivity per spore of *M. graminicola* on seedling leaves of wheat cultivar riband, calculated as in Osbourn *et al.* (1987), against number of spores applied in a 5 μl point inoculation. The line is the best-fitting regression, with slope 0.72 $\pm$ 0.16 (2 SEM). $R^2 = 0.83$. Open symbols: infectivity per spore of *P. nodorum*, calculated from the number of spores applied to a leaf and the number of lesions observed. Note the remarkable parallelism of the two sets of data.
(Data from Eyal and Scharen, 1977).

It is of considerable interest to test the extent to which spores of different species substitute for one another in attacking a leaf. This has been attempted in a number of pathogens (Harrower, 1978; Adee *et al.*, 1990; Montesinos and Bonaterra, 1996; Nolan *et al.*, 1999) but in some cases the results are hard to interpret because the density dependence of each pathogen has not been determined independently, but has been assumed to be inversely linear. This is not the case for *M. graminicola* or *P. nodorum* (Fig. 2), and is only true for rusts at extremely low densities (Fig. 1). Zelikovitch and Eyal (1991) showed that the severity of infections resulting from simultaneous attack by two isolates of *M. graminicola*, at high density, could be much lower than that resulting from either alone. This lends support to the idea that active plant responses are an important component of the density dependence of *M. graminicola*, because they suggest that one isolate may make resources unavailable to another, without itself benefiting.

Infection is of course only a single phase of the life cycle. Newton *et al.* (1997), working on rust, showed that competitive inferiority in infection efficiency was compensated for by increased spore production in a mixed infection. How the *per capita* reproductive rate of a population depends on its biotic environment will depend on all the stresses experienced at all phases of the life cycle. In this sense, a physiologist's view of stress and an ecologist's are rather different: if the organism, for example, makes enzymes which successfully destroy a xenobiotic without discernable cost, it would be of great physiological interest to know how the organism had responded to this stress; but there would be little for a population biologist to study. Various scenarios help to understand the consequences of different types of competition, typified by the scramble and contest extremes. If each individual requires a unit of resource to live, and can find it infinitely efficiently, the rate of growth is a constant until the environment is full, when the growth rate falls immediately to zero—scramble competition. On the other hand, if we assume that the available resource is equally available everywhere, then it makes sense to assume that individual reproduction rates fall linearly in proportion to the local population density. The resulting model is a standard logistic.

However, if infection success rates and spore output are linearly related to competition, as in the rust example above, there will not be a linear relation between population density and *per capita* growth rate, r. The product of the proportion of spores produced which successfully infect—including losses in dispersal as well as failure to infect—and the number of spores each produces if it is successful is often denoted by R. The average time between infection and new spore production is approximately the latent period p of the pathogen. Making the approximation $r \approx \ln(R)/p$ (Segarra *et al.*, 2001), we predict that the growth rate of a population which is dispersing spores roughly at random will decline as the logarithm of the population density. This gives rise to the well-known Gompertz model for growth of a single population (Campbell and Madden, 1990, p. 175). This is much the commonest observed growth curve for plant disease epidemics (Berger, 1981), but there are of course many other ways in which such an asymmetric curve, approximately fit by a Gompertz can be produced. In particular, the nutritional value of plant hosts for fungal pathogens often declines towards the end of the growing season—systemic defences become steadily more activated as disease accumulates—and pathogen epidemics are likely to start during periods when the environment is most suitable, so the environment will often be deteriorating at the finish of an epidemic.

Since pathogens almost all go through periods of great rarity, it is likely that they will be selected so that single spores can infect naïve hosts very efficiently, with the least danger of triggering host defences. They will be relatively weakly selected when populations are dense. The evolutionary pressures on a plant will not necessarily counter this: a single infection is not very important, but it is very important to stop the next attacks and prevent the pathogen population building up. Given that plant defences may be locally destructive (e.g. active oxygen), costly (e.g. changes to wall structure) or effective against one species while facilitating another (Govrin and Levine, 2000), it makes sense that the primary result of attack by single spores would be to prime defences for the next attack. This makes sense of the regulated nature of plant defences (e.g. Quirino and Bent,

2003), and perhaps explains some of the problems experienced in improving plant resistance by altering defences (Schlaich *et al.*, 2006).

3. SYNERGISM

There are several situations in which shortage of resources or an unfavourable environment inducing stress may trigger changes which in themselves have nothing to do with overcoming or surviving the stress, but with longer-term survival. These include altered foraging strategies, sexual reproduction and production of long-term survival structures.

For necrotrophic pathogens one may speculate that density dependence may be inverse up to a certain point: one individual supply of degradative enzymes might not be enough to overcome host defences and achieve self-sustaining growth, but many would. This is presumably the evolutionary force behind quorum sensing and the local density dependence of many bacterial pathogens (Dong *et al.*, 2001); it seems surprising that fungi do not use the same trick, though quorum sensing does not yet appear to have been reported in fungi, plant pathogenic or not. *Botrytis cinerea* seems to be a possible organism in which such a phenomenon might occur: excretion of cell wall degrading enzymes has been shown to be greater when the local environment is at low pH, and individual hyphae excrete oxalic acid (Manteau *et al.*, 2003; Prusky and Yakoby, 2003; Akimitsu *et al.*, 2004). Thus when hyphae are abundant, they will excrete much enzyme involved in necrotrophic nutrition, sufficient to avoid damage from local plant defenses simply because of the rate of the attack; when they are rare, they might adopt a scavenger strategy, simply growing at the rate possible by passively absorbing local nutrients. This would explain both the undoubted success of *B. cinerea* as a necrotrophic pathogen (Govrin and Levine, 2000), and its ability to grow as a systemic endophyte in a number of hosts, without apparently damaging its host at all (*Primula*: Barnes and Shaw, 2003. *Lactuca, Taraxacum, Cyclamen, Euphorbia*: Unpublished data).

In dense populations of heterothallic fungi, nutrient stress is often a physiological signal to trigger sexual reproduction. This makes evolutionary and ecological sense: if your host is healthy and nutrient rich, then it is both unlikely that you will find a sexual partner, and likely that asexual reproduction will be successful, so your best chance of finding a sexual partner may be to multiply asexually. On the other hand, if resources are limited, this limits the chances for successful asexual reproduction and may be because you are surrounded by conspecifics. In this case, sex is both possible (because there is a good chance of finding an individual of opposite mating type) and desirable (because the environment is changing). Thus in heterothallic fungi the more dense the population, sexual reproduction will tend to occur more often and more successfully. Eriksen and Munk (2003) found evidence for this in *M. graminicola*. This inverse density dependence could give rise to Allee effects, as discussed by Garrett and Bowden (2002). In the case of *Tilletia indica*, this Allee effect was seen as a barrier to invasion; the argument should apply to all smuts.

In the heterothallic *M. graminicola*, where ascospores are the mechanism of spread between crops in successive years (Shaw and Royle, 1989; Hunter *et al.*, 1999), long-term trends in abundance could be influenced by similar Allee effects. In northern Europe, for example, hundreds of ascospores may infect each square metre of crop early in the season at present, and it will be rare for a plant, at least on its older leaves, not to have several infections of distinct genotype and opposite mating type. But if the fungus ever became rare, for example because good host resistance was common or because wheat were an uncommon crop, it could be rare for a wheat leaf to be infected by two spores of opposite mating type within a few centimetres of each other. In this case few matings would occur, and the multiplication rate across fields would be low. In fact, with relatively simple assumptions there will be a threshold of resistance and cropping intensity below which disease will tend to disappear year-to-year and above which it will increase. This could lead to unexpectedly large changes in disease prevalence with relatively small changes in cropping area, and hysteresis, in which the prevalence of disease in the system depends on past history (Fig. 3).

Based on estimates of the quantity of pathogen-specific DNA in preserved samples of wheat from the Broadbalk experiment, Bearchell *et al.* (2005) showed that *M. graminicola* appears to have been common in the mid 19th century, becoming rarer by the early 20th century and re-appearing as a serious problem in the mid 1980s. This was about 10 years after the introduction and widespread use of extremely susceptible but high-yielding cultivars such as Longbow. Bearchell *et al.* (2005) noted a very close correlation of these changes with atmospheric sulphur deposition, but alternative explanations remain possible. The patterns of change in area and susceptibility in Fig. 3 have been chosen to be moderately realistic, but the parameters and initial conditions are not based on data.

4. MULTI-TROPHIC INTERACTIONS

Pathogenic fungi are no less subject to infection and predation than other organisms. They presumably have natural defence mechanisms, which will inevitably carry costs and are likely to be activated conditionally, when the appropriate stress occurs. There appears to have been little study of this, but the extreme specialisation of *Sphaerellopsis filum* and *Ampelomyces quisqualis* races to particular genera or even species of rust or mildew (Yuan *et al.*, 1999; Kiss *et al.*, 2004; Szentivanyi *et al.*, 2005) suggests a rather specific host defence against attack. When we are able to elucidate the mechanism of this, it will be a most interesting example to compare with mammalian and plant defence mechanisms.

Whenever a natural enemy is a passively dispersing specialist, it will tend to be more successful in a dense population of its host than a sparse one; this has consequences for both the spatial distribution of a pathogen and for the temporal dynamics. They differ between vertically infectious hyperparasites, such as mycoviruses, and horizontally infectious hyperparasites and predators. I shall examine the two cases separately.

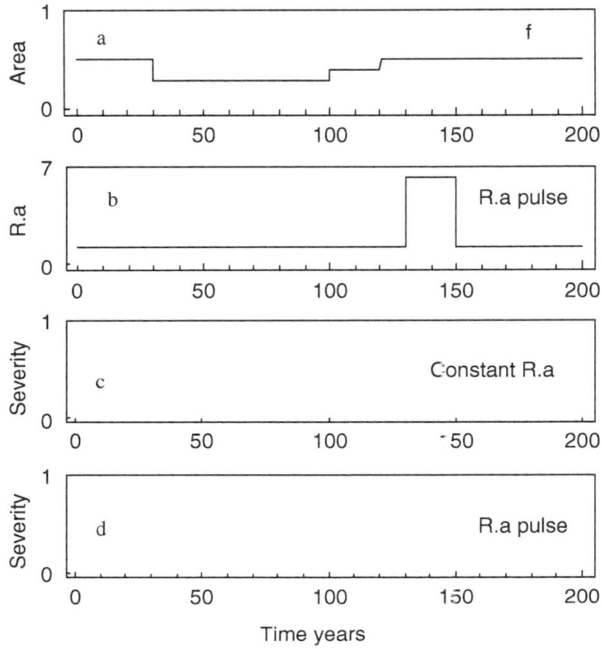

Figure 3 Output of a model showing peak severity of a disease like *M. graminicola* over a period where area of host (wheat for *M. graminicola*) changes over a threshold. The model governing the figure is as follows. Let s_i be the peak severity in year i. This severity increases from the initial severity by a factor R. The initial disease arises from the previous year's peak severity by random infection, so the density of infection is proportional to the cropping intensity φ. Because infection with two mating types within a reasonable range of each other is necessary to produce ascospores (Eriksen *et al.*, 2001), the initial density of infection is also proportional to s^2. The combined constant of proportionality for the dependence on φ and severity is a. Allowing for logistic density dependence, the recurrence between years is therefore $s_{i+1} = 1/\{1 + \exp(R_{i+1}a_i\phi_i s_i^2)\}$ where the indices on a symbol refer to a particular year. There is a threshold at $s_i = (R_i a_i \phi_i)^{-1}$: If s_i is less than this then the population decreases, if more, it increases. (a) The pattern of change in ϕ over time; (b) the pattern of change in varietal susceptibility assumed in panel d; (c) the consequent pattern of change in s if susceptibility is constant; (d) the consequent pattern of change in s if susceptibility changes briefly as in panel b.

First, it is instructive to look at a very simplified model of how hyperparasites might interact with a pathogen, in a seasonal environment. If they are completely specialised they will be able to multiply only in proportion to host density, and the host fungus will become rare out of the season of its host plant. As discussed more fully in Shaw (1994) this can lead to oscillations, irregular in detail despite the fact that no environmental variation is modelled, but with a general pattern of alternating years of abundance and rarity (Fig. 4). Uncorrelated environmental noise has surprisingly little effect on these patterns (Shaw, 1994), but stochastic effects at low population sizes tend to interrupt the regular alternation of severity. These models are clearly too simple to be fit to real data, but they emphasise

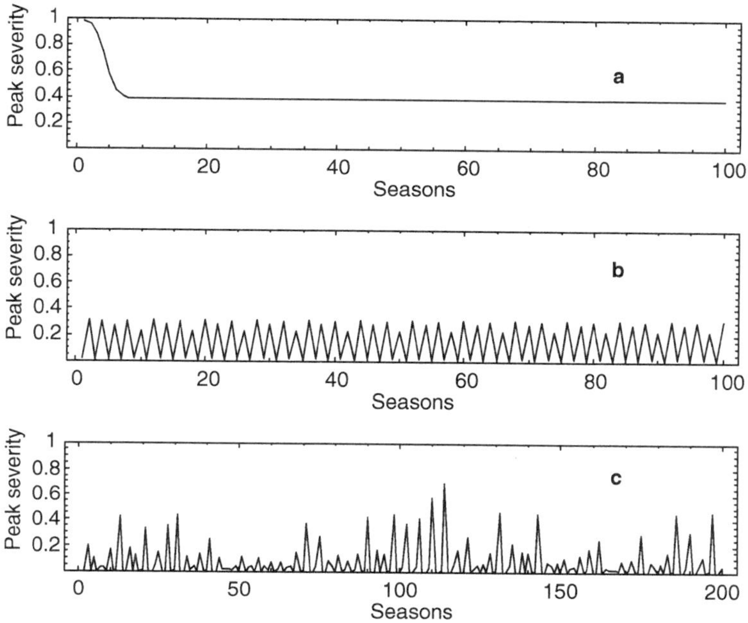

Figure 4 Output of a model showing peak severity of a pathogen with an annual cycle of increase and crash. The model governing the figure is as follows. A pathogen, at density $h(t)$ at time t, increases logistically during each season at a rate r_h, which declines linearly during the season as conditions become unfavourable. It is also subject to destruction by a hyperparasite with density $s(t)$, which infects at a rate proportional to density of both pathogen and hyperparasite. So within a season, from $t = 0$ to $t = 1$ the governing equations are $\frac{dh}{dt} = r_h h(1 - h)(1 - t) - bhs$; $\frac{ds}{dt} = bhs$ at the turn of each year, the host dies or disappears and both h and s are set to a small fraction a of their value at the end of the year. For some ranges of a, b and r_h the end-of-year severity is the same every year. In other ranges it fluctuates regularly; in yet others, it fluctuates chaotically. Parameters for each panel are: (a) $b = 7.2$/Year, $r_h = 25.2$/Year, $a = 0.01$; (b) $b = 18$/Year, $r_h = 18.4$/Year, $a = 0.01$; (c) $b = 24$/Year, $r_h = 18.4$/Year, $a = 0.01$.

some surprising features which do occur—for example, *Puccinia coronata* on *Holcus* in the Reading area has been extremely scarce this year (2006), following a summer in 2005 when infection incidence of *Sphaerellopsis filum* reached 90% of sori (unpublished data). The feature of these models which seems least to correspond with reality is the more-or-less regular occurrence of years in which the host fungus is almost absent at the end of the year.

Mycoviruses share some of the characteristics discussed in the previous section, as they usually spread only when compatible (but in this case vegetatively compatible, rather than opposite mating type) hyphae come in contact. Once infected, a lineage will often tend to reproduce less effectively than uninfected individuals and disappear. This may allow immune individuals that carry some fitness penalty to remain in the population long-term. The results of such a situation, if modelled, may be a stable equilibrium, but are also likely to be quite short period oscillations in peak host density, which will be high when infection is low, exactly the conditions required to increase infection. These may have a regular period, but may also be irregular, either because environmental variation interacts with the period or because the process is intrinsically chaotic or has a complex long-period pattern. Useful examples of control of disease by mycoviruses must presumably involve equilibria; a function of the modelling exercise is simply to remind researchers that this does not have to be the outcome of a deleterious virus spreading efficiently in a pathogen population. More guidance can be obtained from more complicated models. In particular, significant evolutionary co-adaptation can occur over short time scales, with the virus evolving to be better spread and less deleterious, and the pathogen evolving to reduce rates of spread and deleteriousness. Predictive, as opposed to illustrative, models probably need to take this into account, but would require careful characterisation of the variation present in populations (Swinton and Gilligan, 1996, 1999).

5. POPULATION STRESS

As the theme of this conference is stress, it seems appropriate to consider what we might mean by stress acting at a population level. For example, we can ask what sets the limits of distribution of a pathogen. These will usually be narrower than the limits apparent from laboratory measurements of conditions in which the pathogen can multiply, because what matters is the integration of these, and pressures from similar competing species and natural enemies, into a population multiplication rate. In turn, the pathogen will be found where this multiplication rate is greater than one from season to season, and will not be found where the rate is less than one. This can produce a sharp boundary, outside which little physiological adaptation by individuals is actually necessary for them to survive: the stress on the population is very different from the stress on any particular individual (Fig. 3).

Where a population is mildly stressed, one would expect evolution which would reduce the stress; this would lead to indefinite expansion of species ranges, which we do not commonly observe. An obvious hypothesis as to the reason for this is competition: the niche into which a species might evolve is full. Some evidence in favour of this is the extremely subtle differentiation of niches that appears to allow similar species to co-exist with superficially similar lifestyles. Examples here include *Oculimacula yallundae* and *O. acuformis* which despite both attacking the stem base of cereals with similar invasion strategies appear scarcely to compete (Bierman *et al.*, 2002); *Leptosphaeria maculans* and *L. biglobosa* are superficially very similar fungi on brassicas but with enough small differences in environmental relations that in the UK one is a serious one a minor pathogen (Huang *et al.*, 2003). Models of this differentiation have been discussed by Gudelj *et al.* (2004).

However, it is equally plausible that many niches remain unfilled, simply because there is no sequential set of niches through which a sequence of evolutionary changes can take place to arrive at the necessary phenotype. Thus, a pathogen able to attack one host, because it has lost a series of functions required by a free-living fungus and therefore does not trigger an R-gene response will usually be unable to cross to another host because it would have to reacquire lost functions and lose a new set. Thus, at the population level, the analogy of a stress which an organism's physiology is insufficiently flexible to deal with is the inability to evolve to deal with an altered environment. At this point a link between the two levels of organisation may emerge through an extension of the argument put for stress-induction of sexual reproduction earlier. A side effect of stress is often to reduce the fidelity of DNA replication and repair—either because of diversion of metabolic effort away from these proteins or because the stressor actually increases the rate of damage of the DNA. This will increase the mutation rate, which could increase the extent of subsequent adaptation (Poole *et al.*, 2003) if variation was limiting the extent of selective change.

REFERENCES

Adee, S.R., Pfender, W.F. and Hartnett, D.C. (1990). Competition between *Pyrenophora tritici-repentis* and *Septoria nodorum* in the wheat leaf as measured with de Wit replacement series. *Phytopathology,* **80**, 1177–1181.

Akimitsu, K., Isshiki, A., Ohtani, K., Yammoto, H., Eshel, D. and Prusky, D. (2004). Sugars and pH: A clue to the regulation of fungal cell wall-degrading enzymes in plants. *Physiological and Molecular Plant Pathology,* **65**, 271–275.

Barnes, S.E. and Shaw, M.W. (2003). Infection of commercial hybrid primula seeds by *Botrytis cinerea* and latent disease spread through the plants. *Phytopathology,* **93**, 573–578.

Bearchell, S.J., Fraaije, B.A., Shaw, M.W. and Fitt, B.D.L. (2005). Wheat archive links long-term fungal pathogen population dynamics to air pollution. *Proceeding of the National Academy of Sciences of the USA,* **102**, 5301–5636.

Berger, R.D. (1981). Comparison of the Gompertz and logistic equations to describe plant disease progress. *Phytopathology,* **71**, 716–719.

Bierman, S.M., Fitt, B.D.L., van den Bosch, F., Bateman, G.L., Jenkyn, J.F. and Welham, S.J. (2002). Changes in populations of the eyespot fungi *Tapesia yallundae* and *T. acuformis* under different fungicide regimes in successive crops of winter wheat, 1984-2000. *Plant Pathology,* **51**, 191–201.

Campbell, C.L. and Madden, L.V. (1990). *Introduction to Plant Disease Epidemiology.* Wiley, New York.

Dong, Y.H., Wang, L.H., Xu, J.L., Zhang, H.B., Zhang, X.F. and Zhang, L.H. (2001). Quenching quorum-sensing-dependent bacterial infection by an N-acyl homoserine lactonase. *Nature,* **411**, 813–817.

Eriksen, L. and Munk, L. (2003). The occurrence of *Mycosphaerella graminicola* and its anamorph *Septoria tritici* in winter wheat during the growing season. *European Journal of Plant Pathology,* **109**, 253–259.

Eriksen, L., Shaw, M.W. and Østergård, H. (2001). A model of the effect of pseudothecia on genetic recombination and epidemic development in populations of *Mycosphaerella graminicola. Phytopathology,* **91**, 240–248.

Eyal, Z. and Scharen, A.L. (1977). A Quantitative method for inoculation of wheat seedlings with pycnidiospores of *Septoria nodorum. Phytopathology,* **67**, 712–714.

Fokkema, N.J. (1981). Fungal leaf saprophytes, beneficial or detrimental? In: *Microbial Ecology of the Phylloplane* (J.P. Blakeman, ed.), pp. 433–454. Academic Press, London.

Garrett, K.A. and Bowden, R.L. (2002). An Allee effect reduces the invasive potential of *Tilletia indica. Phytopathology,* **92**, 1152–1159.

Govrin, E.M. and Levine, A. (2000). The hypersensitive response facilitates plant infection by the necrotrophic pathogen *Botrytis cinerea. Current Biology,* **10**, 751–757.

Gudelj, I., Fitt, B.D.L. and van den Bosch, F. (2004). Evolution of sibling fungal pathogens in relation to host specialisation. *Phytopathology*, **94**, 789–795.

Harrower, K.M. (1978). Effects of mixed inocula of *Leptosphaeria nodorum* and *Septoria tritici* on wheat seedlings. *Transactions of the British Mycological Society*, **70**, 41–45.

Huang, Y.J., Toscano-Underwood, C., Fitt, B.D.L., Hu, X.J. and Hall, A.M. (2003). Effects of temperature on ascospore germination and penetration of oilseed rape (*Brassica napus*) leaves by A- or B-group *Leptosphaeria maculans* (phoma stem canker). *Plant Pathology*, **52**, 245–255.

Hunter, T., Coker, R.R. and Royle, D.J. (1999). The teleomorph stage, *Mycosphaerella graminicola*, in epidemics of *Septoria tritici* blotch on winter wheat in the UK. *Plant Pathology*, **48**, 51–57.

Kema, G.H.J., Yu, D., Rijkenberg, F.H.J., Shaw, M.W. and Baayen, R.P. (1996). Histology of the pathogenesis of *Septoria tritici* in wheat. *Phytopathology*, **86**, 777–786.

Keon, J., Rudd, J.J., Antoniw, J., Skinner, W., Hargreaves, J. and Hammond-Kosack, K. (2005). Metabolic and stress adaptation by *Mycosphaerella graminicola* during sporulation in its host revealed through microarray transcription profiling. *Molecular Plant Pathology*, **6**, 527–540.

Kessel, G.J.T., Köhl, J., Powell, J.A., Rabbinge, R. and van der Werf, W. (2005). Modeling spatial characteristics in the biological control of fungi at leaf scale: Competitive substrate colonisation by *Botrytis cinerea* and the saprophyte antagonist *Ulocladium atrum*. *Phythopathology*, **95**, 439–448.

Kiss, L., Russell, J.C., Szentivanyi, O., Xu, X. and Jeffries, P. (2004). Biology and biocontrol potential of Ampelomyces mycoparasites, natural antagonists of powdery mildew fungi. *Biocontrol Science and Technology*, **14**, 635–651.

Magan, N. and Lacey, J. (1986). The phylloplane microflora of ripening wheat and effect of late fungicide applications. *Annals of Applied Biology*, **109**, 117–128.

Manteau, S., Abouna, S., Lambert, B. and Legendre, L. (2003). Differential regulation by ambient pH of putative virulence factor secretion by the phytopathogenic fungus *Botrytis cinerea*. *FEMS Microbiology Ecology*, **43**, 359–366.

Montesinos, E. and Bonaterra, A. (1996). Dose-response models in biological control of plant pathogens: An empirical verification. Phytopathology, **86**, 464–472.

Newton, M.R., Kinkel, L.L. and Leonard, K.J. (1997). Competition and density-dependent fitness in a plant parasitic fungus. *Ecology*, **78**, 1774–1784.

Nolan, S., Cooke, B.M. and Monahan, F.J. (1999). Studies on the interaction between *Septoria tritici* and *Stagonospora nodorum* in wheat. *European Journal of Plant Pathology*, **105**, 917–925.

Osbourn, A.E., Caten, C.E., and Scott, P.R. (1987). Adaptation of *Septoria nodorum* to wheat or barley on detached leaves. *Plant Pathology*, **36**, 565–576.

Poole, A.M., Phillips, M.J. and Penny, D. (2003). Prokaryote and eukaryote evolvability. *Biosystems*, **69**, 163–185.

Prusky, D. and Yakoby, N. (2003). Pathogenic fungi: Leading or led by ambient pH? *Molecular Plant Pathology*, **4**, 509–516.

Quirino, B.F. and Bent, A.F. (2003). Deciphering host resistance and pathogen virulence: The *Arabidopsis/Pseudomonas* interaction as a model. *Molecular Plant Pathology*, **4**, 517–530.

Schlaich, T., Urbaniak, B.M., Malgras, N., Ehler, E., Birrer, C., Meier, L. and Sautter, C. (2006). Increased field resistance to Tilletia caries provided by a specific antifungal virus gene in genetically engineered wheat. *Plant Biotechnology Journal*, **4**, 63–75.

Segarra, J., Jeger, M.J. and van den Bosch, F. (2001). Epidemic dynamics and patterns of plant diseases. *Phytopathology*, **91**, 1001–1010.

Shaw, M.W. (1994). Seasonally induced chaotic dynamics and their implications in models of plant disease. *Plant Pathology*, **43**, 790–801.

Shaw, M.W. and Royle, D.J. (1989). Airborne inoculum as a major source of *Septoria tritici* (*Mycosphaerella graminicola*) infections in winter wheat crops in the UK. *Plant Pathology*, **38**, 35–43.

Silvertown, J. and Lovett-Doust, J. (1993). *Introduction to Plant Population Biology*. Blackwell, Oxford.

Southwell, R.J., Brown, J.F. and Welsby, S.M. (1999). Microbial interactions on the phylloplane of wheat and barley after applications of mancozeb and triadimefon. *Australasian Plant Pathology*, **28**, 139–148.

Swinton, J. and Gilligan, C.A. (1996). Dutch Elm Disease and the Future of the Elm in the UK — A quantitative analysis. *Philosophical Transactions of the Royal Society of London B*, **351**, 605–615.

Swinton, J. and Gilligan, C.A. (1999). Selecting hyperparasites for biocontrol of Dutch elm disease. *Proceedings of the Royal Society B*, **266**, 437–445.

Szentivanyi, O., Kiss, L., Russell, J.C., Kovacs, G.M., Varga, K., Jankovics, T., Lesemann, S., Xu, X.M. and Jeffries, P. (2005). Ampelomyces mycoparasites from apple powdery mildew identified as a distinct group based on single-stranded conformation polymorphism analysis of the rDNA ITS region. *Mycological Research*, **109**, 429–438.

Yuan, Z.W., Pei, M.H., Hunter, T., Ruiz, C. and Royle, D.J. (1999). Pathogenicity to willow rust, *Melampsora epitea*, of the mycoparasite *Sphaerellopsis filum* from different sources. *Mycological Research*, **103**, 509–512.

Zelikovitch, N. and Eyal, Z. (1991). Reduction in pycnidial coverage after inoculation of wheat with mixtures of isolates of *Septoria tritici*. *Plant Disease*, **75**, 907–910.

CHAPTER **5**

Differences in Stress Responses Between Model and Pathogenic Fungi

Janet Quinn

Contents

1. Introduction 68
2. Stress Responses and Virulence 68
3. Stress Responses in the Host 69
4. Stress Responses *In Vitro* 70
   4.1 General-Stress Responses 70
   4.2 Stress-Specific Responses 73
5. Regulation of Stress-Induced Gene Expression 74
   5.1 The Hog1 SAPK Pathway 74
   5.2 Role of Hog1 in the Core-Stress Response in *C. albicans* 76
   5.3 Role of Hog1 in the Regulation of Stress-Specific Responses in *C. albicans* 77
   5.4 Comparison of SAPK-Regulated Genes 78
   5.5 Stress-Responsive Transcription Factors 78
6. Stress Responses in *C. glabrata* 80
7. Conclusions: Specialised Stress Responses in *Candida* 81
Acknowledgements 82
References 82

Abstract

The virulence of the human fungal pathogens *Candida albicans* and *Candida glabrata* is dependent upon their ability to mount stress responses. This reflects the importance of these responses in protecting these fungal pathogens against host defences. Recent molecular and genomic studies have contributed significantly to our understanding of *C. albicans* stress responses and how they are regulated. Interestingly, it is now apparent that *C. albicans* has diverged significantly from the benign model yeasts *Saccharomyces cerevisiae* and *Schizosaccharomyces pombe* with respect to the nature and regulation of its stress responses. For example, recent studies, from us and others, have revealed that while key regulatory molecules are conserved in *C. albicans*, their contributions to the regulation of stress responses have diverged. While much less is known about stress responses in *C. glabrata*, initial findings suggest that the sensitivity and regulation of stress responses

British Mycological Society Symposia Series
Published by Elsevier Ltd.

in this fungal pathogen also differs from that in *S. cerevisiae* and *S. pombe*. In this chapter I will highlight such differences and discuss how pathogenic *Candida* species may exploit specialised stress responses to protect themselves during disease progression in the human host.

1. INTRODUCTION

Fungal diseases represent a current and increasing threat to the health of the UK population. *Candida* species cause extremely common and often recurrent infections of the oral, gastrointestinal and urogenital tracts in even mildly immunocompromised individuals (Calderone and Fonzi, 2001). Furthermore, in immunocompromised hosts these fungi can invade the bloodstream and infect internal organs leading to life-threatening systemic disease (Pfaller *et al.*, 2000). Invasive *Candida* infections are rapidly increasing in incidence and have a high-attributable mortality (Pfaller *et al.*, 1999a). *Candida albicans* is the major agent isolated during systemic-fungal infections whereas *Candida glabrata* is now the second most common causative agent of candidiasis, having been isolated from 20% of invasive *Candida* infections (Pfaller *et al.*, 1999b, 2000). *C. glabrata* infections have a relatively high mortality compared with those caused by other *Candida* species (Gleason *et al.*, 1997; Viscoli *et al.*, 1999), as this species is relatively resistant to azole-based anti-fungal treatments.

2. STRESS RESPONSES AND VIRULENCE

A critical feature of *C. albicans* and *C. glabrata* is their versatility. Both can survive as a commensal in several anatomically distinct sites, each associated with a unique set of environmental pressures. Furthermore, under appropriate conditions these opportunistic pathogens cause systemic infections, which require adaptation to additional microenvironments. Therefore, it is perhaps not surprising that an increasing body of evidence indicates that stress responses are critical for survival of *C. albicans* in the diverse microenvironments encountered within the mammalian host. For example, inactivation of stress-protective enzymes such as Tps1 and Tps2 involved in trehalose biosynthesis, the superoxide dismutases Sod1 and Sod5, catalase Cta1, and the nitric oxide dioxygenase Yhb1, all result in attenuated virulence of *C. albicans* in the mouse model of systemic candidiasis (Wysong *et al.*, 1998; Zaragoza *et al.*, 1998; Hwang *et al.*, 2002; Van Dijck *et al.*, 2002; Martchenko *et al.*, 2004; Hromatka *et al.*, 2005). Similar findings have also been reported with strains carrying deletions in genes encoding stress-regulatory proteins such as the Hog1 stress-activated protein kinase (SAPK) and the Ssk1 response regulator in *C. albicans* (Alonso-Monge *et al.*, 1999; Calera *et al.*, 2000). Therefore, stress-regulatory pathways and the downstream targets of these pathways are required for the full virulence of *C. albicans*. While much less is known about stress responses in *C. glabrata*, a recent study illustrated that cells deleted for the Ste11 MAPKKK displayed an osmosensitive phenotype and attenuated virulence in a mouse model of system candidiasis (Calcagno *et al.*, 2005). This indicates that stress responses are also important for *C. glabrata* to survive in the mammalian host.

3. STRESS RESPONSES IN THE HOST

What stresses do *Candida* species encounter in the human host that requires the function of the stress-protective proteins, listed above, to facilitate survival? A striking feature of *C. albicans* and *C. glabrata* is their ability to colonise a broad range of niches within the human host. The exact nature of these various microenvironments has yet to be defined. However, recent genomics experiments, which have examined the global responses of *C. albicans* to host cells, have provided valuable clues in identifying which stress responses are activated following contact with the host.

As the dissemination of *C. albicans* in the bloodstream is an essential step in systemic candidiasis, a number of studies have examined the global transcriptional response of *C. albicans* following exposure to whole human blood or specific blood fractions. In response to whole blood, a number of stress-protective genes were found to be up-regulated in *C. albicans*, in particular genes involved in protection against oxidative stress (Fradin *et al.*, 2003). In a separate study, the transcript profile of *C. albicans* cells following phagocytosis by human neutrophils revealed the induction of an anti-oxidant transcriptional response and an amino-acid starvation response (Rubin-Bejerano *et al.*, 2003). Similarly, Lorenz and colleagues (2004) observed that oxidative-stress genes were induced in *C. albicans* upon internalisation by mammalian macrophages. Genes were also induced that encode proteins associated with DNA damage repair, presumably due to DNA damage caused by the oxidative burst of the macrophages. These gene responses were accompanied with a major metabolic response indicative of carbohydrate starvation involving the activation of the glyoxylate cycle and gluconeogenesis (Lorenz *et al.*, 2004).

A recent paper has further dissected the genomic-response of *C. albicans* to human blood by examining the transcript profiles of *C. albicans* to different blood fractions (Fradin *et al.*, 2005). This analysis convincingly demonstrated that it is the neutrophil sub-population that initiates stress-responses in *C. albicans* upon blood exposure. Exposure of *C. albicans* to neutrophils resulted in growth arrest and the expression of many genes known to encode factors involved in the detoxification of H_2O_2 and free radicals, such as enzymes involved in the thioredoxin and glutaredoxin systems and catalase (Fradin *et al.*, 2005). Genes involved in the regulation, biosynthesis and transport of amino acids were also found to be up-regulated, consistent with the amino-acid deprivation response previously observed by Rubin-Bejerano and colleagues (2003). Furthermore, genes of *C. albicans* involved in the recruitment of internal nitrogen sources and those encoding ammonium transporters were also significantly up-regulated in neutrophils, illustrating that *C. albicans* actually mounts a general nitrogen-deprivation response upon phagocytosis. In addition, exposure of *C. albicans* to neutrophils triggers a carbohydrate-starvation response, as genes encoding key enzymes of the glyoxylate cycle are strongly induced (Fradin *et al.*, 2005).

Collectively therefore, data from several transcript profiling experiments have strongly reinforced the idea that stress responses, and in particular the oxidative-stress response, are important for the survival of *C. albicans* following exposure to the host's innate immune system. In addition, carbohydrate- and nitrogen-starvation responses are evoked upon phagocytosis, indicating that the micro-environments of the neutrophil and macrophage are relatively nutrient poor. Strikingly, comparative studies examining the transcriptional response of *C. albicans* and the non-pathogenic yeast *Saccharomyces cerevisiae* to neutrophil exposure, found that the oxidative-stress response evoked in *C. albicans* was largely absent in *S. cerevisiae* (Rubin-Bejerano *et al.*, 2003). Similarly, the transcriptional reprogramming that was observed in *C. albicans* upon internalisation by macrophages was not seen in the non-pathogenic yeast *S. cerevisiae* (Lorenz *et al.*, 2004). Furthermore, a feature common to all of the profiling studies outlined above is that a large number of *C. albicans* genes induced in response to host-cell contact have no known homologues. Together, these studies highlight (i) the transcriptional plasticity of *C. albicans* and, (ii) the substantial adaptations that have evolved to suit the environmental niche of the pathogenic fungus. Both of these factors are likely to be critical for the survival of *C. albicans* upon exposure to the diverse environments within the host.

It is conceivable that stress conditions, in addition to those encountered upon blood cell exposure will be encountered in the many other microenvironments occupied by *C. albicans* during the various stages of infection. For example, it is possible that the fungus will be exposed to osmotic stress in the kidney, and pH stress in the gastrointestinal or urogenital tracts. However, further experimentation to examine the global transcriptional responses of *C. albicans* within these specific tissues is needed before we can be certain of the stresses encountered by *C. albicans* in these distinct niches (Hube, 2004).

4. STRESS RESPONSES *IN VITRO*

In this section I discuss transcript-profiling data, which examines the response of *C. albicans* to artificial stresses imposed *in vitro*, with particular emphasis on our recent work. I also highlight the differences and similarities that exist in transcriptional responses to stress between this fungal pathogen and *S. cerevisiae* and *Schizosaccharomyces pombe*. While such *in vitro* imposed stresses cannot replicate the complex microenvironments of the host, such investigations are worthwhile as they allow a detailed analysis of specific responses under well-defined conditions, and the information gained can provide insight into the specific-stress responses evoked in *C. albicans in vivo*.

4.1 General-Stress Responses

Global analysis of stress responses in the budding yeast *S. cerevisiae*, and the distantly related fission yeast *S. pombe*, revealed that hundreds of genes respond in a stereotypical manner to a range of unrelated stresses (Gasch *et al.*, 2000;

Causton *et al.*, 2001; Chen *et al.*, 2003). Significantly, there is considerable overlap between the genes of budding and fission yeast that are regulated in response to different stresses (Chen *et al.*, 2003). In general three main classes of stress-responsive genes are commonly induced in *S. cerevisiae* and *S. pombe* which include (i) those that encode heat-shock proteins or chaperones that protect against protein unfolding, (ii) those with anti-oxidant functions or with roles in repair of oxidative-stress induced damage and (iii) those which encode proteins involved in carbohydrate metabolism, including those involved in the synthesis of the osmo-protectants glycerol and trehalose, and those which generate the metabolic energy required for stress adaptation (sugar transporters, glycolytic enzymes). In addition, a large class of genes are stereotypically repressed in response to stress in budding and fission yeast; these are largely associated with protein synthesis and cellular growth which indicates that the requirement to preserve energy is a conserved feature of all stress responses. These stereotypical responses have been termed the 'environmental stress response' (ESR) or 'common environmental response' (CER) in budding yeast (Gasch *et al.*, 2000; Causton *et al.*, 2001), and the 'core environmental stress response' (CESR) in fission yeast (Chen *et al.*, 2003). Such stereotypical responses are thought to underline the 'general-stress response' previously identified in *S. cerevisiae* and the phenomenon of 'stress-cross protection' (Siderius and Mager, 1997), in which exposure to one type of stress can protect the cell against subsequent exposure to an apparently unrelated stress.

The first indication that stress-responses in *C. albicans* may differ significantly from those in model yeasts came from a transcript-profiling study by the Whiteway group, which examined the transcriptional response of *C. albicans* to similar oxidative (0.4 mM H_2O_2), osmotic (0.3 M NaCl) and heat-stress (23–37 °C) conditions that induced the ESR/CER in *S. cerevisiae* and CESR in *S. pombe* (Enjalbert *et al.*, 2003). Strikingly, no *C. albicans* genes were commonly induced by all three of these stresses, which is in stark contrast to that seen in budding and fission yeasts (Gasch *et al.*, 2000; Causton *et al.*, 2001; Chen *et al.*, 2003). Furthermore, stress-cross protection, a classic hallmark of a 'general-stress response', was not detected in *C. albicans* under these conditions. For example, pre-treatment of *C. albicans* cells with a mild heat shock (23–37 °C) did not confer protection against a subsequent exposure to oxidative stress (Enjalbert *et al.*, 2003).

It is possible, however, that different stresses may be needed to evoke a core-stress response in a pathogenic commensal fungus compared to non-pathogenic environmental yeasts. Indeed, *C. albicans* is relatively resistant to certain stress conditions, such as oxidative stress and heat stress, compared with *S. cerevisiae* and *S. pombe* (Jamieson *et al.*, 1996; Smith *et al.*, 2004). In addition, while a number of studies have highlighted the importance of the Hog1 SAPK in *C. albicans* stress responses (San Jose *et al.*, 1996; Alonso-Monge *et al.*, 2003; Smith *et al.*, 2004), and stress-cross protection is observed if *C. albicans* cells are exposed to stresses that activate Hog1 (Smith *et al.*, 2004), only one of the three stress conditions (0.3 M NaCl) employed in the previous transcript-profiling study (Enjalbert *et al.*, 2003) activates the Hog1 SAPK (Smith *et al.*, 2004, see below).

Therefore to reinvestigate the possible presence of a core transcriptional response to stress, the global responses of *C. albicans* to three stresses that activate Hog1 was examined (0.3 M NaCl; 5 mM H_2O_2; 0.5 mM $CdSO_4$: Enjalbert *et al.*, 2006). Excitingly, subsets of genes were identified that were significantly induced (24) or repressed (37) in response to all three stresses, illustrating that *C. albicans* does mount a core stress response. However, the subset of 24 induced core stress genes identified in *C. albicans* is significantly smaller than the number of induced core stress genes identified in *S. cerevisiae* (300 genes; Gasch *et al.*, 2000: 216 genes; Causton *et al.*, 2001) and in *S. pombe* (140 genes; Chen *et al.*, 2003). The actual number of core stress genes defined in each study is, however, heavily influenced by experimental design and the criteria used to define such genes. Therefore, in an attempt to directly compare the magnitude of the induced core-stress responses in *C. albicans* with that in *S. cerevisiae* and *S. pombe*, to osmotic, oxidative and heavy metal stress, equivalent transcript-profiling datasets were analysed using equivalent analytical tools. Using this approach it is likely that the proportion of core-stress genes in *S. cerevisiae* will be an underestimate, as datasets had to be assembled from three different laboratories to accumulate the necessary experimental conditions. Consequently, the level of consistency between these datasets was lower than that for the corresponding *C. albicans* and *S. pombe* datasets, each of which was generated by a single laboratory. Nevertheless, this comparison revealed that induced core-stress genes represented 3.7% of all stress genes in *C. albicans*, 6.0% of all stress genes in *S. cerevisiae* and 22% of all stress genes in *S. pombe* (Enjalbert *et al.*, 2006). This systematic analysis in which comparable transcript-profiling datasets were analysed using equivalent analytical tools confirmed that the set of induced core-stress genes is genuinely smaller in *C. albicans* compared with budding and fission yeasts.

To compare the global roles of core-stress genes in the three yeast species, functional categories were identified that are significantly enriched in the sets of core-stress genes from *C. albicans*, *S. cerevisiae* and *S. pombe*. In this analysis, the assignment of genes to specific functional categories was done using the gene ontology (GO) resources at SGD (www.yeastgenome.org/GOContents.shtml). Although fewer induced core-stress genes were identified in *C. albicans*, those that were identified belong to some of the same functional categories as budding and fission yeasts. Response to stress and carbohydrate-metabolism genes were significantly over-represented in the core-stress genes from *C. albicans*, *S. cerevisiae* and *S. pombe*. However, other functional categories that were enriched in the core-stress genes of budding and fission yeasts were not over-represented in *C. albicans* core-stress genes. For example, while individual oxidative- and osmotic-stress genes were part of the *C. albicans* core-stress response, these functional categories were not significantly enriched in the core-stress genes. Therefore, there is some overlap, but overall a much narrower set of functional categories were enriched in the core-stress genes in *C. albicans* than in *S. cerevisiae* and *S. pombe* (Enjalbert *et al.*, 2006).

As seen with the induced core-stress genes, the subset of 34 repressed core-stress genes identified in *C. albicans* is relatively small compared to those in *S. cerevisiae* (Gasch *et al.*, 2000: 400 genes; Causton *et al.*, 2001: 283 genes) and

S. pombe (Chen *et al.*, 2003: 106 genes). However, irrespective of the number of genes involved, a dominant feature common to the sets of repressed core-stress genes in *C. albicans*, *S. cerevisiae* and *S. pombe* is their role in protein synthesis. This illustrates that, despite the differences in ecological niches, the requirement to preserve energy in response to stress is conserved between the model yeasts *S. cerevisiae* and *S. pombe* and the pathogenic yeast *C. albicans*.

4.2 Stress-Specific Responses

To compare the specific global transcriptional responses of *C. albicans* to oxidative (H_2O_2), osmotic (NaCl) and heavy metal (cadmium) stresses, the profiles obtained for *C. albicans* were compared with similar transcript-profiling datasets from *S. cerevisiae* and *S. pombe* (Gasch *et al.*, 2000; Fauchon *et al.*, 2002; Chen *et al.*, 2003; O'Rourke and Herskowitz, 2004). Stress-induced genes were assigned to specific functional categories using the GO resources at SGD. This analysis revealed that these three yeasts respond to oxidative, osmotic and heavy metal stress using similar strategies (Enjalbert *et al.*, 2006). For example, in *C. albicans*, *S. cerevisiae* and *S. pombe*, the following functional categories response to oxidative stress, glutathione metabolism, carbohydrate metabolism and energy-reserve metabolism were significantly enriched in the sets of oxidative stress-induced genes. In response to osmotic stress, genes involved in the hyperosmotic response, the response to oxidative stress, carbohydrate metabolism and energy reserve metabolism were induced in all three yeasts. Finally, genes involved in the response to oxidative stress, glutathione metabolism and protein folding were induced by heavy metal stress in all three yeasts. Therefore, there is a high degree of functional overlap between the global oxidative, osmotic and heavy metal stress responses in these three yeasts (Enjalbert *et al.*, 2006). Significantly, many of the genes induced by H_2O_2 treatment of *C. albicans* strains under laboratory growth conditions (Enjalbert *et al.*, 2003, 2006) are also activated upon neutrophil exposure (Fradin *et al.*, 2005), suggesting that the protective functions of these genes are significant *in vivo* against the oxidative burst of host neutrophils. It is also interesting to note that, similar to previous findings in *S. pombe* (Quinn *et al.*, 2002), the transcriptional response to oxidative stress in *C. albicans* is dependent on the dose of H_2O_2, as some sets of genes are differentially induced in response to 0.4–5 mM H_2O_2 (Enjalbert *et al.*, 2003, 2006). For example, while genes involved in peroxide detoxification (*CAP1*, *CTA1*, *GPX1*, *GST3*, *TRR1*, *TRX1*) are induced at both low and high levels of peroxide stress, genes implicated in carbohydrate metabolism (*ICL1*, *GPM2*, *GSY1*, *MLS1*, *NTH1*, *PCK1*) are only induced in response to high levels H_2O_2, whereas genes involved in the DNA-damage response were activated specifically in response to low levels of H_2O_2 (*HNT2*, *IPF4708*, *IPF4356*, *RGA2*). Comparison of the transcript profiles to low and high doses of H_2O_2 (Enjalbert *et al.*, 2003, 2006) with the profile of the *C. albicans* response to neutrophils (Fradin *et al.*, 2005) indicates that *C. albicans* cells may be exposed to relatively high levels of reactive species following neutrophil exposure. For example, many of the carbohydrate-metabolism genes specifically induced following high doses of peroxide stress were also induced upon exposure to neutrophils. Thus, this fungal pathogen appears to have evolved specific mechanisms to defend itself against the potent oxidative burst of mammalian neutrophils (Fang, 2004).

Comparisons were also made with specific-stress genes in *C. albicans*, *S. cerevisiae* and *S. pombe*. To allow such analysis *C. albicans* genes were identified that have orthologues in both budding and fission yeasts. This was achieved by selecting the most significant bidirectional hit for each *C. albicans* protein in *S. cerevisiae* and *S. pombe*. A comparison of the transcriptional responses of hallmark stress genes present in these yeasts revealed that only a small number of genes are regulated in a similar fashion in all three yeasts (Enjalbert *et al.*, 2006). For example, *HSP12* was induced in all three yeasts in response to all three of the stresses, and orthologues of *C. albicans HSP90* were induced in response to heavy metal stress in all three yeasts. However, the majority of genes displayed different regulatory profiles. For example, orthologues of the *C. albicans* oxidative-stress responsive transcription factor Cap1 were strongly induced in *C. albicans* and *S. cerevisiae*, but not in *S. pombe*, in response to oxidative stress. Also, genes involved in trehalose metabolism were significantly induced in *S. cerevisiae* and *S. pombe* in response to osmotic and heavy metal stresses, but with the exception of a small induction of *TPS2*, this was not observed in *C. albicans* (Enjalbert *et al.*, 2006). Collectively this analysis indicates that while *C. albicans*, *S. cerevisiae* and *S. pombe* exploit similar strategies to protect themselves from oxidative, osmotic and heavy metal stresses, the specific genes employed to execute these strategies show a level of divergence.

5. REGULATION OF STRESS-INDUCED GENE EXPRESSION

Studies in the model yeasts *S. cerevisiae* and *S. pombe* have provided significant insight into the stress-signalling pathways and transcription factors that regulate the transcriptional response to stress in unicellular eukaryotes. While similar pathways are present in *C. albicans*, it is becoming increasingly apparent that clear differences exist between the stress responses of this fungal pathogen and those of budding and fission yeasts. In this section I discuss the recent advances that have been made in the molecular responses to stress in *C. albicans*, with particular emphasis on our recent work, and highlight differences in such responses with those in *S. cerevisiae* and *S. pombe*.

5.1 The Hog1 SAPK Pathway

The *S. pombe* Sty1 and the *S. cerevisiae* Hog1 SAPK pathways are among the best characterised eukaryotic stress-signalling systems. The *S. pombe* Sty1 SAPK pathway is activated in response to many diverse stress stimuli including oxidative stress, osmotic stress, temperature up shift, cold stress, nutrient limitation, DNA damaging agents and heavy metals (reviewed in Toone and Jones, 1998), and regulates the CESR in this yeast (Chen *et al.*, 2003). In contrast, the primary function attributed to the *S. cerevisiae* Hog1 SAPK is to respond to osmotic stress (reviewed in Hohmann, 2002), although recent studies have also implicated this SAPK in other stress responses.

Whilst considerably less is known about the regulation of stress responses in *C. albicans* compared to that in the model yeasts *S. cerevisiae* and *S. pombe*, a number of studies have highlighted the importance of the Hog1 SAPK in *C. albicans* stress responses (San Jose *et al.*, 1996; Alonso-Monge *et al.*, 2003; Smith *et al.*, 2004; Arana *et al.*, 2005). The *C. albicans* Hog1 SAPK was originally cloned by functional complementation of the osmosensitive phenotype associated with the *S. cerevisiae hog1* mutant (San Jose *et al.*, 1996). Furthermore, similar to *S. cerevisiae hog1* cells, *C. albicans hog1* cells are sensitive to osmotic stress and do not accumulate the osmo-protectant glycerol on high-osmolarity media (San Jose *et al.*, 1996; Kayingo and Wong, 2005). However, subsequent studies have revealed that the *C. albicans* Hog1 SAPK may be functionally more similar to the Sty1 SAPK pathway in *S. pombe* rather than the Hog1 pathway in *S. cerevisiae*. For example, like Sty1, *C. albicans* Hog1 is phosphorylated (and therefore activated), in response to a diverse range of environmental signals in addition to osmotic stress including various oxidative-stress agents (Alonso-Monge *et al.*, 2003; Smith *et al.*, 2004), the heavy metals cadmium and arsenic, the drug staurosporine, the purine analogue caffeine, and the quorum sensing molecule farnesol (Smith *et al.*, 2004). Furthermore, ectopic expression of *C. albicans HOG1* in *S. pombe*, can rescue all the phenotypes associated with deletion of the *sty1*$^{+}$ SAPK gene (Smith *et al.*, 2004). Significantly, *C. albicans hog1* cells display increased sensitivity compared with wild-type cells when challenged with many of the stresses which activate the Hog1 SAPK, thus illustrating that Hog1 signalling is crucial for the cellular response to these stresses in this pathogen (San Jose *et al.*, 1996; Alonso-Monge *et al.*, 2003; Smith *et al.*, 2004). Hog1 translocates to the nucleus following exposure to stress conditions that activate the kinase (Smith *et al.*, 2004; Arana *et al.*, 2005). Furthermore, Hog1 mediates stress-cross protection (in which exposure to one type of stress can protect the cell against subsequent exposure to an unrelated stress) in *C. albicans* (Smith *et al.*, 2004).

Although the *C. albicans* Hog1 pathway is similar to the *S. pombe* Sty1 pathway in that both respond to diverse stimuli, the activation profiles of the *C. albicans* and *S. pombe* SAPKs are not identical (Smith *et al.*, 2004). In general, the *S. pombe* SAPK pathway responds to more stimuli than the *C. albicans* Hog1 pathway (D. Smith and J. Quinn, unpublished). An analysis of some of the differences indicates that the *C. albicans*-specific SAPK activation profile may reflect an adaptation of this pathway to respond appropriately to the specific microenvironments encountered within the host. For example, while the *S. pombe* Sty1 SAPK is activated following exposure to temperatures of 37 °C or above, the opposite is seen in *C. albicans* where a 'reduction' in the basal levels of Hog1-phosphorylation following temperature increase is observed (Smith *et al.*, 2004). *C. albicans* has evolved to thrive in warm-blooded mammalian hosts where temperatures fluctuations are uncommon and even in the febrile host temperature changes are relatively slow. Therefore, it would not be logical for this human pathogen to perceive temperatures of 37 °C as a stress. Furthermore, higher levels of oxidative stress are required to activate the *C. albicans* Hog1 SAPK than the Sty1 SAPK in *S. pombe*, and Hog1 only translocates to the nucleus in response to high but not low doses of peroxide stress (Smith *et al.*, 2004). Activation of Hog1 specifically to high levels of oxidative stress may be central to the mechanism(s) employed by *C. albicans* to distinguish between the low levels of oxidative stress generated intracellularly, and the high levels of reactive oxygen species generated by the host immune system.

5.2 Role of Hog1 in the Core-Stress Response in *C. albicans*

As discussed above, the *S. pombe* Sty1 SAPK pathway is robustly activated in response to many diverse stress stimuli and co-ordinates a general stress response in this yeast (Chen *et al.*, 2003). As *C. albicans* Hog1 is also activated in response to diverse stress stimuli, it is plausible that Hog1 would regulate a core transcriptional response to stress in this pathogen. To test this prediction we defined the global transcriptional responses in *C. albicans* to three stress stimuli that activate the Hog1 SAPK and, as described above, found that a small but statistically significant core-stress response was evoked under such conditions (Enjalbert *et al.*, 2006). However, the prediction that the Hog1 SAPK would regulate a core transcriptional response to stress in *C. albicans* under conditions that activate this SAPK proved to be only partially correct. The transcript profiles from *hog1* cells revealed that only three core-stress genes (*GPD2, IPF18207* and *RHR2*) displayed some dependency upon Hog1 for their activation in response to all stress conditions tested. While inactivation of *HOG1* significantly attenuated the transcriptional response of core-stress genes to osmotic and heavy metal stress, the activation of the majority of core-stress genes in response to oxidative stress was much less dependent on Hog1. Our subsequent analyses revealed that Cap1 plays a major role in the regulation of core-stress genes in response to oxidative stress (see below). Hence, the Hog1 SAPK pathway functions alone, or in parallel with other pathways, to regulate the core transcriptional response to stress in *C. albicans* (Enjalbert *et al.*, 2006).

This is different to the situation in *S. pombe* in which the Sty1 SAPK is the major regulator of core-stress genes in response to diverse stress stimuli including osmotic, oxidative and heavy metal stresses (Chen *et al.*, 2003). The role of the Hog1 SAPK in the regulation of core-stress responses in *C. albicans* also differs from that of the Hog1 SAPK in *S. cerevisiae*. In *S. cerevisiae*, the induction of core-stress genes in response to osmotic stress requires both Hog1 and the general-stress transcription factors Msn2 and Msn4 (O'Rourke and Herskowitz, 2004). In contrast, *C. albicans* Hog1 alone plays a major role in osmotic-stress induction of the core-stress response in this fungal pathogen. This difference is possibly due to the reassignment of function of Msn2 and Msn4-related proteins in *C. albicans* (Nicholls *et al.*, 2004). Hence, although *C. albicans, S. cerevisiae* and *S. pombe* mount a core transcriptional response to stress, the mechanisms employed to regulate such responses is different in the three fungi.

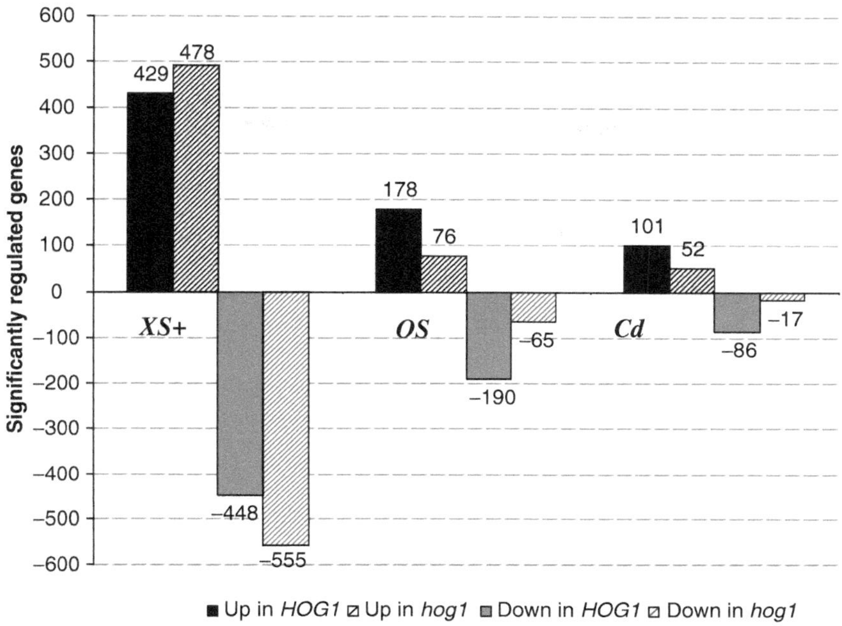

Figure 1 Deletion of *HOG1* has significant effects upon the *C. albicans* transcriptome in response to osmotic and heavy metal stress, but not oxidative stress. The number of significantly regulated genes in *C. albicans* in wild type and *hog1* cells, in response to oxidative (XS+), osmotic (OS), and heavy metal (Cd) stress are shown.

5.3 Role of Hog1 in the Regulation of Stress-Specific Responses in *C. albicans*

Transcript profiles from *hog1* cells revealed that Hog1 plays a key role in the global transcriptional response to osmotic and heavy metal stress, consistent with the stress-sensitive phenotypes associated with *hog1* cells (Figure 1). For example, *C. albicans* genes involved in sulphur amino-acid biosynthesis, which are required for heavy metal tolerance, and genes involved in glycerol accumulation, which are required for osmotic stress tolerance, were deregulated in *hog1* cells (Enjalbert *et al.*, 2006). The transcript-profiling data also provided a molecular basis for the previously observed stress-cross protection in *C. albicans* in which pre-treatment of cells with an osmotic stress conferred significant cross protection against oxidative stress (Smith *et al.*, 2004). Hog1 regulates the expression of genes encoding key anti-oxidant enzymes (*CTA1, IPF6629*) following osmotic stress, which can account for the Hog1-dependency for this stress-cross protection (Enjalbert *et al.*, 2006).

However, the transcript-profiling data revealed that, similar to that seen with the core-stress genes, Hog1 does not play a major role in the overall transcriptional response to oxidative stress in *C. albicans* (Figure 1). These results are surprising as deletion of *HOG1* results in cells with impaired tolerance to oxidative stress and Hog1 is robustly phosphorylated and translocates to the nucleus in response to oxidative stress (Alonso-Monge *et al.*, 2003; Smith *et al.*,

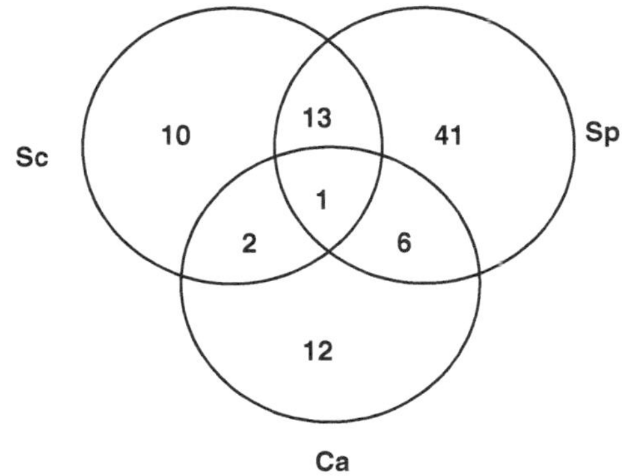

Figure 2 Venn diagram showing SAPK-dependence of stress-regulated orthologues in *C. albicans* (Ca), *S. cerevisiae* (Sc) and *S. pombe* (Sp).

2004). This indicates that the role of Hog1 in the oxidative-stress response in *C. albicans* occurs independently of gene transcription.

5.4 Comparison of SAPK-Regulated Genes

To investigate the role of *C. albicans, S. cerevisiae* and *S. pombe* SAPKs in the regulation of gene expression, a systematic analysis of the SAPK-regulation of orthologues that exist in the three fungi was performed (Figure 2, Enjalbert *et al.*, 2006). Of the 85 orthologues that displayed stress regulation in at least one of the three yeasts, only one such gene (*C. albicans IPF5981, S. cerevisiae MSC1, S. pombe ish1*$^{+}$) was found to be under SAPK regulation in all three yeast. Further cross-comparisons revealed an additional 2 or 6 orthologues to display SAPK-dependency in *C. albicans* and *S. cerevisiae* or *C. albicans* and *S. pombe*, respectively. Interestingly, more orthologues were commonly regulated by SAPKs in the benign yeasts, *S. pombe* and *S. cerevisiae.*

A limitation of such an approach is that SAPK-regulated genes, which have similar functions, but were not defined as orthologues, are disregarded. For example, genes encoding catalase and glyceraldehyde dehydrogenase enzymes are regulated by SAPKs in all three yeast. However, this was not highlighted in the analysis, as the defined orthologues encoding these enzymes were either not regulated by stress or were not under SAPK-regulation, in all three yeasts. However, the systematic approach undertaken in this study was necessary due to the large size of the datasets that were compared, and overall this analysis further highlights the differences in the roles of the SAPKs in gene regulation in the three yeast, *C. albicans, S. cerevisiae* and *S. pombe.*

5.5 Stress-Responsive Transcription Factors

5.5.1 Cap1

A key regulator of oxidative-stress responses in *C. albicans* is the Cap1 transcription factor (Alarco and Raymond, 1999; Zhang *et al.*, 2000; Alonso-Monge *et al.*,

2003; Enjalbert *et al.*, 2006). Cap1 is a bZip transcription factor of the AP-1 family, and is highly similar to the *S. cerevisiae* Yap1 and *S. pombe* Pap1 proteins which have well-studied roles in oxidative-stress responses (see Toone *et al.*, 2001 for a review). Although there is no data regarding the virulence of *C. albicans cap1* strains, it is likely that Cap1 will be essential to survive host responses. For example, *CAP1* is strongly induced upon neutrophil exposure, and *cap1* cells display reduced viability when exposed to whole blood or neutrophils (Fradin *et al.*, 2005).

AP-1-like factors activate the transcription of their target genes via the Yap response element (YRE) (Fernandes *et al.*, 1997; Cohen *et al.*, 2002). The YRE is significantly over-represented in the promoters of genes induced by oxidative stress, compared with other stresses in *C. albicans* (Enjalbert *et al.*, 2006). Consistent with this finding, the oxidative-stress induction of many *C. albicans* genes is significantly attenuated upon inactivation of *CAP1* (Enjalbert *et al.*, 2006; Wang *et al.*, 2006) confirming the key role of Cap1 in oxidative-stress gene transcription.

All of the fungal AP-1-like factors accumulate in the nucleus in response to H_2O_2 (Toone *et al.*, 1998; Delaunay *et al.*, 2000; Zhang *et al.*, 2000). Studies in *S. cerevisiae* first illustrated that this is linked to changes in the redox status of two cysteine-rich domains (n-CRD and c-CRD) that prevent the interaction of Yap1 with the Crm1 nuclear export factor (Delaunay *et al.*, 2000). Similar mechanisms are in place in *S. pombe* and *C. albicans* (Toone *et al.*, 1998; Zhang *et al.*, 2000). However, recent studies, from us and others, have defined an important difference in the regulation of *C. albicans* Cap1 and *S. pombe* Pap1. In *S. pombe*, the nuclear accumulation of Pap1 is dependent upon the Sty1 SAPK, and is inhibited by increasing peroxide concentrations (Toone *et al.*, 1998; Quinn *et al.*, 2002; Vivancos *et al.*, 2004; Bozonet *et al.*, 2005). However, in *C. albicans*, Cap1 rapidly accumulates in the nucleus at both low and high levels of H_2O_2 in a Hog1-independent manner (Alonso-Monge *et al.*, 2003; Enjalbert *et al.*, 2006). The differing relationships between the SAPK and AP-1 signalling pathways in *S. pombe* and *C. albicans* might account for the different roles of the Sty1 and Hog1 SAPKs in oxidative-stress signalling in these fungi. These findings may explain why Sty1 plays a major role in oxidative-stress responses in *S. pombe*, while Hog1 plays a less central role in the global transcriptional response of *C. albicans* to oxidative stress.

5.5.2 Msn2/Msn4

The general transcriptional response to stress in *S. cerevisiae* is partially dependent upon the transcription factors Msn2 and Msn4 (Schmitt and McEntee, 1996; Gasch *et al.*, 2000; Causton *et al.*, 2001). Msn2 and Msn4 activate target genes by binding to specific stress responsive elements (STRE) in their promoters (CCCCT: Mager and de Kruijff, 1995; Martinez-Pastor *et al.*, 1996; Moskvina *et al.*, 1998). The promoters of many stress genes in *C. albicans* contain STRE-like sequences, and two genes encoding Msn2/4-like proteins are present in the *C. albicans* genome (*MSN4* and *MNL1*, for Msn2/4-like protein) (Nicholls *et al.*, 2004). However, a detailed analysis in *C. albicans* revealed that the two Msn2/4-like proteins play no obvious roles in the cellular responses of this pathogen to stress (Nicholls *et al.*, 2004). These observations are consistent with transcript-profiling studies described above, which have shown that the core transcriptional response to stress in *C. albicans* has diverged significantly from the corresponding response in *S. cerevisiae* (Enjalbert *et al.*, 2003, 2006).

6. STRESS RESPONSES IN *C. glabrata*

Virtually nothing is known about 'classical' stress responses in *C. glabrata*. However, it is well documented that this species is relatively resistant to anti-fungal agents (such as azoles), exposure to which can be viewed as a stress (Gleason *et al.*, 1997; Kaur *et al.*, 2004). This resistance does not seem to be confined to anti-fungal toxins as we have found that *C. glabrata* displays a remarkable resistance to extremely high levels of peroxide stress (Figure 3; J. Quinn and K. Haynes, unpublished). Current experiments are aimed at dissecting the functional significance, and the molecular mechanisms, underlying this remarkable peroxide resistance.

Recent studies also suggest that distinct differences in osmotic stress-signalling pathways exist in *C. glabrata* compared with *S. cerevisiae*. In *S. cerevisiae*, osmotic-stress signals are relayed to Hog1 by two redundant pathways. The osmotic stress sensors at the top of these pathways (Sln1 and Sho1) relay signals to the redundant MAPKKKs Ste11 and Ssk2/22, which regulate the activity of Hog1. Deletion of both upstream branches is needed before any effect on osmotic-stress signalling to Hog1 is seen (Maeda *et al.*, 1995; Posas and Saito, 1997). However, *C. glabrata ste11* cells display significant sensitivity to osmotic stress compared to wild-type cells (Calcagno *et al.*, 2005). Preliminary data suggests that Ste11 is the sole MAPKKK, which relays osmotic stress signals to the *C. glabrata* Hog1 SAPK (J. Quinn and K. Haynes, unpublished). This is in direct contrast to that seen in *S. cerevisiae*, in which three MAPKKKs, Stell, Ssk2

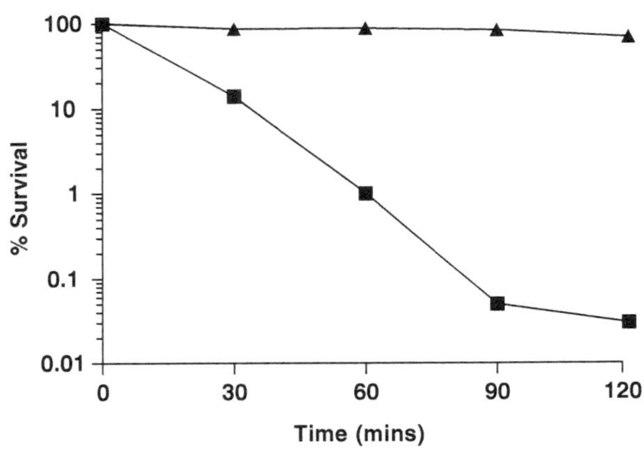

Figure 3 *C. glabrata* is more resistant to oxidative stress than *C. albicans*. Graph illustrating % survival (colony forming units) of *C. albicans* (■) and *C. glabrata* (▲) following treatment with 25 mM H_2O_2.

and Ssk22 need to be inactivated before effects on Hog1 phosphorylation, or sensitivity to osmotic stress, is seen (Maeda *et al.*, 1995; Posas and Saito, 1997). This is different again to the situation in *C. albicans* in which inactivation of both the Sho1 and Sln1 osmotic stress-signalling pathways does not prevent osmotic-stress induced activation of Hog1 (Roman *et al.*, 2005). The functional significance of the different strategies employed to relay osmotic-stress signals to the Hog1 SAPK in *S. cerevisiae, C. albicans* and *C. glabrata* awaits further experimentation.

7. CONCLUSIONS: SPECIALISED STRESS RESPONSES IN *Candida*

Recent data indicate that stress responses are required for the full virulence of *Candida* species. As discussed above, *Candida* has retained many of the stress-response genes and stress-signalling molecules that are present in *S. cerevisiae* and *S. pombe*. However, it is now becoming abundantly clear that there is considerable specialisation in the nature and sensitivity of stress responses in these human fungal pathogens compared with the benign model yeasts *S. cerevisiae* and *S. pombe*. For example, *C. albicans* and *C. glabrata* are significantly more resistant to oxidative stress than these budding and fission yeasts, and whilst *S. cerevisiae* and *S. pombe* struggle to grow at temperatures of 37 °C, *C. albicans* and *C. glabrata* thrive under such conditions. Clearly these differences represent adaptations of these human fungal pathogens to survive the immune defence mechanisms and body temperature of the human host. Consequently, *C. albicans* does not mount a general-stress response to low doses of peroxide stress or a mild heat stress, which stimulate such responses in *S. cerevisiae* and *S. pombe* (Enjalbert *et al.*, 2003). Furthermore, although *C. albicans* has many of the stress-regulatory proteins

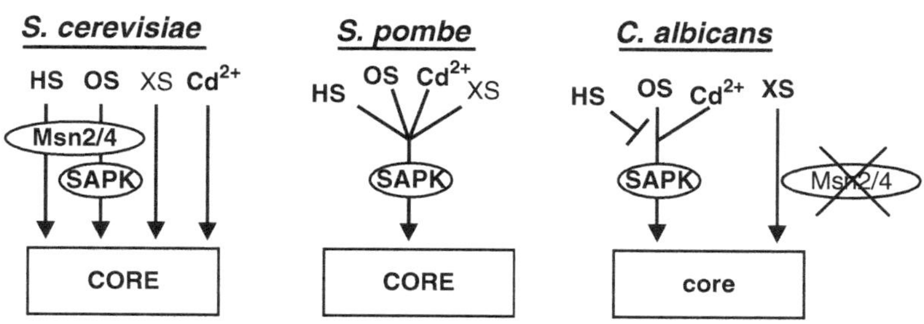

Figure 4 Regulation of core stress responses in *S. cerevisiae, S. pombe* and *C. albicans*. The SAPK pathways in *S. pombe* and *C. albicans* play a central role in regulating core stress responses. however, these pathways have adapted to respond in a niche-specific manner, reflecting the different growth environments of these two yeasts. In contrast, core stress responses in *S. cerevisiae* and *S. pombe* are stimulated by similar conditions, presumably as these organisms inhabit very similar environments. In *S. cerevisiae*, however, core stress responses are regulated by stress-specific pathways which include the SAPK pathway and the Msn2 and Msn4 transcription factors. In contrast, homologues of Msn2 and Msn4 in *C. albicans* do not regulate stress-induced gene Expression. Key, **HS** = heat stress, **OS** = osmotic stress, **XS** = high doses of oxidative stress, XS = low doses of oxidative stress, **Cd^{2+}** = heavy metal stress.

found in *S. cerevisiae* and/or *S. pombe*, the specific molecular mechanisms employed by *C. albicans* to respond to stress have diverged from those observed in these model yeasts. This is exemplified by the different mechanisms in place to regulate core transcriptional responses to stress in *C. albicans*, *S. cerevisiae* and *S. pombe* (Figure 4). A major challenge remaining is to understand the molecular mechanisms underlying the differences in the sensitivity and regulation of stress responses that have evolved in *Candida* species to allow adaptation to environmental stresses that are relevant in the host.

ACKNOWLEDGEMENTS

I gratefully acknowledge my friends and colleagues for many stimulating discussions on fungal stress responses, in particular Al Brown, Ken Haynes and Brian Morgan. I also thank Brice Enjalbert for preparing Figure 1, and Debbie Smith, Jill Cheetham and Iswarya Dantaluru for their continuing hard work at the bench. In addition I also thank the Medical Research Council, the Biotechnology and Biological Sciences Research Council, and the Wellcome Trust for their financial support.

REFERENCES

Alarco, A.M. and Raymond, M. (1999). The bZip transcription factor Cap1p is involved in multidrug resistance and oxidative stress response in *Candida albicans*. *Journal of Bacteriology*, **181**, 700–708.

Alonso-Monge, R., Navarro-Garcia, F., Molero, G., Diez-Orejas, R., Gustin, M., Pla, J., Sanchez, M. and Nombela, C. (1999). Role of the mitogen-activated protein kinase Hog1p in morphogenesis and virulence of *Candida albicans*. *Journal of Bacteriology*, **181**, 3058–3068.

Alonso-Monge, R., Navarro-Garcia, F., Roman, E., Negredo, A.I., Eisman, B., Nombela, C. and Pla, J. (2003). The Hog1 mitogen-activated protein kinase is essential in the oxidative stress response and chlamydospore formation in *Candida albicans*. *Eukaryotic Cell*, **2**, 351–361.

Arana, D.M., Nombela, C., Alonso-Monge, R. and Pla, J. (2005). The Pbs2 MAP kinase kinase is essential for the oxidative-stress response in the fungal pathogen *Candida albicans*. *Microbiology*, **151**, 1033–1049.

Bozonet, S.M., Findlay, V.J., Day, A.M., Cameron, J., Veal, E.A. and Morgan, B.A. (2005). Oxidation of a Eucaryoticic 2-Cys peroxiredoxin is a molecular switch controlling the transcriptional response to increasing levels of hydrogen peroxide. *Journal of Biological Chemistry*, **280**, 23319–23327.

Calcagno, A.M., Bignell, E., Rogers, T.R., Jones, M.D., Muhlschlegel, F.A. and Haynes, K. (2005). Candida glabrata Ste11 is involved in adaptation to hypertonic stress, maintenance of wild-type levels of filamentation and plays a role in virulence. *Medical Mycology*, **43**, 355–364.

Calderone, R.A. and Fonzi, W.A. (2001). Virulence factors of *Candida albicans*. *Trends in Microbiology*, **9**, 327–335.

Calera, J.A., Zhao, X.J. and Calderone, R. (2000). Defective hyphal development and avirulence caused by a deletion of the *SSK1* response regulator gene in *Candida albicans*. *Infection and Immunity*, **68**, 518–525.

Causton, H.C., Ren, B., Koh, S.S., Harbison, C.T., Kanin, E., Jennings, E.G., Lee, T.I., True, H.L., Lander, E.S. and Young, R.A. (2001). Remodeling of yeast genome expression in response to environmental changes. *Molecular Biology of the Cell*, **12**, 323–337.

Chen, D., Toone, W.M., Mata, J., Lyne, R., Burns, G., Kivinen, K., Brazma, A., Jones, N. and Bahler, J. (2003). Global transcriptional responses of fission yeast to environmental stress. *Molecular Biology of the Cell*, **14**, 214–229.

Cohen, B.A., Pilpel, Y., Mitra, R.D. and Church, G.M. (2002). Disrimination between paralogs using microarray analysis: Application to the Yap1p and Yap2p transcriptional networks. *Molecular Biology of the Cell*, **13**, 1608–1614.

Delaunay, A., Isnard, A.D. and Toledano, M.B. (2000). H_2O_2 sensing through oxidation of the Yap1 transcription factor. *The EMBO Journal*, **19**, 5157–5166.

Enjalbert, B., Nantel, A. and Whiteway, M. (2003). Stress-induced gene expression in *Candida albicans*: Absence of a general stress response. *Molecular Biology of the Cell*, **14**, 1460–1467.

Enjalbert, B., Smith, D.A., Cornell, M.J., Alam, I., Nicholls, S., Brown, A.J.P. and Quinn, J. (2006). Role of the Hog1 stress-activated protein kinase in the global transcriptional response to stress in the fungal pathogen *Candida albicans*. *Molecular Biology of the Cell*, **17**, 1018–1032.

Fang, F.C. (2004). Antimicrobial reactive oxygen and nitrogen species: Concepts and controversies. *Nature Reviews in Microbiology*, **2**, 820–832.

Fauchon, M., Lagniel, G., Aude, J.-C., Lombardia, L., Soularue, P., Petat, C., Marguerie, G., Sentenac, A., Werner, M. and Labarre, J. (2002). Sulfur sparing in the yeast proteome in response to sulfur demand. *Molecular Cell*, **9**, 713–723.

Fernandes, L., Rodrigues-Pousada, C. and Struhl, K. (1997). Yap, a novel family of eight bZIP proteins in *Saccharomyces cerevisiae* with distinct biological functions. *Molecular Cell Biology*, **17**, 6982–6993.

Fradin, C., De Groot, P., MacCallum, D., Schaller, M., Klis, F., Odds, F.C. and Hube, B. (2005). Granulocytes govern the transcriptional response, morphology and proliferation of *Candida albicans* in human blood. *Molecular Microbiology*, **56**, 397–415.

Fradin, C., Kretschmar, M., Nichterlein, T., Gaillardin, C., d'Enfert, C. and Hube, B. (2003). Stage-specific gene expression of *Candida albicans* in human blood. *Molecular Microbiology*, **47**, 1523–1543.

Gasch, A.P., Spellman, P.T., Kao, C.M., Carmel-Harel, O., Eisen, M.B., Storz, G., Botstein, D. and Brown, P.O. (2000). Genomic expression programs in the response of yeast cells to environmental changes. *Molecular Biology of the Cell*, **11**, 4241–4257.

Gleason, T.G., May, A.K., Caparelli, D., Farr, B.M., Sawyer, R.G. (1997). Emerging evidence of selection of fluconazole-tolerant fungi in surgical intensive care units. *Archives of Surgery*, **132**, 1197–1201.

Hohmann, S. (2002). Osmotic stress signaling and osmoadaptation in yeasts. *Microbiology and Molecular Biology Reviews*, **66**, 300–372.

Hromatka, B.S., Noble, S.M. and Johnson, A.D. (2005). Transcriptional response of *Candida albicans* to nitric oxide and the role of the *YHB1* gene in nitrosative stress and virulence. *Molecular Biology of the Cell*, **16**, 4814–4826.

Hube, B. (2004). From commensal to pathogen: Stage- and tissue-specific gene expression of *Candida albicans*. *Current Opinion in Microbiology*, **7**, 336–341.

Hwang, C.S., Rhie, G.E., Oh, J.H., Huh, W.K., Yim, H.S. and Kang, S.O. (2002). Copper- and zinc-containing superoxide dismutase (Cu/ZnSOD) is required for the protection of *Candida albicans* against oxidative stresses and the expression of its full virulence. *Microbiology*, **148**, 3705–3713.

Jamieson, D.J., Stephen, D.W. and Terriere, E.C. (1996). Analysis of the adaptive oxidative stress response of *Candida albicans*. *FEMS Microbiology Letters*, **138**, 83–88.

Kaur, R., Castano. I. and Cormack, B.P. (2004). Functional genomic analysis of fluconazole susceptibility in the pathogenic yeast *Candida glabrata*: Roles of calcium signaling and mitochondria. *Antimicrobial Agents and Chemotherapy*, **48**, 1600–1613.

Kayingo, G. and Wong, B. (2005). The MAP kinase Hog1p differentially regulates stress-induced production and accumulation of glycerol and D-arabitol in *Candida albicans*. *Microbiology*, **151**, 2987–2999.

Lorenz, M.C., Bender, J.A. and Fink, G.R. (2004). Transcriptional response of *Candida albicans* upon internalization by macrophages. *Eukaryotic Cell*, **3**, 1076–1087.

Maeda, T., Takekawa, M. and Saito, H. (1995). Activation of yeast PBS2 MAPKK by MAPKKKs or by binding of an SH3-containing osmosensor. *Science*, **269**, 554–558.

Mager, W.H. and de Kruijff, A.J.J. (1995). Stress-induced transcriptional activation. *Microbiology Reviews*, **59**, 506–531.

Martchenko, M., Alarco, A.M., Harcus, D. and Whiteway, M. (2004). Superoxide dismutases in *Candida albicans*: Transcriptional regulation and functional characterization of the hyphal-induced SOD5 gene. *Molecular Biology of the Cell*, **15**, 456–467.

Martinez-Pastor, M.T., Marchler, G., Schuller, C., Marchler-Bauer, A., Ruis, H. and Estruch, F. (1996). The *Saccharomyces cerevisiae* zinc finger proteins Msn2p and Msn4p are required for transcriptional induction through the stress-response element. *EMBO Journal*, **15**, 2227–2235.

Moskvina, E., Schuller, C., Maurer, C.T.C., Mager, W.H. and Ruis, H. (1998). A search in the genome of *Saccharomyces cerevisiae* for genes regulated via stress response elements. *Yeast*, **14**, 1041–1050.

Nicholls, S., Straffon, M., Enjalbert, B., Nantel, A., Macaskill, S., Whiteway, M. and Brown, A.J.P. (2004). Msn2/4-like transcription factors play no obvious roles in the stress responses of the fungal pathogen, *Candida albicans*. *Eukaryotic Cell*, **5**, 1111–1123.

O'Rourke, S.M. and Herskowitz, I. (2004). Unique and redundant roles for HOG MAPK pathway components as revealed by whole-genome expression analysis. *Molecular Biology of the Cell*, **15**, 532–542.

Pfaller, M.A., Jones, R.N., Doern, G.V., Fluit, A.C., Verhoef, J., Sader, H.S., Messer, S.A., Houston, A., Coffman, S. and Hollis, R.J. (1999a). International surveillance of blood stream infections due to *Candida* species in the European SENTRY Program: Species distribution and antifungal susceptibility including the investigational triazole and echinocandin agents. SENTRY participant group (Europe). *Diagnostic Microbiology and Infectious Disease*, **35**, 19–25.

Pfaller, M.A., Jones, R.N., Doern, G.V., Sader, H.S., Messer, S.A., Houston, A., Coffman, S. and Hollis, R.J. (2000). Bloodstream infections due to Candida species: SENTRY antimicrobial surveillance program in North America and Latin America, 1997–1998. *Antimicrobial Agents and Chemotherapy*, **44**, 747–751.

Pfaller, M.A., Messer, S.A., Hollis, R.J., Jones, R.N., Doern, G.V., Brandt, M.E. and Hajjeh, R.A. (1999b). Trends in species distribution and susceptibility to fluconazole among blood stream isolates of *Candida* species in the United States. *Diagnostic Microbiology and Infectious Disease*, **33**, 217–222.

Posas, F. and Saito, H. (1997). Osmotic activation of the HOG MAPK pathway via Ste11p MAPKKK: Scaffold role of Pbs2p MAPKK. *Science*, **276**, 1702–1705.

Quinn, J., Findlay, V.J., Dawson, K., Millar, J.B., Jones, N., Morgan, B.A. and Toone, W.M. (2002). Distinct regulatory proteins control the graded transcriptional response to increasing H_2O_2 levels in fission yeast *Schizosaccharomyces pombe*. *Molecular Biology of the Cell*, **13**, 805–816.

Roman, E., Nombela, C. and Pla, J. (2005). The Sho1 adaptor protein links oxidative stress to morphogenesis and cell wall biosynthesis in the fungal pathogen *Candida albicans*. *Molecular and Cell Biology*, **25**, 10611–10627.

Rubin-Bejerano, I., Fraser, I., Grisafi, P. and Fink, G.R. (2003). Phagocytosis by neutrophils induces an amino acid deprivation response in *Saccharomyces cerevisiae* and *Candida albicans*. *Proceedings of the National Academy of Sciences United States of America*, **100**, 11007–11012.

San-Jose, C., Monge, R.A., Perez-Diaz, R., Pla, J. and Nombela, C. (1996). The mitogen-activated protein kinase homolog *HOG1* gene controls glycerol accumulation in the pathogenic fungus *Candida albicans*. *Journal of Bacteriology*, **178**, 5850–5852.

Schmitt, A.P. and McEntee, K. (1996). Msn2p, a zinc finger DNA-binding protein, is the transcriptional activator of the multistress response in *Saccharomyces cerevisiae*. *Proceedings of the National Academy of Sciences United States of America*, **93**, 5777–5782.

Siderius, M. and Mager, W.H. (1997). *General Stress Response*. Springer-Verlag, New York.

Smith, D.A., Nicholls, S., Morgan, B.A., Brown, A.J. and Quinn, J. (2004). A conserved stress-activated protein kinase regulates a core stress response in the human pathogen *Candida albicans*. *Molecular Biology of the Cell*, **15**, 4179–4190.

Toone, W.M. and Jones, N. (1998). Stress-activated signalling pathways in yeast. *Genes to Cells*, **3**, 485–498.

Toone, W.M., Kuge, S., Samuels, M., Morgan, B.A., Toda, T. and Jones, N. (1998). Regulation of the fission yeast transcription factor Pap1 by oxidative stress: Requirement for the nuclear export factor Crm1 (Exportin) and the stress-activated MAP kinase Sty1/Spc1. *Genes and Development*, **12**, 1453–1463.

Toone, W.M., Morgan, B.A. and Jones, N. (2001). Redox control of AP-1-like factors in yeast and beyond. *Oncogene*, **20**, 2336–2346.

Van Dijck, P., De Rop, L., Szlufcik, K., Van Ael, E. and Thevelein, J.M. (2002). Disruption of the *Candida albicans TPS2* gene encoding trehalose-6-phosphate phosphatase decreases infectivity without affecting hypha formation. *Infection and Immunity*, **70**, 1772–1782.

Viscoli, C., Girmenia, C., Marinus, A., Collette, L., Martino, P., Vandercam, B., Doyen, C., Lebeau, B., Spence, D., Krcmery, V., De Pauw, B. and Meunier, F. (1999). Candidemia in cancer patients:

A prospective, multicenter surveillance study by the invasive fungal infection group (IFIG) of the European organization for research and treatment of cancer (EORTC). *Clinical Infectious Diseases*, **28**, 1071–1079.

Vivancos, A.P., Castillo, E.A., Jones, N., Ayte, J. and Hidalgo, E. (2004). Activation of the redox sensor Pap1 by hydrogen peroxide requires modulation of the intracellular oxidant concentration. *Molecular Microbiology*, **52**, 1427–1435.

Wang, Y., Cao, Y.Y., Jia, X.M., Cao, Y.B., Gao, P.H., Fu, X.P., Ying, K., Chen, W.S. and Jiang, Y.Y. (2006). Cap1p is involved in multiple pathways of oxidative stress response in *Candida albicans*. *Free Radical Biology and Medicine*, **40**, 1201–1209.

Wysong, D.R., Christin, L., Sugar, A.M., Robbins, P.W. and Diamond, R.D. (1998). Cloning and sequencing of a *Candida albicans* catalase gene and effects of disruption of this gene. *Infection and Immunity*, **66**, 1953–1961.

Zaragoza, O., Blasquez, M.A. and Gancedo, C. (1998). Disruption of the *Candida albicans TPS1* gene encoding trehalose-6-phosphate synthase impairs formation of hyphae and decreases virulence. *Journal of Bacteriology*, **180**, 3809–3815.

Zhang, X., De Micheli, M., Coleman, S.T., Sanglard, D. and Moye-Rowley, W.S. (2000). Analysis of the oxidative stress regulation of the *Candida albicans* transcription factor, Cap1p. *Molecular Microbiology*, **36**, 618–629.

CHAPTER 6

The Remarkably Diverse Pathogenicity Effectors of the Obligate Oomycete *Hyaloperonospora parasitica*

Jim Beynon

Contents

1. Plants have Several Layers of Resistance Mechanisms 88
   1.1 RPP13/ATR13 90
   1.2 RPP1/ATR1 92
2. Oomycete and Malarial Pathogencity Effectors Reveal a Conserved Motif 93
3. Oomycete Genomics will Revolutionise Analysis of Plant Immune Systems 94

Acknowledgements 94
References 94

Abstract

Oomycete pathogens of plants, such as potato blight, cause massive yield losses every year but they are also proving to be important to our understanding of host defence mechanisms. *Hyaloperonospora parasitica* (*Hpat*) causes downy mildew disease of Arabidopsis and has been used in many studies that analyse the genes and signalling mechanisms of the host plant involved in disease resistance. We have cloned two resistance genes, *RPP1* and *RPP13*, from Arabidopsis that specify isolate-specific resistance to *Hpat* and shown them to have diverse recognition capabilities and high levels of allelic diversity. Recently, we have cloned the matching pathogen genes, *ATR1* and *ATR13*, that trigger the *RPP1* and *RPP13* specified resistance, respectively. These genes show no homology to other known proteins and reveal amazing levels of allelic diversity. They demonstrate clear evidence of an "arms race" between host immunity and pathogen-effector proteins. Both genes contain a signal peptide followed by an RXLR motif that we predict is involved in targeting the effectors to the host plant cytoplasm as it shares common features with the RXL motif found in malarial parasite effectors. We predict that these pathogen effectors are involved in suppressing host basal or race-specific immune response mechanisms. The advent of the *Hpat*

British Mycological Society Symposia Series
Published by Elsevier Ltd.

genome sequence will, therefore, provide new tools with which to unravel the Arabidopsis immune system.

1. PLANTS HAVE SEVERAL LAYERS OF RESISTANCE MECHANISMS

Plants have developed systems to recognise invading pathogens and respond to that invasion by preventing pathogen growth. A basal defence system responds to signals of a potential pathogen's presence (pathogen-associated molecular patterns — PAMPS) and prevents invasion. PAMPS are usually highly conserved components of the pathogen such as glucans, flagellin or chitin that act as general signals to the host immune systems preventing the invasion of a large range of organisms (Chisholm *et al.*, 2006). Potentially, therefore, most organisms are not pathogens of a particular plant species because their PAMPS activate a basal immune response. In order to become a host-specific pathogen some organisms are able to interfere with this basal system and establish a successful infection. To achieve this suppression of immunity, a pathogen has to develop a range of proteins, pathogenicity effectors, specifically targeted to host cells, which interfere with basal immunity and allow the organism to grow and reproduce on host tissues (e.g. He *et al.*, 2006).

To counteract this suppression of basal immunity, plants have developed another line of defence represented by the plant resistance (R) genes. The largest class of R genes contain leucine-rich repeats (LRR); the model plant Arabidopsis contains over 200 predicted open reading frames that could represent R genes. These can be divided into those with extracellular LRR domain and those that are likely to be located in the cytoplasm (approximately 150). The presence of LRRs implies some sort of interaction. This is consistent with the role of R genes being the detection of pathogen proteins (or metabolites) that are exposed to the plant cell on invasion. The LRR domains of R genes are often under diversifying selection implying that the pathogen component of the interaction is being changed to avoid detection. It is logical, therefore, that the plant R protein and the pathogen protein responds to interact directly. Although this has been shown in a few cases (Tang *et al.*, 1996; Deslandes *et al.*, 2003; Dodds *et al.*, 2006), usually such direct interaction has not been demonstrated. Indeed, it appears that the pathogen proteins themselves have other plant-protein targets and it is that interaction that is detected by the R protein, triggering a resistance response. The pathogen genes involved in these interactions have confusingly been called avirulence genes as their presence triggers a resistance gene resulting in the pathogen being avirulent (fails to grow) on the host plant. However, it is more likely that such proteins are in fact pathogenicity effectors targeted to plant proteins involved in the basal immune system to eliminate their ability to co-ordinate an effective resistance response. The R proteins may have then evolved to initiate a resistance response when the basal immune system has been or is under threat of being suppressed or bypassed. A good example of such an interaction is that mediated via the Arabidopsis protein RIN4. *RPM1*-mediated resistance is triggered by the *Pseudomonas AvrRpm1* avirulence gene product. RPM1 was shown by yeast two hybrid (Y2H) to interact with a host protein called RIN4, which is required for RPM1 function (Mackey *et al.*, 2002) and is likely to, itself, play a role in innate immunity (Kim *et al.*, 2005). The AvrRPM1 protein has been shown to phosphorylate and so inactivate RIN4 and, hence, it is RPM1's role to detect this activity. Another avirulence protein from *Pseudomonas*, AvrRpt2, degrades RIN4 resulting in loss of RPM1-mediated resistance (Mackey *et al.*, 2003). To counteract this, the plant has developed the *RPS2* resistance gene, which triggers resistance by detecting loss of RIN4 due to AvrRpt2. Hence, the pathogen has developed two attack proteins to inactivate RIN4, an important defence protein. RPM1 also interacts with RIN13, an Arabidopsis protein that specifically enhances the RPM1 resistance response (Al-Daoude *et al.*, 2005). Overexpression of RIN13 results in loss of the physical manifestation of plant-resistance responses but increased resistance to a bacterial pathogen. This may imply that resistance and plant cell death are not coupled. Another example is that of the pathogen protein AvrPphB that cleaves a protein kinase (PBS1) resulting in the triggering of resistance mediated by the R protein RPS5 (Shao *et al.*, 2003). These fascinating stories are just two of the many potential complex interactions between Arabidopsis and its pathogens. Similarly, in tomato the AvrPto protein interacts with the Pto kinase, an interaction that is detected by the Prf R protein.

Importantly, it appears that pathogens have developed proteins that interfere with host proteins important to basal immune responses. It is also possible to imagine that under selective pressure caused by the presence of the R proteins that the pathogen would evolve proteins targeted at suppressing this secondary host-resistance mechanism. Therefore, such pathogen effector proteins are potentially valuable tools that can be used to identify host proteins that may be fundamental to plant defence processes.

Arabidopsis is the model molecular plant system, consequently many tools are available for studying the effects of pathogen invasion and to elucidate host-resistance mechanisms. The oomycete pathogen *Hpat* causes downy mildew disease of Arabidopsis and can easily be isolated from wild populations of Arabidopsis. This suggests that these organisms have co-evolved and may, therefore, be good models for studying host–pathogen interactions. When a range of Arabidopsis accessions are challenged with different *Hpat* isolates a diverse set of interactions are revealed from heavy sporulation, through various restricted sporulation phenotypes to clear host responses manifested through localised cell death restricting pathogen invasion (Holub and Beynon, 1997). These phenotypes suggest that there is a dynamic interaction between host and pathogen. The key question is whether this is mediated via a complex interaction between pathogen effector proteins and host innate and R gene-mediated resistance mechanisms. Over the last decade, several R genes have been cloned or genetically defined in Arabidopsis and the majority of these have come from the Arabidopsis-*Hpat* interaction. The *R* genes *RPP1* (recognition of *Hpat*), *RPP2a*, *RPP2b*, *RPP4*, *RPP5*, *RPP8*, *RPP13* and *RPP26* have all been cloned from Arabidopsis and code for R proteins (Parker *et al.*, 1997; Botella *et al.*, 1998; McDowell *et al.*, 1998; Bittner-Eddy *et al.*, 2000; van der Biezen *et al.*, 2002; Sinnapidou *et al.*, 2004; Sepahvand and Beynon, unpublished). All other code for proteins predicted to contain LRR domains and to be cytoplasmically located. As Arabidopsis has evolved mechanisms to detect *Hpat* intra-cellularly, this would imply that *Hpat* effectors are delivered to the host cytoplasm, consistent with a role in suppressing host immune responses through a range of targets. The commitment by Arabidopsis to the development of many *R* genes to detect *Hpat* implies that the host regards this organism as a significant pathogen, and that the pathogen is required to deploy a diverse collection of effector proteins to successfully invade host tissue without eliciting a defence response.

Recent cloning of pathogen effector and resistance gene pairs from the *Hpat*/Arabidopsis interaction has revealed a fascinating insight into what is to come from studying this host–pathogen interaction:

1.1 RPP13/ATR13

The *RPP13* gene codes for a protein that contains a coiled coil (CC) domain, a nucleotide-binding site (NB) and a leucine-rich repeat domain and, hence, falls into the CC:NB:LRR class of cytoplasmically located R proteins (Bittner-Eddy *et al.*, 2000). RPP13Nd (the two-letter code represents the Arabidopsis accession containing the particular *RPP13* allele) recognises the presence of the *Hpat* isolates Maks9 and Emco5, whereas RPP13-Rld recognises the isolate Wela3. This suggests that there is diversity of recognition capability at the *RPP13* locus and that it plays an active role in preventing invasion by *Hpat*. *RPP13* is present as a single gene in Arabidopsis allowing easy analysis of the gene from a range of Arabidopsis accessions (Bittner-Eddy *et al.*, 2000). Extensive allele sequencing has revealed that *RPP13* is a highly variable gene: the level of nucleotide change (*K*(s)) is one of the highest in genes so far studied in Arabidopsis but what is most startling is the frequency (*K*(a)) with which these result in amino acid changes in the protein (Rose *et al.*, 2004). The *K*(a) score is an order of magnitude greater than any gene analysed so far in Arabidopsis. The *K*(a)/*K*(s) ratio is a measure of the nature of selective pressure on an open reading frame: a ratio of one implies neutral selection, less than one implies selection for amino acid conservation and greater than one that the gene is under diversifying selection. The *K*(a)/*K*(s) ratio for *RPP13* is approximately one, implying that the gene is under neutral selection. Although, this is misleading as if the analysis is carried out on the CC:NB and LRR domains separately, it is clear that the CC:NB is under purifying selection whereas the LRR region is under very high levels of diversifying selection (Figure 1). This implies that the LRR domain is under the influence of a diversifying factor and as it is possible that this region is involved in protein–protein interactions, it would seem likely that it is involved in detecting the presence of *Hpat*-effector protein. As the CC:NB region is under selection for conservation of amino acid sequence, it is likely that this is involved in a conserved role of the protein, such as signal transduction in response to detection of the *Hpat* effector. When cloned and transformed into a susceptible Arabidopsis accession, the *RPP13-Nd* and *RPP13-Rld* genes only recognised Maks9 and Emco5 or Wela3, respectively. This suggests that RPP13-Nd and RPP13-Rld either recognise different alleles of the same *Hpat* effector or different effector proteins altogether. If it

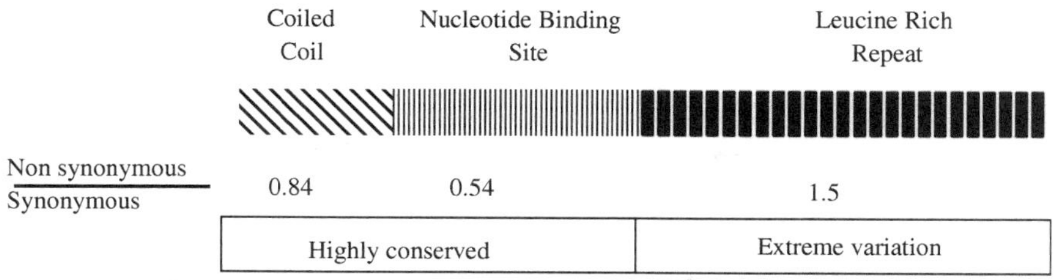

Figure 1 Ratio of non-synonymous to synonymous mutation rates *K*(a)/*K*(s) provides a measure of nature of the selective pressures experienced by the gene.

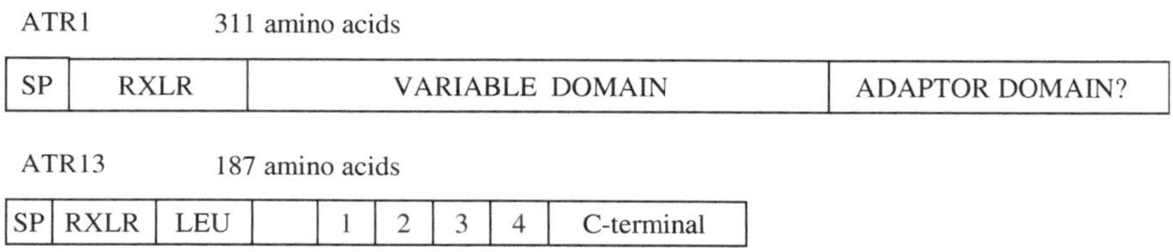

Figure 2 SP: signal peptide. RXLR: conserved amino acid motif probably involved in transport of effector into host cells. LEU: a heptad leucine repeat. In ATR13 numbers represent 11 amino acid repeats.

were alleles of the same effector then the pathogen gene could also be expected to be under diversifying selection to match that seen in *RPP13*.

ATR13, the *Hpat* effector that triggers isolate-specific resistance specified by RPP13, codes for a protein of 187 amino acids (Figure 2) (Allen *et al.*, 2004). At the N-terminus is a predicted signal peptide (SP), implying that the protein is secreted by the pathogen into the apoplastic space allowing the protein to come into contact with the host cell. The rest of the protein can be divided into three regions: following the SP is a heptad leucine repeat, reminiscent of a CC domain but not associated with a predicted alpha helix, which precedes four copies of an imperfect eleven amino acid repeat, that shows no homology to known protein motifs, and terminates with a region containing no clear patterns in the amino acid sequence. As with *RPP13*, *ATR13* is present as a single copy gene in *Hpat* and allelic sequencing revealed very high levels of sequence diversity. Determining *K*(a)/*K*(s) ratios in these examples was not meaningful as all alleles contained many nucleotide changes and in some cases all of them resulted in amino acid changes. Therefore, ATR13 is under very high levels of diversifying selection, the only region of the gene to reveal any level of conservation was the SP, which would not be expected to be exposed to the host-resistance mechanisms and, hence, would not be anticipated to be under selective pressure. Allelic sequencing also implied that recognition specificity by RPP13 was coded in the C-terminal region of ATR13. Domain swap experiments have confirmed this observation (Rebecca Allen, unpublished). However, alleles recognised or not by RPP13, all showed extensive levels of sequence diversity suggesting that other genes in Arabidopsis were exerting selective pressure on ATR13. This raises the possibility that other resistance genes may recognise ATR13.

1.2 RPP1/ATR1

RPP1 codes for a resistance protein that is predicted to be cytoplasmically located and contains a toll/interleukin receptor (TIR)-like domain, a nucleotide-binding site and a leucine-rich repeat (TIR:NB:LRR) (Botella *et al.*, 1998). Unlike *RPP13*, *RPP1* lies within a complex locus, the structure of which can vary greatly between Arabidopsis accessions. For example accession Ws-0 contains four open reading frames within the *RPP1* locus, three of which have been cloned and shown to be active resistance genes. Each gene was shown to be capable of recognising and initiating a resistance response to a different set of *Hpat* isolates. For example, RPP1-WsA recognised *Hpat* isolates Noks1, Maks9, Emoy2 and Cala2 whereas RPP1-WsB recognised Noks1, Maks9 and Emoy2 but not Cala2. This raised two possibilities, that RPP1-WsA recognised the same pathogen protein in Noks1, Maks9, Emoy2 and Cala2 and RPP1-WsB recognised a different protein from Noks1, Maks9 and Emoy2. Alternatively, RPP1-WsA and RPP1-WsB recognised the same protein in Noks1, Maks9 and Emoy2 and RPP1-WsA responded to a different protein from Cala2. In contrast, from Southern hybridisation data, accession Nd-1 appears to contain many potential *R* genes at the *RPP1* locus, however sequence analysis reveals only one complete open reading frame amidst many pseudo-genes (Gordon, 2002). This single open reading frame, *RPP1-Nd*, was cloned and shown to be capable of recognising all the isolates (Emoy2, Hiks1 and Waco5) for which isolate-specific recognition had been mapped to the *RPP1* locus (Gordon, 2002; Rehmany *et al.*, 2005). The simplest explanation of this phenotype was that RPP1-Nd recognised the same protein in each pathogen isolate. However, RPP1-Nd and RPP1-WsA both recognise isolate Emoy2 but no other similar isolates, hinting at even more complex potential interactions.

ATR1, the *Hpat* effector that triggers isolate-specific resistance specified by RPP1, codes for a protein of 311 amino acids (Figure 2) (Rehmany *et al.*, 2005). As with ATR13, a SP is present at the N-terminus of the protein suggesting export into the apoplastic space, however, no other motifs could be identified in the rest of the protein. Sequencing revealed that *Hpat* isolates Emoy2, Hiks1 and Waco5 contained the same allele, consistent with recognition by RPP1-Nd. However, *ATR1* alleles from Emoy2, Noks1, Maks9 and Cala2 showed high levels of DNA and amino acid sequence diversity. RPP1-WsB recognises ATR1-Emoy2, ATR1-Maks9 and ATR1-Noks1 but not ATR1-Cala2 consistent with the isolate-specific recognition capability of the resistance gene. This demonstrated that, as with RPP13, Arabidopsis resistance genes can recognise highly variable pathogen proteins. However, RPP1-WsA did not recognise ATR1-Cala2 suggesting that this resistance protein can interact with a different pathogen protein. The pathogen isolate Emco5 is not recognised by Arabidopsis accession Ws-0 and would not, therefore, be expected to contain an *ATR1* allele capable of triggering RPP1-mediated resistance. However, RPP1-WsB does trigger a resistance response when challenged with ATR1-Emco5. *ATR1*-Emco5 is expressed *in planta* suggesting that Arabidopsis plants challenged with Emco5 are exposed to ATR1. Therefore, it is possible that another pathogen protein acts as a suppressor of RPP1 recognition of ATR1. This suggests the possibility that *Hpat* has evolved another class of effector to suppress the *R* gene-mediated resistance mechanisms of Arabidopsis.

This analysis of two host–pathogen interacting gene pairs has already revealed a wealth of tantalising results that demonstrate an "arms race" between the two organisms. In this battle, the pathogen effector designed to overcome host-resistance mechanisms, is under constant selection to generate novel protein structure to evade detection by R proteins while maintaining function. In response to rapidly changing effector structure the R proteins themselves generate novel forms. However, this battle is complex involving the interactions of multiple pathogen effectors and host-resistance proteins, possibly innate immunity and *R* gene-dependent mechanisms, to determine a successful invasion pathogen or its exclusion by host defence mechanisms.

2. OOMYCETE AND MALARIAL PATHOGENCITY EFFECTORS REVEAL A CONSERVED MOTIF

Comparisons of proven effector proteins from oomycete plant pathogens (ATR1 and ATR13 from *Hpat*, Avr3a from *Phytophthora infestans* and Avr1b from *Phytophthora sojae*) show no overall homology (Allen *et al.*, 2004; Shan *et al.*, 2004; Armstrong *et al.*, 2005; Rehmany *et al.*, 2005). This would be consistent with them interacting with host-specific immune systems and evolving separately under high levels of diversifying selection resulting in diverged protein sequences. However, a region following the signal peptide, present in each protein, showed significant levels of sequence conservation (Rehmany *et al.*, 2005). All contained an RXLR (arginine, any amino acid, leucine, arginine) sequence followed in all but ATR13 by an acidic region often containing the motif DEER (aspartate, glutamate, glutamate, arginine). Spatial and sequence conservation of this motif in effector proteins from such diverse pathogens implies a fundamental role in pathogenicity. All four effectors are recognised by cytoplasmically located R proteins implying that they are all targeted to the host-plant cytoplasm. Therefore, as the RXLR motif follows the SP, it would be in the correct position to be involved in transferring the effector protein across the host membrane to enter the plant cell (Rehmany *et al.*, 2005). This possibility was enhanced by the observation that the amino acid sequence RXL located, following an SP in pathogenicity effectors from the malarial parasite *Plasmodium*, was required for transferring the pathogen proteins into human erythrocytes (Hiller *et al.*, 2004; Marti *et al.*, 2004; Bhattacharjee *et al.*, 2006).

Although oomycetes phenotypically resemble fungi, in that they produce hyphae, produce similar invagination structures to obtain intimate contact with host cell membranes and produce sporulation structures and vegetative and sexual spores for propagation, they are only very distantly related. Oomycetes belong to the Stramenopiles that are closely related to the Alveolates in which lie the Apicomplexans containing *Plasmodium*. Hence, the RXL motif may have evolved in an ancient group that gave rise to the Stramenopiles and Alveolates and it will be interesting to determine how wide spread this motif is in exported proteins from organisms in these regions of the evolutionary tree. Another key question will be whether the RXL motif is always associated with pathogenic life cycles.

3. OOMYCETE GENOMICS WILL REVOLUTIONISE ANALYSIS OF PLANT IMMUNE SYSTEMS

The genome sequence of *Hpat* isolate Emoy2 is currently being completed and a primary assembly has recently been released. This will allow bioinformatic analyses to be carried out to identify a significant proportion of the effector component of the pathogen. The proteins produced by these genes are presumably designed to overcome host immune systems in a highly adapted host–pathogen interaction. As such these become tools to the scientists to probe and unravel the complex molecular interactions between these two organisms.

Three *Phytophthora* genome sequences (*P.sojae*—http://phytophthora.vbi.vt.edu, *P. ramorum*—http://phytophthora.vbi.vt.edu and *P.infestans*—www.broad.mit.edu/annotation/genome/phytophthora_infestans/Home.html) have either been completed or are in progress. These sequences will enable studies of interaction between pathogens and crop plants and combined with *Hpat* studies allow the design of new methods of protecting crops from disease in the field. They will also allow comparative genomic studies to reveal those genes common between oomycetes and those that under host-specific selective pressures have diverged. Fundamental mechanism of pathogenesis will, therefore, be revealed which could become targets for therapeutic intervention.

ACKNOWLEDGEMENTS

I would like to thank Peter Bittner-Eddy, Anne Rehmany, Rebecca Allen, Sharon Hall, Laura Baxter, Mary Coates, Julia Meitz and Linda Hughes for carrying out much of the work described in this paper and for opening up this research field enabling the transition to a genomic era. I would also like to thank Brett Tyler and John McDowell for their major contributions to the *Hpat* genome-sequencing project. Finally, I must mention Eric Holub, Ian Crute and Paul Birch whose scientific influence has been highly significant in enabling the analysis of *Hpat* at the genetic and genomic levels. The work in my laboratory has been extensively funded by the Biotechnology and Biological Sciences Research Council.

REFERENCES

Al-Daoude, A., de Torres Zabala, M., Ko, J.H. and Grant, M. (2005). RIN13 is a positive regulator of the plant disease resistance protein RPM1. *The Plant Cell*, **17**, 1016–1028.

Allen, R.L., Bittner-Eddy, P.D., Grenville-Briggs, L.J., Meitz, J.C., Rehmany, A.P., Rose, L.E. and Beynon, J.L. (2004). Host-parasite coevolutionary conflict between Arabidopsis and downy mildew. *Science*, **306**, 1957–1960.

Armstrong, M.R., Whisson, S.C., Pritchard, L., Jorunn, I.B. Bos, Venter, E., Avrova, A.O., Rehmany, A.P., Böhme, U., Brooks, K.,Cherevach, I., Hamlin, N., White, B., Fraser, A., Lord, A., Quail, M.A., Churcher, C., Hall, N., Berriman, M., Huang, S., Kamoun, S., Beynon, J.L. and Birch, P.R.J. (2005). An ancestral oomycete locus contains late blight avirulence gene Avr3a, encoding a protein that is recognized in the host cytoplasm. *Proceedings of the National Academy of Science*, **102**, 7766–7771.

Bhattacharjee, S., Hiller, N.L., Liolios, K., Win, J., Kanneganti, T.D., Young, C., Kamoun, S. and Haldar, K. (2006). The malarial host-targeting signal is conserved in the Irish potato famine pathogen. *PLoS Pathogens*, **2**(5), e50 doi:10.1373/journal.ppat.0020050.

Bittner-Eddy, P.D., Crute, I.R., Holub, E.B. and Beynon, J.L. (2000). *RPP13* is a simple locus in *Arabidopsis thaliana* for alleles that specify downy mildew resistance to different avirulence determinants in *Peronospora parasitica*. *The Plant Journal*, **21**, 177–188.

Botella, M.A., Parker, J.E., Frost, L.N., Bittner-Eddy, P.D., Beynon, J.L., Daniels, M.J., Holub, E.B. and Jones, J.D.G. (1998). Three genes of the Arabidopsis *RPP1* complex resistance locus recognise distinct *Peronospora parasitica* avirulence determinants. *The Plant Cell*, **10**, 1847–1860.

Chisholm, S.T., Coaker, G., Day, B. and Staskawicz, B.J. (2006). Host-microbe interactions: Shaping the evolution of the plant immune response. *Cell*, **124**, 803–814.

Deslandes, L., Olivier, J., Peeters, N., Feng, D.X., Khounlotham, M., Boucher, C., Somssich, I., Genin, S. and Marco, Y. (2003). Physical interaction between RRS1-R, a protein conferring resistance to bacterial wilt, and PopP2, a type III effector targeted to the plant nucleus. *Proceedings of the National Academy of Science, USA*, **100**, 8024–8029.

Dodds, P.N., Lawrence, G.J., Catanzariti, A.M., The, T., Wang, C.A., Ayliffe, M.A., Kobe, B. and Ellis, J.G. (2006). Direct protein interaction underlies gene-for-gene specificity and coevolution of the flax resistance genes and flax rust avirulence genes. *Proceedings of the National Academy of Science, USA*, **103**, 8888–8893.

Gordon, A. (2002). Analsysis of the *RPP1* resistance gene cluster in Arabidopsis accession Niederzenz. Ph.D. thesis, University of Birmingham.

He, P., Shan, L., Lin, N. C., Martin, G.B., Kemmerling, B., Nürnberger, T. and Sheen, J. (2006). Specific bacterial suppressors of MAMP signaling upstream of MAPKKK in *Arabidopsis* innate immunity. *Cell*, **125**, 563–575.

Hiller, N.L., Bhattacharjee, S., van Ooij, C., Liolios, K., Harrison, T., Lopez-Estrano, C., and Haldar, K. (2004). A host-targeting signal in virulence proteins reveals a secretome in malarial infection. *Science*, **306**, 1934–1937.

Holub, E.B. and Beynon, J.L. (1997). Symbiology of mouse-ear cress (*Arabidopsis thaliana*) and oomycetes. *Advances in Botanical Research*, **24**, 227–273.

Kim, M.G., da Cunha, L., McFall, A.J., Belkhadir, Y., DebRoy, S., Dangl, J.L. and Mackey, D. (2005) Two *Pseudomonas syringae* Type III effectors inhibit RIN4-regulated basal defense in *Arabidopsis*. *Cell*, **121**, 749–759.

Mackey, D., Belkhadir, Y., Alonso, J.M., Ecker, J.R. and Dangl, J.L. (2003). Arabidopsis RIN4 is a target of the Type III virulence effector AvrRpt2 and modulates RPS2-mediated resistance. *Cell*, **112**, 379–389.

Mackey, D., Holt III, B.F., Wiig, A. and Dangl, J.L. (2002). RIN4 interacts with *Pseudomonas syringae* type III effector molecules and is required for RPM1-mediated resistance in Arabidopsis. *Cell*, **108**, 743–754.

Marti, M., Good, R.T., Rug, M., Knuepfer, E. and Cowman, A.F. (2004). Targeting malaria virulence and remodeling proteins to the host erythrocyte. *Science*, **306**, 1930–1933.

McDowell, J.M., Dhandaydham, M., Long, T.A., Aarts, M.G.M., Goff, S., Holub, E.B. and Dangl, J.L. (1998). Intragenic recombination and diversifying selection contribute to the evolution of downy mildew resistance at the *RPP8* locus of *Arabidopsis*. *The Plant Cell*, **10**, 1861–1874.

Parker, J.E., Coleman, M.J., Szabò, V., Frost, L.N., Schmidt, R., van der Biezen, E.A., Moores, T., Dean, C., Daniels, M.J. and Jones, J.D.G. (1997). The Arabidopsis downy mildew resistance gene *RPP5* shares similarity to the Toll and Interleukin-1 receptors with *N* and *L6*. *The Plant Cell*, **9**, 879–894.

Rehmany, A.P., Gordon, A., Rose, L.E., Allen, R.L., Armstrong, M.R., Whisson, S.C., Kamoun, S., Tyler, B.M., Birch, P.R.J. and Beynon, J.L. (2005). Differential recognition of highly divergent downy mildew avirulence gene alleles by *RPP1* resistance genes from two Arabidopsis lines. *The Plant Cell*, **17**, 1839–1850.

Rose, L.E., Bittner-Eddy, P.D., Langley, C.H., Holub, E.B., Michelmore, R.W. and Beynon, J.L. (2004). The maintenance of extreme amino acid diversity at the disease resistance gene, *RPP13*, in *Arabidopsis thaliana*. *Genetics*, **166**, 1517–1527.

Shan, W., Cao, M., Leung, D. and Tyler, B.M. (2004). The Avr1b locus of Phytophthora sojae encodes an elicitor and a regulator required for avirulence on soybean plants carrying resistance gene Rps1b. *Molecular Plant-Microbe Interactions*, **17**, 394–403.

Shao, F., Goldstein, C., Ade, J., Stoutmeyer, M., Dixon, J.E. and Innes, R.W. (2003). Cleaveage of Arabidopsis PBS1 by a bacterial type III effector. *Science*, **301**, 1230–1233.

Sinappidou, E., Williams, K., Nott, L., Bahkt, S., Tör, M., Crute, I.R., Holub, E.B., Bittner-Eddy, P.D. and Beynon, J.L. (2004). The *RPP2* locus of Arabidopsis accession Col-5 contains at lest two genes required for isolate specific recognition of *Peronospora parasitica* isolate *Cala2*. *The Plant Journal*, **38**, 898–909.

Tang, X., Frederick, R.D., Zhou, J., Halterman, D.A., Jia, Y. and Martin, G.B. (1996). Initiation of plant disease resistance by physical interaction of AvrPto and Pto Kinase. *Science*, **274**, 2060–2063.

Tör, M., Brown, D., Cooper, A., Woods-Tör, A., Sjolander, K., Jones, J.D. and Holub, E.B. (2004). Arabidopsis downy mildew resistance gene RPP27 encodes a receptor-like protein similar to CLAVATA2 and tomato Cf-9. *Plant Physiology*, **135**, 1100–1101.

van der Biezen, E.A., Freddie, C.T., Kahn, K., Parker, J.E. and Jones, J.D. (2002). Arabidopsis RPP4 is a member of the RPP5 multigene family of TIR-NB-LRR genes and confers downy mildew resistance through multiple signalling components. *The Plant Journal*, **29**, 439–451.

CHAPTER 7

Ace2 and Fungal Virulence

Helen Findon and **Ken Haynes**

Contents

1. Introduction to Ace2 97
2. Mitotic Exit and Regulation of Ace2 98
3. Ram: Regulation of *ACE2* and Cellular Morphogenesis 99
4. Ace2 and Fungal Virulence 101
5. Concluding Remarks 103
Acknowledgements 103
References 103

Abstract

The DNA binding protein, Ace2, plays a major role in the control of cell-cycle progression in *Saccharomyces cerevisiae*, being a principal regulator of the M/G1 transition and cell separation in particular. Ace2 also plays a significant role in effecting virulence, though in completely different ways, in the fungal pathogens *Candida albicans* and *Candida glabrata*. Here we will briefly review Ace2 regulation and function in terms of cell-cycle progression, cell separation and fungal virulence.

1. INTRODUCTION TO ACE2

Saccharomyces cerevisiae Ace2 is a zinc-finger transcription factor that is paralogous to another DNA binding protein, Swi5 (Butler and Thiele, 1991), indeed the proteins share 37% identity and have highly conserved zinc-finger motifs (83% identity). Transcription and localisation similarities also exist, the proteins are expressed during the same phase of the cell cycle (G2) and are localised within the cytoplasm until mitosis when they are translocated to the nucleus (Nasmyth *et al.*, 1990; Dohrmann *et al.*, 1992; Tebb *et al.*, 1993).

Swi5 and Ace2 are both capable of expressing a number of G1 specific genes (Doolin et al., 2001).

Swi5 and Ace2 also function independently of each other (Dohrmann *et al.*, 1992). For example, *S. cerevisiae* Ace2 is the key regulator of cell separation and specifically controls the expression of *SCW11, YHR143W, YER124C* (Doolin *et al.*, 2001) and most notably *CTS1* that encodes an endochitinase required for

British Mycological Society Symposia Series
Published by Elsevier Ltd.

the degradation of the chitin ring that exists at the septum between mother and daughter cells (Dohrmann *et al.*, 1992). *S. cerevisiae* cells that are mutated in either *CTS1* or *ACE2* are defective in cell separation, form clumps of cells (Kuranda and Robbins, 1991; King and Butler, 1998). Furthermore, disruption of *ACE2* gene dramatically reduces expression of *CTS1* (Dohrmann *et al.*, 1992). Cell-cycle-dependent nuclear localisation of Ace2 is required for expression of its target genes and Ace2 that is restricted to the nucleus throughout the cell cycle causes up-regulation of *CTS1* expression (O'Conallain *et al.*, 1999). Scw11 has similarities to endoglucanases (Cappellaro *et al.*, 1998) and is thought to be involved in mating (Zeitlinger *et al.*, 2003). Yhr143w (Dse2) and Yer124c (Dse1) may have roles in cell-wall metabolism (Doolin *et al.*, 2001).

Swi5, on the other hand, specifically activates *HO*, *PIR1* and *YPL158C* (Doolin *et al.*, 2001). *HO* encodes an endonuclease involved in mating-type switching in *S. cerevisiae* (reviewed by Nasmyth, 1993). Pir1 is a cell-wall protein (Toh-e *et al.*, 1993; Mrsa *et al.*, 1997) while Ypl158c is a protein of unknown function.

The similarities in the ways in which Ace2 and Swi5 are regulated in terms of their cellular location, their ability to jointly regulate a subset of genes and the fact that they both recognise the same core-binding sequence in the promoters of their target genes might suggest that each is capable of activating the target genes specific to the other. This is not the case, although Ace2 and Swi5 can bind to the same sequences in vitro, mechanisms exist to ensure the activation of expression of their specific target genes *in vivo* (Dohrmann *et al.*, 1996).

In the case of Swi5, this regulation is exerted *via* the action of the homeodomain protein Pho2 that interacts specifically with Swi5 at the *HO* promoter (Brazas and Stillman, 1993a, 1993b; Dohrmann *et al.*, 1996). In addition, mutations in *SIN5* in the presence of a functional *ACE2* allele lead to rescue of *HO* expression in a *swi5* null mutant suggesting that Sin5 functions to prevent inappropriate expression of *HO* by Ace2 (Stillman *et al.*, 1994).

Within the upstream region of the *CTS1* promoter there is a 258-base pair region, which requires Ace2 but not Swi5 for activation of expression. Within this region is a 46-base pair segment of DNA that acts as a negative regulatory sequence, which blocks Swi5 activation of *CTS1* (Dohrmann *et al.*, 1996). Furthermore, mutations in the *NCE3* (negative regulator of *CTS1* expression) gene result in Swi5 activation of *CTS1* in the absence of Ace2 (Dohrmann *et al.*, 1996).

To summarise, despite the fact that Ace2 and Swi5 can bind to the same DNA sequences in vitro, elegant regulatory mechanisms have evolved to ensure that each transcription factor binds to the correct DNA sequences *in vivo*, thus ensuring appropriate transcription of target genes (Dohrmann *et al.*, 1996).

2. MITOTIC EXIT AND REGULATION OF ACE2

In order to be transcriptionally active, Ace2 itself requires activation by the kinase Cbk1 in daughter cells only (Racki *et al.*, 2000; Colman-Lerner *et al.*, 2001; Weiss *et al.*, 2002). Cbk1 mediated control of Ace2 also involves Mob2. Cbk1 and Mob2 are very similar to the protein kinases Dbf2 and Mob1, respectively; both pairs of proteins have been shown to interact via two hybrid experiments (Luca and Winey, 1998; Racki *et al.*, 2000). Dbf2 and Mob1 are part of a regulatory cascade that controls mitotic exit. This cascade is known as the mitotic exit network (MEN) and is preceeded by the Cdc14 early-anaphase release (FEAR) network (see D'Amours and Amon, 2004 for a review of FEAR). Within the MEN, activation of a GTPase DNA binding protein, Tem1, (Shirayama *et al.*, 1994) is controlled by a Bub2 Bfa1 GTPase activating complex (GAP) (Pereira *et al.*, 2000) and by a guanine nucleotide exchange factor (GEF) Lte1 (Keng *et al.*, 1994; Shou *et al.*, 1999; Bardin *et al.*, 2000). This control is dependent on the position of the spindle pole and the phase of the cell cycle (Bardin *et al.*, 2000; Pereira *et al.*, 2000). Once active, Tem1 is able to trigger a signalling cascade via the kinases Cdc15 and Dbf2 (which functions with Mob1) (Komarnitsky *et al.*, 1998; Luca and Winey, 1998; Lee *et al.*, 2001; Mah *et al.*, 2001) and a number of accessory proteins including Nud1, which anchors Bfa1 to the spindle pole body (SPB) thus facilitating the appropraite localisation of the Tem1 Bub2 complex (Gruneberg *et al.*, 2000), the kinase Cdc5 (Lee *et al.*, 2001) and Cdc55, a regulatory subunit of protein phosphatase 2A (Yellman and Burke, 2006). The ultimate function of each pathway is to promote and maintain Cdc14 release from the nucleolus throughout anaphase and telophase such that this key protein phosphatase can counteract the effects of cyclin-dependent kinases (specifically Cdc28) and promote termination of mitosis (Visintin *et al.*, 1998, 1999). For a complete review of MEN see McCollum and Gould (2001). Interestingly, it has recently been shown that *Candida albicans* Cdc14 is required for cell separation, activation and localisation of Ace2 and mitotic exit (Clemente-Blanco *et al.*, 2006), suggesting that, as in *S. cerevisiae*, *C. albicans* Cdc14 is an important regulator of late cell-cycle events (Clemente-Blanco *et al.*, 2006).

All of the observations described above have lead to the suggestion that Cbk1 and Mob2 may be controlled by, and act within, a pathway similar to the MEN (Weiss *et al.*, 2002; Nelson *et al.*, 2003). This pathway has been named RAM for regulation of Ace2 activity and cellular morphogenesis (Nelson *et al.*, 2003) and is described below.

3. RAM: REGULATION OF *ACE2* AND CELLULAR MORPHOGENESIS

The RAM network, which controls cell polarity and Ace2 activity, contains the Cbk1 Mob2 protein kinase complex, Tao3, a 270 kDa protein of unknown molecular function that interacts with Cbk1 and Kic1 (Du and Novick, 2002), Hym1, a homologue of the *Aspergillus nidulans* hyphal growth protein, HymA (Karos and Fischer, 1999; Nelson *et al.*, 2003), Sog2 a novel leucine-rich repeat-containing protein and the kinase Kic1 (Nelson *et al.*, 2003). Loss of function mutations in most of the genes encoding these proteins were found to have round-cell morphologies, indicative of defects in cell polarisation, and had cell separation defects (Nelson *et al.*, 2003).

Furthermore mutants lacking Tao3, Hym1 and Kic1 were found to have altered Ace2-GFP localisation. Ace2 was not restricted to the daughter cell but localised to both the mother and daughter, suggesting that these proteins have a role in Ace2 daughter cell specific localisation (Nelson *et al.*, 2003).

Two-hybrid and epitope tagged co-immunoprecipitation analysis experiments revealed interactions that exist between the components of RAM and led to the suggestion of a model for RAM in which Hym1, Sog2 and Kic1 function as a protein kinase complex. Indeed Hym1 binds to both Sog2 and Kic1 and requires both proteins for appropriate localisation (Nelson *et al.*, 2003). The kinase activity of the Hym1–Sog2–Kic1 complex may phosphorylate and activate Cbk1 and Mob2, perhaps via interactions with Tao3 and Ssd1, a protein involved with maintaining cellular integrity (Kaeberlein and Guarente, 2002), to activate Ace2 activity and thus expression of daughter cell specific genes, cell integrity and polarised growth.

Cbk1 and Ssd1 have been shown to interact suggesting that they may both function in RAM (Racki *et al.*, 2000). Ssd1 functions in conjunction with Zrg8 and Srl1, two proteins of unknown molecular function, as part of the RAM network to control cell integrity and this forms a third, novel branch of the RAM network (Kurischko *et al.*, 2005). It is noteworthy that deletion of *SSD1* in *S. cerevisiae* caused increased virulence in a murine model of infection and resulted in up-regulation of pro-inflammatory cytokines suggesting that the RAM network may be critical for virulence (Wheeler *et al.*, 2003).

Recently, investigations have been undertaken to establish the role of the RAM proteins in cell-cycle progression (Bogomolnaya *et al.*, 2006). If such a role existed it would represent a fourth function of RAM; however, the authors found no role for the RAM network as a whole in the timing of cell-cycle initiation, rather that the different members of RAM had positive or negative regulatory roles in promoting START. For example, Hym1 promotes cell-cycle progression while Mob2, Sog2, Tao3, and Kic1 delay cell-cycle progression (Bogomolnaya *et al.*, 2006).

Following mitotic exit and activation of Ace2, the mother and daughter cell separate so that a new cell cycle can begin. Ace2 is one of the principle effectors of cell separation (a subject covered at length in several excellent reviews; Bi, 2001; Walther and Wendland, 2003; Yeong, 2005). The final act of cell separation following successful cytokinesis is the degradation of the septum that exists between the mother and daughter cell. This is achieved via the action of the chitinase Cts1 (Kuranda and Robbins, 1991) and the endoglucanases Eng1 (Baladron *et al.*, 2002) and Scw11 (Cappellaro *et al.*, 1998), the expression of the genes, which encode these proteins, is controlled by the transcription factor Ace2 in a daughter cell-specific manner (Dohrmann *et al.*, 1992; Doolin *et al.*, 2001). When mother and daughter cells are separated both re-enter S phase and are able to begin a new cell cycle.

In conclusion, cell division is a complex process which is initiated at the start of the cell cycle when the presumptive bud site is selected and ends with a separate mother and daughter cell, each of which are able to be begin a new cycle; it requires the co-ordinated orchestration of a great number of proteins (reviewed by Yeong, 2005). One of the ultimate effectors of cell separation is Ace2 since it acts following cytokinesis to activate proteins essential for the destruction of the primary and secondary septa (Dohrmann *et al.*, 1992; Doolin *et al.*, 2001) and therefore the final aspect of cell division, the separation of mother cell from daughter cell.

4. ACE2 AND FUNGAL VIRULENCE

Inactivation of Ace2 in both *C. albicans* and *Candida glabrata* results in a failure of cell separation. *C. albicans* cells remain attached at the mother–daughter junction, exhibit increased pseudohyphal growth, even under conditions where growth in the yeast phase is favoured, and are hyper-invasive in agar (Kelly *et al.*, 2004) *C. glabrata ace2* cells form massive clumps and have significant changes in their proteome (Kamran *et al.*, 2004; Stead *et al.*, 2005). However, the role that Ace2 plays in the virulence of these two pathogens is strikingly different. *C. albicans ace2* null mutants are completely attenuated in a DBA2 murine model and slightly attenuated in a BALB-c model (Kelly *et al.*, 2004; MacCallum *et al.*, 2006). In stark contrast, infection of either CD-1 or BALB-c mice, with *C. glabrata ace2* null cells, results in an increased ability to cause disease (Kamran *et al.*, 2004; MacCallum *et al.*, 2006). In the CD-1, model infection resulted in 100% mortality at 18 h post-infection, this was accompanied by significant invasion of the tissue parenchyma and massive over-stimulation of the host innate immune response, most notably 16- and 38-fold increases of IL6 and TNFα levels, respectively, in mice infected with *ace2* null mutant compared to wild-type cells. These cytokines are known to be vital in the host defence against *Candida* infections (Netea *et al.*, 2002) and this response is reminiscent of septic shock. Septic shock results from an inappropriately elevated host response to infection, the presence of excess pro-inflamatory cytokines causes a decrease, both in blood volume and in blood pressure; the ultimate effect of which can be, in extreme cases, multiple organ failure and death (reviewed by Sriskandan and Cohen, 1995). It is possible that similar events are occurring in mice infected with *C. glabrata ace2* null mutant cells, leading to a septic shock-like response which is likely contributing to, if not underpinning, the increased virulence phenotype.

Why does deletion of *C. glabrata ACE2* cause such a dramatic phenotype? It is possible that deletion of *ACE2* has resulted in exposure of a previously hidden molecule, over or under expression of a cell surface molecule or extracellular secretion of a toxin or similar molecule that has stimulated the pro-inflammatory arm of the innate immune system such that a septic shock response has been elicited. Indeed, analysis of the *C. glabrata* proteome has shown the regulation of 61 proteins to be altered in the absence of *C. glabrata ACE2* (Stead *et al.*, 2005). One or more of these proteins may be altering the composition of the cell wall resulting in the septic shock-like response in the host. Indeed, the majority of proteins up-regulated in *C. glabrata ace2* null cells are involved in energy and metabolism, particularly central carbon metabolism, a cellular process previously implicated in the virulence of *C. albicans* (Lorenz and Fink, 2001).

It can be argued that in order for an organism to retain a gene, which when absent increases its effectiveness as a pathogen, the gene product must offer some evolutionary benefit. It has been suggested that such benefit might be related to transmission and therefore dissemination (Foreman-Wykert and Miller, 2003). Unlike *C. albicans* and *S. cerevisiae*, *C. glabrata* is not able to undergo yeast to hyphal transition, although it can produce pseudohyphae *in vitro* (Csank and Haynes, 2000). Morphogenic transition allows hyphal cells to switch to yeast cells and therefore allow dissemination throughout the host *via* the blood stream. The ability to undergo morphological switching has long been considered a virulence factor and although there is a lack of definite demonstration of this (Calderone and Fonzi, 2001). Separation of mother and daughter cells is likely to be the only mechanism of dissemination in *C. glabrata*, certainly no others are known at present. Retention of *C. glabrata ACE2* allows efficient cell separation resulting in a single cell phenotype that benefits dissemination through the host and therefore provides an evolutionary advantage.

C. glabrata ACE2 appears to be the first hypervirulence gene to be identified in any *Candida* species. As discussed previously *C. glabrata ace2* null cells have a cell separation defect in addition to the hypervirulence phenotype. It is reasonable to hypothesise that the cell separation defect may contribute towards the increased virulence observed. It is, however, unlikely to be the cause of the hypervirulence as *C. glabrata ace2* cells are avirulent in immunocompetent mice (MacCallum *et al.*, 2006) and other clumpy mutants have been demonstrated to be avirulent *in vivo*. Such mutants include *C. albicans tup1* strains. *TUP1* encodes a negative regulator of filamentous growth (Braun and Johnson, 1997). *C albicans tup1* mutants exist only in the filamentous form, have a large fuzzy phenotype (Braun and Johnson, 1997; Zhao *et al.*, 2002) and are attenuated for virulence in a murine model of candidiasis (Braun *et al.*, 2000). A further example of a mutant with cell separation defects is *C. albicans och1*, defective in a mannosyltransferase (Bates *et al.*, 2006). This mutant is also attenuated for virulence in a murine model of candidiasis (Bates *et al.*, 2006) suggesting that mutants with cell separation defects do not, *per se*, cause hypervirulence *in vivo*. But the most persuasive evidence that hypervirulence and clumpy growth are separable comes from the observation that *C. albicans ace2* cells have a clumpy growth phenotype, similar to that observed in the *C. glabrata ace2* null mutant, but are completely attenuated. Mice infected with the *C. albicans ace2* null strain exhibit 100% survival over 28 days (Kelly *et al.*, 2004).

The question then arises why should inactivation of orthologues of the same gene from two *Candida* species give such markedly different virulent profiles despite having similar clumpy growth phenotypes? *C. glabrata* is more closely related to *S. cerevisiae* than *C. albicans* (Barns *et al.*, 1991). The evolutionary distance between *C. albicans* and *C. glabrata* is a possible reason for the stark difference in virulence seen. Ace2 may have different functions in *C. glabrata* and *C. albicans*, as well as different activators and downstream targets. In evidence, there is only a single orthologue of *S. cerevisiae ACE2* and *SWI5* in *C. albicans* (Kelly *et al.*, 2004), but in *C. glabrata* orthologues of both genes exist (Kamran *et al.*,

2004). This is likely due to a genome duplication in the *S. cerevisiae–C. glabrata* lineage subsequent to their divergence from the *C. albicans* lineage (Wolfe and Shields, 1997; Wong *et al.*, 2002). Furthermore, *C. albicans* has four chitin synthase genes (Au-Young and Robbins, 1990; Sudoh *et al.*, 1993; Gow *et al.*, 1994; Munro *et al.*, 2003) and three chitinase genes (McCreath *et al.*, 1995) of which *CHT3* but not *CHT2* appears to be under the transcriptional control of *C. albicans* Ace2 (Kelly *et al.*, 2004). The transcriptional regulation of *CHT1* remains to be established. In *C. glabrata*, the expression of the single chitinase-encoding gene *CTS1* is controlled by Ace2 (Findon and Haynes, unpublished data). These differences suggest that the mechanisms by which *C. albicans* and *C. glabrata* control cell separation are different and may involve different proteins in each organism. Differences such as these and others, which remain to be identified, may provide insights into the marked disparity in virulence between *C. glabrata* and *C. albicans ace2* mutants.

5. CONCLUDING REMARKS

Ace2 plays a vital role in regulating progression through the cell cycle and is the primary regulator of cell separation in *S. cerevisiae*. This is in part, at least, mediated through its control of the expression of a number of genes encoding hydrolytic enzymes that breakdown the septum that links mother and daughter cells post cytokinesis. Remarkably it also plays a fundamental, but distinct, role in the virulence of *C. albicans* and *C. glabrata*. However, many questions remain. How is Ace2 transported into and out of the nucleus? What are the co-factors of Ace2? What is the basis for the increased virulence of *C. glabrata ace2* cells? Why are *C. albicans ace2* cells attenuated? The elucidation of these, and other questions, provides a significant and exciting challenge.

ACKNOWLEDGEMENTS

We are grateful to all members of the Imperial Molecular Mycology groups, past and present, for helpful discussions. This work was funded by the BBSRC (S02/P001, BBS/S/A/2004/11098, BBS/B/10331) and Wellcome Trust (072420).

REFERENCES

Au-Young, J. and Robbins, P.W. (1990). Isolation of a chitin synthase gene (*CHS1*) from *Candida albicans* by expression in *Saccharomyces cerevisiae*. *Molecular Microbiology*, **4**, 197–207.

Baladron,V., Ufano, S., Duenas, E., Martin-Cuadrado, A.B., Del Rey, F. and Vazquez de Aldana, C.R. (2002). Eng1p, an endo-1,3-beta-glucanase localized at the daughter side of the septum, is involved in cell separation in *Saccharomyces cerevisiae*. *Eukaryotic Cell*, **1**, 774–786.

Bardin, A.J., Visintin, R. and Amon, A. (2000). A mechanism for coupling exit from mitosis to partitioning of the nucleus. *Cell*, **102**, 21–31.

Barns, S.M., Lane, D.J., Sogin, M.L., Bibeau, C. and Weisburg, W.G. (1991). Evolutionary relationships among pathogenic *Candida* species and relatives. *Journal of Bacteriology*, **173**, 2250–2255.

Bates, S., Hughes, H.B., Munro, C.A., Thomas, W.P., MacCallum, D.M., Bertram, G. *et al.* (2006). Outer chain N-glycans are required for cell wall integrity and virulence of *Candida albicans. The Journal of Biological Chemistry*, **281**, 90–98.

Bi, E. (2001). Cytokinesis in budding yeast: The relationship between actomyosin ring function and septum formation. *Cell Structure and Function*, **26**, 529–537.

Bogomolnaya, L.M., Pathak, R., Guo, J. and Polymenis, M. (2006). Roles of the RAM signaling network in cell cycle progression in *Saccharomyces cerevisiae. Current Genetics*, **49**(6), 384–392.

Braun, B.R., Head, W.S., Wang, M.X. and Johnson, A.D. (2000). Identification and characterization of *TUP1*-regulated genes in *Candida albicans. Genetics*, **156**, 31–44.

Braun, B.R. and Johnson, A.D. (1997). Control of filament formation in *Candida albicans* by the transcriptional repressor *TUP1. Science*, **277**, 105–109.

Brazas, R.M. and Stillman, D.J. (1993a). Identification and purification of a protein that binds DNA cooperatively with the yeast Swi5 protein. *Molecular and Cellular Biology*, **13**, 5524–5537.

Brazas, R.M. and Stillman, D.J. (1993b). The Swi5 zinc-finger and Grf10 homeodomain proteins bind DNA cooperatively at the yeast *HO* promoter. *Proceedings of the National Academy of Science of the United States of America*, **90**, 11237–11241.

Butler, G. and Thiele, D.J. (1991). *ACE2*, an activator of yeast metallothionein expression which is homologous to *SWI5. Molecular and Cellular Biology*, **11**, 476–485.

Calderone, R.A. and Fonzi, W.A. (2001). Virulence factors of *Candida albicans. Trends in Microbiology*, **9**, 327–335.

Cappellaro, C., Mrsa, V. and Tanner, W. (1998). New potential cell wall glucanases of Saccharomyces cerevisiae and their involvement in mating. *Journal of Bacteriology*, **180**, 5030–5037.

Clemente-Blanco, A., Gonzalez-Novo, A., Machin, F., Caballero-Lima, D., Aragon, L., Sanchez, M. *et al.* (2006). The Cdc14p phosphatase affects late cell-cycle events and morphogenesis in *Candida albicans. Journal of Cell Science*, **119**, 1130–1143.

Colman-Lerner, A., Chin, T.E. and Brent, R. (2001). Yeast Cbk1 and Mob2 activate daughter-specific genetic programs to induce asymmetric cell fates. *Cell*, **107**, 739–750.

Csank, C. and Haynes, K. (2000). *Candida glabrata* displays pseudohyphal growth. *FEMS Microbiology Letters*, **189**, 115–120.

D'Amours, D. and Amon, A. (2004). At the interface between signaling and executing anaphase-Cdc14 and the FEAR network. *Genes and Development*, **18**, 2581–2595.

Dohrmann, P.R., Butler, G., Tamai, K., Dorland, S., Greene, J.R., Thiele, D.J. and Stillman, D.J. (1992). Parallel pathways of gene regulation: Homologous regulators Swi5 and Ace2 differentially control transcription of *HO* and chitinase. *Genes and Development*, **6**, 93–104.

Dohrmann, P.R., Voth, W.P. and Stillman, D.J. (1996). Role of negative regulation in promoter specificity of the homologous transcriptional activators Ace2p and Swi5p. *Molecular and Cellular Biology*, **16**, 1746–1758.

Doolin, M.T., Johnson, A.L., Johnston, L.H. and Butler, G. (2001). Overlapping and distinct roles of the duplicated yeast transcription factors Ace2p and Swi5p. *Molecular Microbiology*, **40**, 422–432.

Du, L.L. and Novick, P. (2002). Pag1p, a novel protein associated with protein kinase Cbk1p, is required for cell morphogenesis and proliferation in *Saccharomyces cerevisiae. Molecular Biology of the Cell*, **13**, 503–514.

Foreman-Wykert, A.K. and Miller, J.F. (2003) Hypervirulence and pathogen fitness. *Trends in Microbiology*, **11**, 105–108.

Gow, N.A., Robbins, P.W., Lester, J.W., Brown, A.J., Fonzi, W.A., Chapman, T. and Kinsman, O.S. (1994). A hyphal-specific chitin synthase gene (*CHS2*) is not essential for growth, dimorphism, or virulence of *Candida albicans. Proceedings of the National Academy of Science of the United States of America*, **91**, 6216–6220.

Gruneberg, U., Campbell, K., Simpson, C., Grindlay, J. and Schiebel, E. (2000). Nud1p links astral microtubule organization and the control of exit from mitosis. *The EMBO Journal*, **19**, 6475–6488.

Kaeberlein, M. and Guarente, L. (2002). *Saccharomyces cerevisiae MPT5* and *SSD1* function in parallel pathways to promote cell wall integrity. *Genetics*, **160**, 83–95.

Kamran, M., Calcagno, A.M., Findon, H., Bignell, E., Jones, M.D., Warn, P. *et al.* (2004). Inactivation of transcription factor gene *ACE2* in the fungal pathogen *Candida glabrata* results in hypervirulence. *Eukaryotic Cell*, **3**, 546–552.

Karos, M. and Fischer, R. (1999). Molecular characterization of HymA, an evolutionarily highly conserved and highly expressed protein of *Aspergillus nidulans. Molecular & General Genetics*, **260**, 510–521.

Kelly, M.T., MacCallum, D.M., Clancy, S.D., Odds, F.C., Brown, A.J. and Butler, G. (2004). The *Candida albicans CaACE2* gene affects morphogenesis, adherence and virulence. *Molecular Microbiology*, **53**, 969–983.

Keng, T., Clark, M.W., Storms, R.K., Fortin, N., Zhong, W., Ouellette, B.F. *et al.* (1994). *LTE1* of *Saccharomyces cerevisiae* is a 1435 codon open reading frame that has sequence similarities to guanine nucleotide releasing factors. *Yeast*, **10**, 953–958.

King, L. and Butler, G. (1998). Ace2p, a regulator of *CTS1* (chitinase) expression, affects pseudohyphal production in *Saccharomyces cerevisiae. Current Genetics*, **34**, 183–191.

Komarnitsky, S.I., Chiang, Y.C., Luca, F.C., Chen, J., Toyn, J.H., Winey, M. *et al.* (1998). Dbf2 protein kinase binds to and acts through the cell cycle-regulated Mob1 protein. *Molecular and Cellular Biology*, **18**, 2100–2107.

Kuranda, M.J. and Robbins, P.W. (1991). Chitinase is required for cell separation during growth of *Saccharomyces cerevisiae.* The Journal of Biological Chemistry, **266**, 19758–19767.

Kurischko, C., Weiss, G., Ottey, M. and Luca, F.C. (2005). A role for the *Saccharomyces cerevisiae* regulation of Ace2 and polarized morphogenesis signaling network in cell integrity. *Genetics*, **171**, 443–455.

Lee, S.E., Frenz, L.M., Wells, N.J., Johnson, A.L. and Johnston, L.H. (2001). Order of function of the budding-yeast mitotic exit-network proteins Tem1, Cdc15, Mob1, Dbf2, and Cdc5. *Current Biology*, **11**, 784–788.

Lorenz, M.C. and Fink, G.R. (2001). The glyoxylate cycle is required for fungal virulence. *Nature*, **412**, 83–86.

Luca, F.C. and Winey, M. (1998). *MOB1*, an essential yeast gene required for completion of mitosis and maintenance of ploidy. *Molecular Biology of the Cell*, **9**, 29–46.

MacCallum, D., Findon, H., Kenny, C., Butler, G., Haynes, K. and Odds, F.C. (2006). Different Pathogenic Outcomes of *ACE2* and *SWI5* Gene Disruptions in Three Yeast Species. *Infection and Immunity*, **74**, 5344–5248.

Mah, A.S., Jang, J. and Deshaies, R.J. (2001). Protein kinase Cdc15 activates the Dbf2-Mob1 kinase complex. *Proceedings of the National Academy of Science of the United States of America*, **98**, 7325–7330.

McCollum, D. and Gould, K.L. (2001). Timing is everything: Regulation of mitotic exit and cytokinesis by the MEN and SIN. *Trends in Cell Biology*, **11**, 89–95.

McCreath, K.J., Specht, C.A. and Robbins, P.W. (1995). Molecular cloning and characterization of chitinase genes from *Candida albicans. Proceedings of the National Academy of Science of the United States of America*, **92**, 2544–2548.

Mrsa, V., Seidl, T., Gentzsch, M. and Tanner, W. (1997). Specific labelling of cell wall proteins by biotinylation. Identification of four covalently linked O-mannosylated proteins of *Saccharomyces cerevisiae. Yeast*, **13**, 1145–1154.

Munro, C.A., Whitton, R.K., Hughes, H.B., Rella, M., Selvaggini, S. and Gow, N.A. (2003). *CHS8*-a fourth chitin synthase gene of *Candida albicans* contributes to *in vitro* chitin synthase activity, but is dispensable for growth. *Fungal Genetics and Biology*, **40**, 146–158.

Nasmyth, K. (1993). Regulating the HO endonuclease in yeast. *Current Opinion in Genetics & Development*, **3**, 286–294.

Nasmyth, K., Adolf, G., Lydall, D. and Seddon, A. (1990). The identification of a second cell cycle control on the *HO* promoter in yeast: Cell cycle regulation of Swi5 nuclear entry. *Cell*, **62**, 631–647.

Netea, M.G., Van Der Graaf, C.A., Vonk, A.G., Verschueren, I., Van Der Meer, J.W. and Kullberg, B.J. (2002). The role of toll-like receptor (*TLR*) 2 and *TLR4* in the host defense against disseminated candidiasis. *The Journal of Infectious Diseases*, **185**, 1483–1489.

Nelson, B., Kurischko, C., Horecka, J., Mody, M., Nair, P., Pratt, L. *et al.* (2003). RAM: A conserved signaling network that regulates Ace2p transcriptional activity and polarized morphogenesis. *Molecular Biology of the Cell*, **14**, 3782–3803.

O'Conallain, C., Doolin, M.T., Taggart, C., Thornton, F. and Butler, G. (1999). Regulated nuclear localisation of the yeast transcription factor Ace2p controls expression of chitinase (*CTS1*) in *Saccharomyces cerevisiae. Molecular & General Genetics*, **262**, 275–282.

Pereira, G., Hofken, T., Grindlay, J., Manson, C. and Schiebel, E. (2000). The Bub2p spindle checkpoint links nuclear migration with mitotic exit. *Molecular Cell*, **6**, 1–10.

Racki, W.J., Becam, A.M., Nasr, F. and Herbert, C.J. (2000). Cbk1p, a protein similar to the human myotonic dystrophy kinase, is essential for normal morphogenesis in *Saccharomyces cerevisiae. The EMBO Journal*, **19**, 4524–4532.

Shirayama, M., Matsui, Y. and Toh, E. (1994). The yeast *TEM1* gene, which encodes a GTP-binding protein, is involved in termination of M phase. *Molecular and Cellular Biology*, **14**, 7476–7482.

Shou, W., Seol, J.H., Shevchenko, A., Baskerville, C., Moazed, D., Chen, Z.W. *et al.* (1999). Exit from mitosis is triggered by Tem1-dependent release of the protein phosphatase Cdc14 from nucleolar RENT complex. *Cell*, **97**, 233–244.

Sriskandan, S. and Cohen, J. (1995). The pathogenesis of septic shock. *Infect*, **30**, 201–206.

Stead, D., Findon, H., Yin, Z., Walker, J., Selway, L., Cash, P. *et al.* (2005). Proteomic changes associated with inactivation of the *Candida glabrata ACE2* virulence-moderating gene. *Proteomics*, **5**, 1838–1848.

Stillman, D.J., Dorland, S. and Yu, Y. (1994). Epistasis analysis of suppressor mutations that allow *HO* expression in the absence of the yeast Swi5 transcriptional activator. *Genetics*, **136**, 781–788.

Sudoh, M., Nagahashi, S., Doi, M., Ohta, A., Takagi, M. and Arisawa, M. (1993). Cloning of the chitin synthase 3 gene from *Candida albicans* and its expression during yeast-hyphal transition. *Molecular & Genereal Genetics*, **241**, 351–358.

Tebb, G., Moll, T., Dowzer, C. and Nasmyth, K. (1993). Swi5 instability may be necessary but is not sufficient for asymmetric *HO* expression in yeast. *Genes & Development*, **7**, 517–528.

Toh-e, A., Yasunaga, S., Nisogi, H., Tanaka, K., Oguchi, T. and Matsui, Y. (1993). Three yeast genes, *PIR1*, *PIR2* and *PIR3*, containing internal tandem repeats, are related to each other, and *PIR1* and *PIR2* are required for tolerance to heat shock. *Yeast*, **9**, 481–494.

Visintin, R., Craig, K., Hwang, E.S., Prinz, S., Tyers, M. and Amon, A. (1998). The phosphatase Cdc14 triggers mitotic exit by reversal of Cdk-dependent phosphorylation. *Molecular Cell*, **2**, 709–718.

Visintin, R., Hwang, E.S. and Amon, A. (1999). Cfi1 prevents premature exit from mitosis by anchoring Cdc14 phosphatase in the nucleolus. *Nature*, **398**, 818–823.

Walther, A. and Wendland, J. (2003). Septation and cytokinesis in fungi. *Fungal Genetic and Biology*, **40**, 187–196.

Weiss, E.L., Kurischko, C., Zhang, C., Shokat, K., Drubin, D.G. and Luca, F.C. (2002). The *Saccharomyces cerevisiae* Mob2p-Cbk1p kinase complex promotes polarized growth and acts with the mitotic exit network to facilitate daughter cell-specific localization of Ace2p transcription factor. *The Journal of Cell Biology*, **158**, 885–900.

Wheeler, R.T., Kupiec, M., Magnelli, P., Abeijon, C. and Fink, G.R. (2003). *A Saccharomyces cerevisiae* mutant with increased virulence. *Proceedings of the National Academy of Science of the United States of America*, **100**, 2766–2770.

Wolfe, K.H. and Shields, D.C. (1997). Molecular evidence for an ancient duplication of the entire yeast genome. *Nature*, **387**, 708–713.

Wong, S., Butler, G., and Wolfe, K.H. (2002). Gene order evolution and paleopolyploidy in hemiascomycete yeasts. *Proceedings of the National Academy of Science of the United States of America*, **99**, 9272–9277.

Yellman, C.M. and Burke, D.J. (2006). The role of Cdc55 in the spindle checkpoint is through regulation of mitotic exit in *Saccharomyces cerevisiae. Molecular Biology of the Cell*, **17**, 658–666.

Yeong, F.M. (2005). Severing all ties between mother and daughter: Cell separation in budding yeast. *Molecular Microbiology*, **55**, 1325–1331.

Zeitlinger, J., Simon, I., Harbison, C.T., Hannett, N.M., Volkert, T.L., Fink, G.R. and Young, R.A. (2003). Program-specific distribution of a transcription factor dependent on partner transcription factor and MAPK signaling. *Cell*, **113**, 395–404.

Zhao, R., Lockhart, S.R., Daniels, K. and Soll, D.R. (2002). Roles of TUP1 in switching, phase maintenance, and phase-specific gene expression in *Candida albicans*. *Eukaryotic Cell*, **1**, 353–365.

CHAPTER 8

Integrative Analysis of Yeast Osmoregulation

Stefan Hohmann

Contents

1. Osmoregulation: Maintaining the Water Balance 110
2. Time Line: Response to Osmotic Shock 110
3. The HOG Signalling System 111
4. Cytosolic Responses 113
5. Fps1 114
6. Transcriptional Responses 117
7. Feedback Control 119
8. Perspectives 123
Acknowledgements 124
References 124

Abstract

Osmoregulation, i.e. the active control of the cellular water balance, encompasses homeostatic mechanisms crucial for life. The osmoregulatory system in the yeast *Saccharomyces cerevisiae* is particularly well understood, although many details of the regulatory mechanisms still remain to be discovered. Central to yeast osmoregulation is the high osmolarity glycerol (HOG) signalling network, a branched mitogen-activated protein kinase (MAPK) pathway that converges on the MAPK Hog1. Active Hog1 controls numerous cellular processes in the cytosol (cell cycle, translation, ion, water and glycerol transport and central metabolism) as well as the expression of numerous genes. In order to achieve a quantitative understanding and to fit different processes in a temporal order, we have generated a comprehensive mathematical model of yeast osmoregulation in collaboration with theoreticians. Using model simulation and quantitative time-course experimentation, we have analysed mechanisms of feedback control of the HOG pathway. These studies illustrate how a signalling pathway combines rigorous feedback control with maintenance of signalling competence, as required for a system controlling cell homeostasis.

British Mycological Society Symposia Series
Published by Elsevier Ltd.

1. OSMOREGULATION: MAINTAINING THE WATER BALANCE

Maintenance of the cellular water balance is fundamental for life. Consequently, all cells, even individual cells in multicellular organisms with an organism-wide osmoregulation, have the ability to actively control their water balance. Many free-living organisms additionally have the ability to survive periods when there is insufficient water in dormant, largely desiccated stages. Control of the water balance is a homeostatic process and is constantly active in living cells to maintain an appropriate intracellular environment for biochemical processes, as well as shape and turgor of cells and organism. In the laboratory, the osmoregulatory system is most conveniently studied as a response to osmotic shock, i.e. a rapid and dramatic change in the extracellular water activity. These rapid changes mediate either water exit from the cell (hyperosmotic shock), causing cell shrinkage, or water entry in the cell (hypo-osmotic shock), causing cell swelling. The yeast *Saccharomyces cerevisiae*, as a free-living organism experiencing both slow and rapid changes in extracellular water activity, has proven a suitable and genetically tractable experimental system to study the underlying signalling pathways governing osmoregulation as well as a number of response and adaptation mechanisms. Although far from complete, the present picture of yeast osmoregulation is both extensive and detailed (De Nadal *et al.*, 2002; Hohmann, 2002). Recently, we have reported the first comprehensive mathematical model based on quantitative and time-resolved data that can correctly reproduce experimental data and has proven to have predictive capability (Klipp *et al.*, 2005).

2. TIME LINE: RESPONSE TO OSMOTIC SHOCK

Following a hyperosmotic shock yeast cells shrink within seconds. Cell shrinkage results in higher cellular ion concentrations, which in turn seems to cause dissociation of many DNA-binding proteins from the DNA template, in this way "resetting" gene expression on a wide scale (Proft and Struhl, 2004). Cells rapidly activate the high osmolarity glycerol (HOG) signal transduction system (De Nadal *et al.*, 2002; Hohmann, 2002) (Figure 1). First responses mediated by active Hog1 appear to encompass stimulation of ion export (Proft and Struhl, 2004), arrest of the cell cycle (Escote *et al.*, 2004; Clotet *et al.*, 2006) and diminished translational capacity (Bilsland-Marchesan *et al.*, 2000; Bilsland *et al.*, 2004). The glycerol export channel Fps1 rapidly closes (Tamás *et al.*, 1999, 2003). Probably, all of these events are crucial for survival of the osmotic shock and they encompass the first adaptive measures of the cell. In addition, it has been reported that active Hog1 phosphorylates PFK2, causing activation of glycolysis to enhance production of the osmolyte glycerol (Dihazi *et al.*, 2004). Glycerol accumulation is essential for adaptation to hyperosmotic conditions (Albertyn *et al.*, 1994; Ansell *et al.*, 1997). A significant portion of active Hog1 accumulates in the nucleus and binds to target promoters where it controls gene expression (Posas *et al.*, 2000; Rep *et al.*, 2000; O'Rourke and Herskowitz, 2004; Pokholok *et al.*, 2006). Genes encoding enzymes in glycerol production as well as an active glycerol-uptake

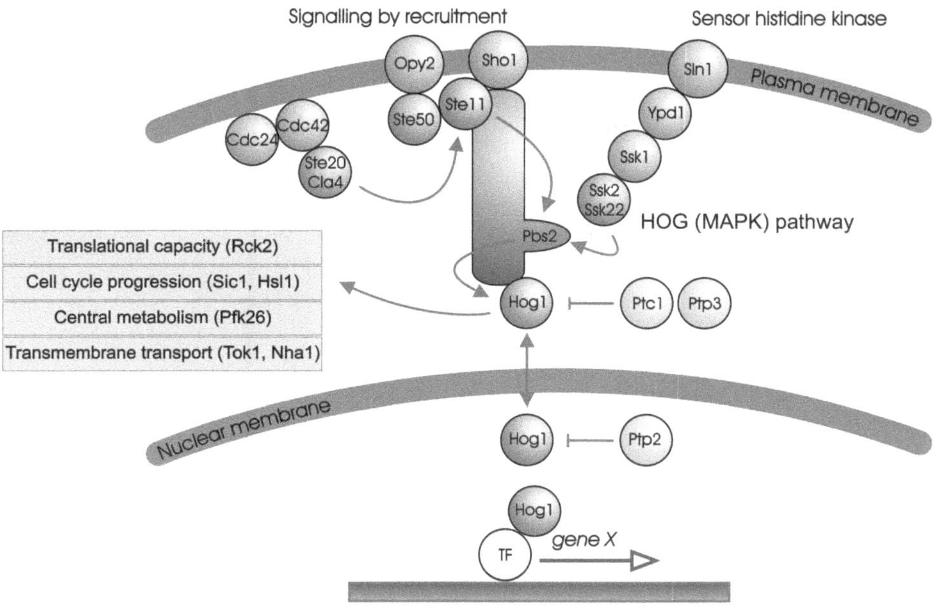

Figure 1 The yeast HOG pathway signalling system and overview of response mechanisms. The pathway consists of two branches: The Sho1 branch with Cdc42/24: G-protein, Ste20: PAK protein kinase, Opy2, Sho1: membrane protein scaffold, Ste11/Ste50: MAPKKK, Pbs2: MAPKK; the Sln1 branch with Sln1-Ypd1-Ssk1: sensing histidine kinase phosphorelay system, Ssk2/22: MAPKKK, Hog1: MAPK, Ptc1, Ptp2/3: protein phosphatases, TF: transcription factor. **(See Colour Section)**

system are among those genes upregulated in a Hog1-dependent manner. Once the cell commences reswelling due to glycerol accumulation, Hog1 becomes dephosphorylated and the osmotic response adapts to a level required for homeostasis at the higher osmolarity (Klipp *et al.*, 2005).

Less is known about the time line of the response to a hypo-osmotic shock, which probably is far simpler. Cell volume increases rapidly following a hypo-osmotic shock. The Fps1 glycerol export channel opens to allow glycerol release and this process is essential for yeast cells to survive such a situation (Luyten *et al.*, 1995; Tamás *et al.*, 1999). In addition, the cell integrity pathway is rapidly and transiently activated (Davenport *et al.*, 1995). Active Slt2 kinase stimulates transcription of genes encoding proteins required for cell wall assembly, which may be important to adapt to the new conditions (Jung and Levin, 1999). In the following sections I will focus mainly on the response to hyperosmotic shock.

3. THE HOG SIGNALLING SYSTEM

The HOG pathway (Figure 1) is one of the best-understood and most intensively studied mitogen-activated protein kinase (MAPK) systems. The first components of the pathway (Hog1, Pbs2) were identified in a genetic screen for osmosensitive mutants (Brewster *et al.*, 1993). Subsequently, numerous additional components were found among suppressor mutations or multi-copy suppressors of the lethality of the *sln1Δ* mutant (Maeda *et al.*, 1994; Posas *et al.*, 1996; Posas and Saito, 1997). Finally, the observation that *ssk1Δ* as well as *ssk2Δ ssk22Δ* mutants are osmotolerant prompted genetic screens that identified components of the Sho1 branch (Maeda *et al.*, 1995). Identification and organisation into the signalling system of the HOG pathway components represent text-book examples of the power of both targeted and global yeast genetics approaches. In addition, significant knowledge has emerged on the flow of information through the pathway (De Nadal *et al.*, 2002; Hohmann, 2002; O'Rourke *et al.*, 2002; Saito and Tatebayashi, 2004).

In yeasts, the HOG signalling system consists of two branches that converge on the MAPKK Pbs2, the Sln1 and the Sho1 branch. The Sho1 branch also plays a role in pseudohyphal development of *S. cerevisiae* (O'Rourke and Herskowitz, 1998). It appears that in many fungi the Sho1 module does not connect to Pbs2 and hence is not involved in osmotic responses (Furukawa *et al.*, 2005; Krantz *et al.*, 2006b). This indicates that the Sho1 module might not primarily have a role in osmosensing but rather perceives signals related to cell shape and/or cell surface conditions. In line with this speculation are observations that Sho1 is specifically located at sites of cell growth and that it does not appear to sense turgor changes (Reiser *et al.*, 2000, 2003).

The Sho1 branch consists almost exclusively of proteins shared with the pseudohyphal development pathway and the pheromone response pathways. Signalling specificity seems to be assured by recruitment to scaffold proteins (Sho1, Opy2, Pbs2) and requires the Hog1 kinase. In *hog1Δ* mutants, exposure to osmotic stress causes "pathway rewiring", i.e. activation of the pseudohyphal and pheromone response pathways and morphological aberrations (O'Rourke and Herskowitz, 1998; Davenport *et al.*, 1999; Rep *et al.*, 2000). The mechanism by which Hog1 prevents such cross talk has not yet been elucidated.

Recently, the mechanisms involved in the activation of the Sho1 branch following osmotic shock have been described in detail using constitutively active Sho1 and Ste11 mutants as well as protein interaction studies (Tatebayashi *et al.*, 2006). Osmoshock somehow converts Cdc42 to its active, GTP-bound form, which then activates the redundant protein kinases Ste20 and Cla4. At the same time, Cdc42-GTP binds to Ste50, which is always associated with Ste11, thereby recruiting Ste20/Cla4 and their substrate Ste11 to the plasma membrane. The recently identified Opy2 membrane protein may assist in recruiting Ste11-Ste50 (Wu *et al.*, 2006). Ste20/Cla4 phosphorylates Ste11 thereby relieving its auto-inhibition. Active Ste11-Ste50 then dissociates from Cdc42. The mechanism that stimulates Sho1 such that it becomes competent to bind the active Ste11–Ste50 complex is presently unclear. However, this binding brings the active MAPKKK Ste11 into vicinity of its substrate, the active MAPKK Pbs2, which also binds to Sho1. It is this step, i.e. binding of Pbs2 to Sho1, which is missing in many fungi such that Sho1/Ste11 does not activate the HOG pathway (Furukawa *et al.*, 2005). Active Pbs2 then phosphorylates the MAPK Hog1. As indicated above, the sensing mechanism of osmotic changes in the Sho1 branch is not understood at this point but must be closely related to Sho1 (Tatebayashi *et al.*, 2006). The observation that Sho1, which is genetically the most upstream component in the pathway, can be replaced by engineered proteins that recruit Pbs2 to the plasma membrane suggests that Sho1 does not function as a sensor itself (Raitt *et al.*, 2000). The protein is overall poorly conserved (Krantz *et al.*, 2006b), and homologues cannot be identified beyond fungi, indicating that Sho1 does not have an enzymatic function.

The above is different to the Sln1 branch. Sln1 is a sensor histidine kinase related to bacterial two-component systems. Such proteins are widespread in fungi and plants (Catlett *et al.*, 2003) and Sln1 has a similar domain organisation as, and sequence similarity to, the bacterial osmosensing histidine kinase EnvZ (Krantz *et al.*, 2006a). Both proteins have two transmembrane domains at their N-terminus, which are connected by a large, poorly conserved extracellular loop, about 300 amino acids in yeasts. It is believed that the extracellular loop and the transmembrane domains sense turgor changes (Reiser *et al.*, 2003), perhaps by responding to movements of the plasma membrane relative to the cell wall. In an unknown manner, the sensing domain transmits such an extracellular or intramembrane signal to the inside of the cell to control the histidine kinase activity of Sln1. In *S. cerevisiae* it appears that the Sln1 histidine kinase is a negative regulator of the downstream MAPK cascade: deletion of Sln1 or inactivation of the kinase by mutation results in lethal Hog1 overactivation (Maeda *et al.*, 1994). Under ambient, low-osmolarity conditions, the Sln1 histidine kinase is active such that phosphate groups are transferred via the Sln1 receiver and response regulator domains as well as the Ypd1 phosphotransfer protein to the Ssk1 response regulator protein, which is inactive in its phosphorylated form. Hyperosmotic shock causes inactivation of Sln1 kinase activity and dephosphorylation of Ssk1 (Posas *et al.*, 1996). Since pathway activation occurs within less than 1 min, phospho-Ssk1 must either be highly unstable (and then constantly phosphorylated to be kept inactive) or be dephosphorylated by a presently unknown phosphatase. Unphosphorylated Ssk1 mediates activation of the redundant MAPKKKs Ssk2 and Ssk22, which in turn activate Pbs2.

The activity and relative contribution of the two pathway branches to Hog1 kinase activity is usually measured in mutants that are blocked in either branch. It is presently not known if such experiments provide a correct impression of the activity of the two branches in wild type cells. In any case, it appears that the Sho1 branch has a higher stress threshold for activation (Maeda *et al.*, 1995; O'Rourke and Herskowitz, 2004) and that the Sho1 branch alone is not sufficient to mediate full pathway activation (unpublished data).

4. CYTOSOLIC RESPONSES

Following its activation, Hog1 visibly accumulates in the nucleus (Ferrigno *et al.*, 1998), an observation that has directed focus on transcriptional targets of Hog1. However, a recent conference report from Jeremy Thorner's lab suggests that important Hog1 targets for survival and adaptation following a hyperosmotic shock must be located in the cytosol—when Hog1 was tethered to the plasma membrane and hence was unable to move to the nucleus to control transcription, cells still achieved osmotolerance. Several such cytosolic targets have been identified over the last years. Activation of the potassium and sodium export pumps Tok1 and Nha1 appears to be required to stabilise the intracellular ion concentration in order to allow for protein–DNA interactions (Proft and Struhl, 2004). It is known that at stronger stress levels (i.e. higher concentration of the osmoticum) yeast cells require a longer period to mediate nuclear accumulation of Hog1 and a transcriptional response. The reason for this might be that the Hog1-dependent ion export first has to reduce cellular ionic strength before Hog1 becomes fully activated, but this has not been tested. In response to arsenite stress (see also below), Hog1 is only moderately activated, does not visibly accumulate in the nucleus and does not seem to mediate a transcriptional response. However, the observed low activation appears to be sufficient to control the cytoplasmic target in acquisition of arsenite tolerance, Fps1 (Thorsen *et al* , 2006). This illustrates that Hog1 can mediate significant responses even if only a fraction of the protein is activated and not visibly recruited to the nucleus.

In addition, it is well established that Hog1 mediates arrest of the cell cycle following osmotic shock in both the G1 as well as the G2 phase by targeting Hsl1 (Clotet *et al.*, 2006) and Sic1 (Escote *et al.*, 2004), respectively. This cell cycle arrest is probably crucial to allow osmotic recovery before cell-division events proceed. Hog1 also controls translation efficiency by targeting the MAPKAP Rck2 (Bilsland-Marchesan *et al.*, 2000; Teige *et al.*, 2001; Bilsland *et al.*, 2004). Furthermore it has been reported that Pfk26 is phosphorylated and activated by Hog1 (Dihazi *et al.*, 2004). Pfk26 mediates production of fructose-2,6-bisphosphate, which in turn activates the glycolytic enzyme phosphofructosekinase. This effect was reported to stimulate the production of glycerol, the essential yeast osmolyte (Figure 2).

5. FPS1

The aquaporin family of transmembrane proteins encompasses water-specific aquaporins as well as aquaglyceroporins with permeability for glycerol and other smaller, usually uncharged compounds (Hohmann *et al.*, 2001; Pettersson *et al.*, 2005). Fps1 is an aquaglyceroporin. As such, it is characterised by two-times three transmembrane domains organised as a direct repeat with certain sequence similarity. Loops B and E dip back into the membrane and meet in the central plane of the membrane bilayer, thereby essentially forming a seventh transmembrane domain. These two loops are part of the transmembrane pore constriction and the best conserved domains of the protein. They contain the two NPA motifs, which are characteristic for the aquaporin family. Fps1, like many other fungal aquaglyceroporins, has long extensions at the N- and the C-terminus (Pettersson *et al.*, 2005).

Fps1 plays a critical role in osmoadaptation by controlling the intracellular glycerol level. It appears that the yeast plasma membrane is rather impermeable

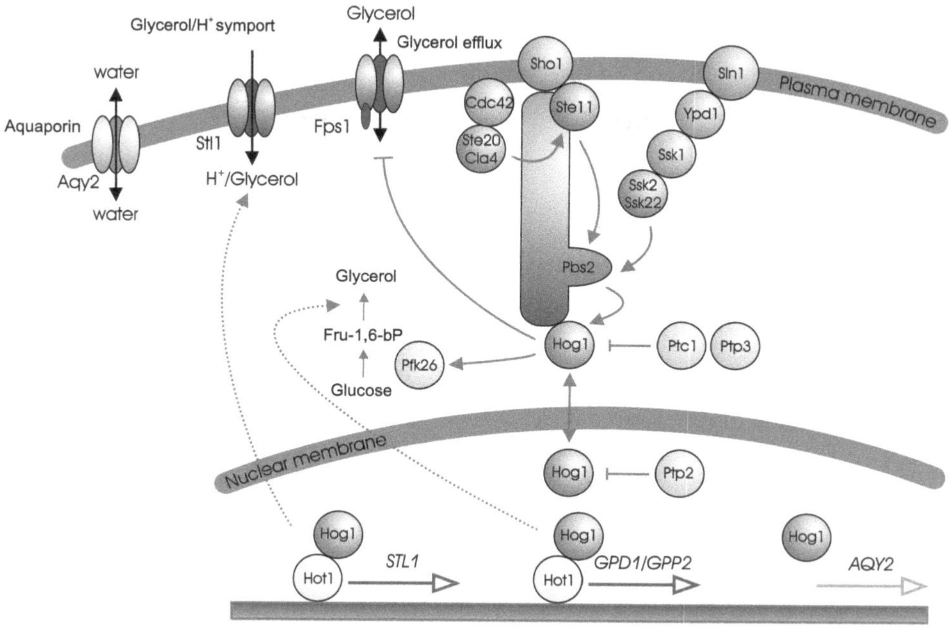

Figure 2 The HOG pathway controls water and Glycerol homeostasis. At the level of gene expression Hog1 mediates reduced expression of *AQY2* encoding a water channel, probably to diminish water loss. Also at gene-expression level, the capacity for glycerol uptake and production are increased by Hog1. Hog1 also appears to stimulate glycolytic flux by activating PFK26 to produce fructose-2,6-bisphosphate. Finally, Hog1 may participate in the control of the glycerol efflux channel Fps1. **(See Colour Section)**

for glycerol (Oliveira *et al.*, 2003) and hence employing a regulated glycerol export channel allows yeast cells rapidly to control and fine-tune their glycerol content. During growth on glucose as a carbon source, glycerol for osmoregulation is produced in the cell via a short branch of glycolysis consisting of two enzymatic steps (Norbeck *et al.*, 1996; Ansell *et al.*, 1997). Mutants lacking Fps1 display a higher intracellular glycerol level (Tao *et al.*, 1999) but under hyperosmotic conditions, they accumulate glycerol with a similar profile as wild type cells (Tamás *et al.*, 1999). When yeast cells are then exposed to a hypo-osmotic shock, i.e. by transfer into distilled water, most of the intracellular glycerol is released to the surrounding medium within less than 3 min.Rapid glycerol export is fully dependent on Fps1 (Tamás *et al.*, 1999). Mutants lacking Fps1 are sensitive to a hypo-osmotic shock and only a fraction of the cells survive such a treatment (Tamás *et al.*, 1999). Deletion of *FPS1* is lethal when combined with mutations that affect the strength of the cell wall, further illustrating the role of Fps1 as an osmotic pressure valve (Tamás *et al.*, 1999; Tong *et al.*, 2004).

The observations that little glycerol leaks out of yeast cells under hyperosmotic conditions while it is rapidly exported upon hypo-osmotic shock clearly illustrates that the transport activity of Fps1 is gated. Constitutively open Fps1 mutants exist and mediate glycerol export even under hyperosmotic conditions.

As a consequence, such yeast cells are unable to properly accumulate glycerol, which they try to compensate for by enhanced glycerol production (Tamás *et al.*, 1999, 2003; Hedfalk *et al.*, 2004; Karlgren *et al.*, 2004). Mutants expressing open Fps1 grow more poorly under hyperosmotic conditions. Hence, mutants lacking Fps1 are hypo-osmosensitive while mutants with overactive Fps1 are hyper-osmosensitive, illustrating the importance of Fps1 for yeast osmoregulation. While proteins with similar functions in regulated osmolyte export exist in probably most organisms, the precise type of aquaglyceroporin represented by *S. cerevisiae* Fps1 (based on conserved regulatory domains) only seems to occur in yeasts, not even in filamentous fungi (Pettersson *et al.*, 2005).

The control mechanism of Fps1 has been investigated by targeted and random mutagenesis. It appears that three regions play a role in Fps1 gating:

(1) The B-loop. A random genetic screen for constitutively open Fps1 (Karlgren *et al.*, 2004) identified residues G348 and H350, both facing the intracellular mouth of the pore in the B-loop. Both residues are located just adjacent to the conserved NPA motif (residues 352–354). It is possible that the mouth of the cytoplasmic side of the pore serves as the target for the closing mechanism.

(2) The region of about 40 amino acids immediately proximal to the first transmembrane domain was first identified following truncation analysis and then by both targeted and random point mutagenesis (Tamás *et al.*, 1999, 2003; Karlgren *et al.*, 2004). Within the core of this domain, 15 out of 18 residues are conserved in all yeast orthologues. Also the spacing of this core region to the first transmembrane domain is conserved (18 residues), and several mutations in this region affecting gating have been identified. We assume that this proximal domain serves as a "lid" that interacts with the B-loop, but direct evidence for such a scenario is presently missing.

(3) The ten amino acids immediately distal of the sixth transmembrane domain are also highly conserved (8 out of 10 identical in all yeast sequences). Truncation of the C-terminus within this sequence, but not further downstream, renders Fps1 constitutively active (Hedfalk *et al.*, 2004; Karlgren *et al.*, 2004). However, deletion of just this part of the C-terminus does not have an effect. This, together with other observations, suggests that the C-terminal extension may have a more indirect role in gating, for instance by supporting a specific configuration of the N-terminal extension and/or oligomerisation Aquaporins usually are tetramers.

The present hypothesis about the mechanism that mediates Fps1 gating focuses on the B-loop plus the N-terminal extension. Interestingly, a specific link between the Fps1 N-terminus and Hog1 signalling was discovered in an entirely different context. Fps1 mediates the uptake of toxic trivalent arsenite into yeast cells (Wysocki *et al.*, 2001), a feature it shares with other aquaglyceroporins. Hydrated arsenite resembles glycerol. Mutants lacking Fps1 are highly tolerant to arsenite while constitutively open Fps1 causes arsenite hypersensitivity (Thorsen *et al.*, 2006). Arsenite seems to activate the HOG pathway in an unknown manner, although only moderately. Mutants lacking Hog1 are sensitive to arsenite. It appears that the only Hog1 target in the acquisition of arsenite tolerance is Fps1. Hog1 phosphorylates T231 in the centre of the highly conserved N-terminal domain and this effect seems to cause inactivation of Fps1 (Thorsen *et al.*, 2006). Mutation of this residue causes constitutively open Fps1, arsenite sensitivity and also hyper-osmosensitivity (Karlgren *et al.*, 2004; Thorsen *et al.*, 2006). However, whether the same Hog1-dependent mechanism controls Fps1 also following a hyper-osmotic shock is presently unclear. The Hog1 effect may represent a long-term control of Fps1 activity, such as at the level of stability, rather than accounting for the very rapid response of Fps1 to osmotic shock. The situation appears to be complex; even in the arsenite response it appears that additional protein kinases can phosphorylate T231 (Thorsen *et al.*, 2006).

6. TRANSCRIPTIONAL RESPONSES

Several studies have reported global gene expression analyses of yeast following a hyperosmotic shock of different intensity (Gasch *et al.*, 2000; Posas *et al.*, 2000; Rep *et al.*, 2000; Causton *et al.*, 2001; Yale and Bohnert, 2001). The reported number of genes up- or downregulated following a hyperosmotic shock varies between studies due to different growth conditions and thresholds applied in the analyses. It appears that about 200–400 genes are upregulated and some 150–250 genes are downregulated. Some important conclusions from these analyses are as follows (Hohmann, 2002):

1. The response is transient. Most genes are up- or downregulated for a certain period of time (depending on the strength of the osmotic treatment from a few minutes to more than an hour) and then mRNA levels fall back to normal.
2. The response has little specificity. Essentially all genes that show robust up- or downregulation following osmotic shock are affected in a similar way by a range of different stress conditions. This general stress response is negatively regulated by protein kinase A.
3. Downregulated genes encode proteins related to growth and biomass formation. In fact most of those genes encode ribosomal proteins, translation factors and glycolytic enzymes. Those genes are positively regulated by protein kinase A.
4. Upregulated genes are not necessary for osmotolerance. Deletion of none of the upregulated genes seems to confer osmosensitivity (Warringer *et al.*, 2003). This appears counter-intuitive at first sight. However, genes encoding functions required to sense and respond to stress need to be present at all times and hence do not need to be regulated. Many proteins whose production is upregulated may rather be relevant for sustaining an osmotolerant state and their deletion may not cause a strong phenotype. In addition, several upregulated genes, such as those encoding enzymes in

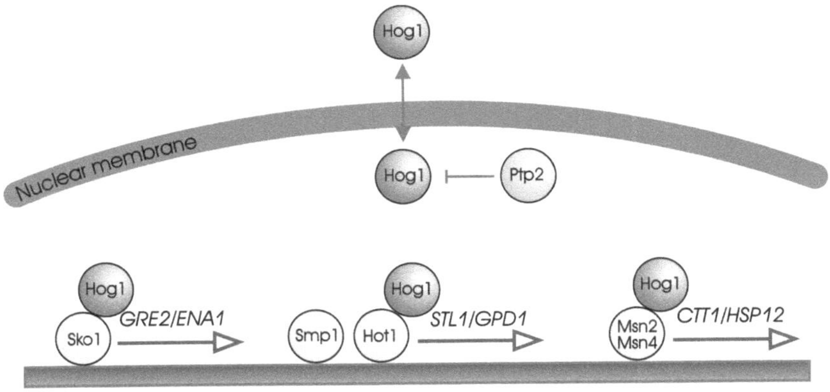

Figure 3 Hog1 and control of gene expression. Hog1 interacts with different transcription factors in the nucleus, such as Sko1, Hot1, Msn2/4 and probably also Smp1 to control different sets of genes (some typical examples are shown). It appears that Hog1 thereby itself is the activating factor. **(See Colour Section)**

glycerol production, are duplicated and only one copy is stress controlled. Its deletion then has only a minor, if any, phenotypic effect.

5. Only a fraction of the genes is HOG dependent. The osmostress-induced upregulation of only about 20% of the genes is fully or largely Hog1 dependent. However, those typically include the most strongly osmoregulated genes.

Several transcription factors have been identified that mediate Hog1-dependent responses: Hot1, Sko1, Msn2/Msn4, Msn1 and Smp1 (Figure 3). The first three of those are best characterised and also seem to play the most significant roles.

Hot1 (Rep *et al.*, 1999b, 2000; Alepuz *et al.*, 2003) is a nuclear protein that seems to control a set of less than ten genes, including those that encode proteins involved in glycerol metabolism and uptake. Hot1 has been demonstrated to recruit Hog1 to target promoters.

Sko1 binds to CRE sites in target promoters, although not all of the about 40 target genes seem to contain such a site (Proft *et al.*, 2001, 2005; Rep *et al.*, 2001; Proft and Struhl, 2002). Active Hog1 converts Sko1, whose repressor function is mediated by Ssn6/Tup1, from a repressor to an activator (Rep *et al.*, 2001; Proft and Struhl, 2002). It appears that Sko1 controls expression of several regulators of the osmoresponse systems, such as the Msn2 transcription factor and the Ptp3 protein phosphatase (Proft *et al.*, 2005). Hence, Sko1 may be central to a transcriptional network important for establishing the adapted state in osmoregulation.

Msn2 and Msn4 are two redundant proteins well known for mediating a general stress response (Ruis and Schüller, 1995). Msn2 and Msn4 are negatively controlled by protein kinase A, which seems to mediate nuclear export of the transcription factor. Although Msn2 and Msn4 are required for transcriptional responses to a range of stress conditions, stimulation of their targets under osmotic stress requires the Hog1 kinase (Rep *et al.*, 2000).

It appears that the Hog1 kinase itself is the activating transcription factor for its targets. Hog1 is recruited to target promoters by Hot1, Sko1 and Msn2/Msn4 (Alepuz *et al.*, 2001, 2003; Proft and Struhl, 2002; De Nadal *et al.*, 2004; Pokholok *et al.*, 2006). The target for the Hog1-mediated transcriptional activation appears to be the Rpd3-Sin3 histone deacetylase complex (De Nadal *et al.*, 2004). In addition, Hog1 is not only required for initiating transcription of target genes, but also for the transcriptional elongation step (Pokholok *et al.*, 2006; Proft *et al.*, 2006). Hence, the perspective of how a protein kinase controls gene expression has changed substantially through studies on the Hog1 MAPK. Rather than phosphorylating and thereby somehow (in)activating a DNA-binding transcription factor, Hog1 (and also other kinases (Pokholok *et al.*, 2006)) associates with DNA binding proteins to mediate gene-specific control of transcription via general components of the gene-regulation machinery.

7. FEEDBACK CONTROL

The response of yeast cells to a hyperosmotic shock is transient. Phosphorylated, active Hog1 becomes detectable within less than 1 min following a moderate osmotic shock with 0.4 M NaCl, but falls back to pre-stress levels within about 30 min.In a similar fashion, the level of mRNAs of Hog1-controlled genes, such as *STL1*, rises rapidly after such an osmotic treatment reaching a peak after about 15 min to fall back to pre-stress levels after 40 min (Figure 4). Hence, HOG pathway activation is controlled by rigorous feedback mechanisms that ensure that the pathway is inactivated during adaptation and the cell cycle block mediated by Hog1 is lifted, allowing proliferation to commence. This also means that the observed, strong HOG pathway activation following a hyperosmotic shock mainly serves to kick on the adaptation process but that maintenance of the

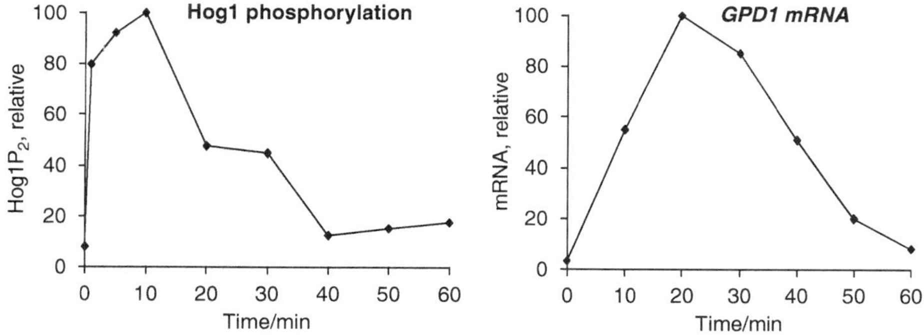

Figure 4 Osmotic responses are transient. The figure shows an example of quantification of the level of phospho-Hog1 and the mRNA of the Hog1-target gene *GPD1*. Cells were treated with 0.5 M NaCl at time 0.

adapted state is either supported by a much smaller level of active Hog1 or by entirely different mechanisms.

Many signalling systems, such as the yeast pheromone response pathway, become desensitised following prolonged or repeated exposure to stimulus. The yeast osmosensing HOG system, on the other hand, controls a homeostatic process and hence needs to remain in a signalling competent state at all times. Indeed, the HOG pathway, and HOG-dependent gene expression can readily be reactivated with essentially the same time course and amplitude by a second osmotic shock at different times following a first shock (e.g. a first increase to 0.5 M NaCl and then a rise to 1 M NaCl) (Klipp *et al.*, 2005). Hence, the mechanisms that control signalling through the HOG pathway must combine rigorous feedback control with maintenance of signalling competence. In fact this property is shared by both branches of the HOG pathway when they individually control Hog1 activation, i.e. when one branch is blocked by mutation (unpublished data).

In order to better understand at a quantitative level, signalling processes in the yeast osmoregulatory system together with the group of Edda Klipp (Max-Planck Institute for Molecular Genetics, Berlin), we generated a mathematical reconstruction of the system (Klipp *et al.*, 2005). The model in its present state encompasses mathematical descriptions of sensing through the Sln1 phosphorelay system, signalling through the MAPK cascade, gene and protein expression for glycerol production, glycolytic metabolism including glycerol production and regulated export through Fps1, as well as changes in cell volume and turgor. These items were initially set up as modules on the basis of literature and large-scale genomics data and tested individually before being connected to an overall model of 35 differential equations and 70 parameters. Some of the parameters were calculated on the basis of standard experiments (osmotic shock with 0.5 M NaCl) and time course data on relative levels of phosphorylated Hog1, levels of mRNA of HOG target genes, specific activity of an enzyme in glycerol production as well as the levels of intracellular and total glycerol. Other parameters were calculated and optimised to best-fit experimental data. The model was then iteratively improved using additional experimental data from perturbations such as different types of stress treatments or genetic changes, including inability to accumulate glycerol. Simulations of the model were then employed to address specific biological questions, such as the mechanism underlying feedback control and maintenance of signalling competence.

Like any other MAPK signalling pathway, the HOG pathway is also negatively controlled by protein phosphatases (Figure 5). It appears that three such phosphatases are physiologically relevant: Ptp2, a nuclear phospho-tyrosine specific phosphatase, Ptp3, a cytosolic phospho-tyrosine specific phosphatase, and Ptc1, a type 2C protein phosphatase (PP2C) located in both the cytosol and the nucleus. Simultaneous deletion of *PTP2* and *PTC1* is lethal due to overactive Hog1 (Maeda *et al.*, 1993), clearly demonstrating the importance of these proteins for controlling signalling through Hog1. Several pieces of evidence appear to suggest that upregulation of protein phosphatases plays a role in adaptation. First, the expression of the genes *PTP2* and *PTP3* is upregulated following osmotic shock (Jacoby *et al.*, 1997; Wurgler-Murphy *et al.*, 1997; Mattison and Ota, 2000).

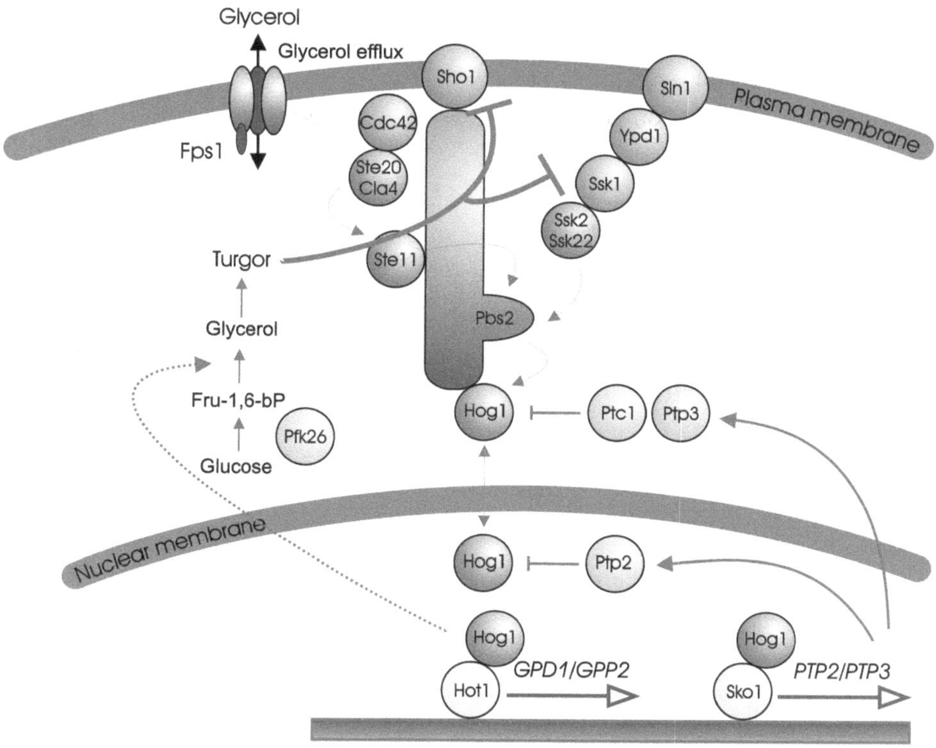

Figure 5 Feedback of the osmotic response. Osmotic adaptation, i.e. Turgor Recovery, seems to determine the time point when HOG Signalling is downregulated. The onset of cell reswelling probably terminates the initial signal. Then, protein phosphatases deactivate the pathway. Upregulation of those phosphatases, indicated here as a transcriptional feedback loop, may play a role in adaptation or, rather, contribute to establish a new steady state in adapted cells. In addition, it appears that Ssk1 and perhaps also Ste11 are destabilised to modulate the response (not depicted here). **(See Colour Section)**

However, it has not been clearly demonstrated that this upregulation is really needed for downregulation of the HOG pathway. Second, deletion of protein phosphatases affects the period of Hog1 phosphorylation (Jacoby *et al.*, 1997; Mattison and Ota, 2000; Warmka *et al.*, 2001), although this seems to depend on the conditions, strains and antibodies employed (Klipp *et al.*, 2005). Third, a kinase-dead Hog1 apparently can be phosphorylated by Pbs2 but is not dephosphorylated anymore suggesting that Hog1-mediated phosphorylation of some target protein is required for dephosphorylation of Hog1 (Wurgler-Murphy *et al.*, 1997). However, this target protein has so far not been identified; while it could well be the nuclear Ptp2 phosphatase, alternative interpretations of this observation exist. For instance, Hog1 could mediate a feedback regulation on an upstream component of the pathway. Such a role is suggested by the observation that Hog1 activity is required to prevent cross talk from the Sho1 branch to the pheromone or pseudohyphal development pathways (O'Rourke and Herskowitz, 1998).

Simulations using the mathematical model of osmoregulation (Klipp *et al.*, 2005) were used to test if Hog1-mediated upregulation of protein phosphatases (transcriptional or otherwise) could account for the observed feedback control on the Hog1 phosphorylation status. This appeared possible. However, it appeared very difficult to fit phosphatase upregulation with HOG pathway signalling competence described above. Hence, it appears that in osmostressed yeast cells downregulation of the HOG pathway, or more specifically the timing of such downregulation, is determined by other mechanisms (genetic upregulation, i.e. by mutations rendering certain pathway components constitutively active, may invoke other feedback mechanisms). We believe that there is a close link between HOG pathway activity and actual osmotic adaptation. In such a scenario, cell shrinking will first activate the HOG pathway and subsequent reswelling will deactivate the pathway. This scenario directly implies that the sensors of the HOG pathway detect osmotic changes rather than an osmotic status. In this picture, the phosphatases would rather play a more passive role and dephosphorylate Hog1 with a more or less constant rate and the Hog1 phosphorylation state would be mainly determined by the activity of the sensing devices. This scenario is backed by simulations and experiments.

Several experimental scenarios support the idea of the Hog1 phosphorylation state being mainly controlled by the osmotic changes caused by external stimuli or cellular adaptation rather than intrinsic feedback mechanisms. First, mutations that prevent glycerol production (*gpd1*Δ *gpd2*Δ) or that delay glycerol accumulation despite glycerol overproduction (constitutively open Fps1) cause sustained Hog1 phosphorylation (Siderius *et al.*, 2000; Klipp *et al.*, 2005). Such cells do not adapt to increased osmolarity, or do so only with a significant delay. Second, cells that accumulate glycerol faster (anaerobically growing cells or cells overproducing Gpd1) display a shorter period of Hog1 phosphorylation (Krantz *et al.*, 2004). Third, it is well known that yeast cells treated with an osmotic shock of higher intensity display a longer period of Hog1 phosphorylation than cells treated with lower stress levels (Van Wuytswinkel *et al.*, 2000). Finally, in an "artificial" osmoadaptation system there is a clear correlation between the apparent period of osmotic adaptation and Hog1 phosphorylation (Karlgren *et al.*, 2005). In this system, yeast cells unable to produce any glycerol (*gpd1*Δ *gpd2*Δ) were transformed with a construct that allows expression of a constitutive aquaglyceroporin. Such cells were then stressed with high levels of glycerol, xylitol and sorbitol. Initially, cells will shrink following such a treatment. However, glycerol, xylitol and sorbitol are transported into cells via the aquaglyceroporin with a rate decreasing from glycerol to sorbitol. In other words, the levels of these compounds will equilibrate across the plasma membrane thereby relieving the osmotic stress, although this will occur at different speeds. Consequently, the period of Hog1 phosphorylation is shortest with glycerol, longer with xylitol and longest with sorbitol. The use of two different aquaglyceroporins mediating transport at different rates confirms this correlation (Karlgren *et al.*, 2005). Taken together, there appears to be a clear link between osmotic adaptation and the period of Hog1 phosphorylation and activity.

8. PERSPECTIVES

By controlling the osmotic response and the timing of its downregulation mainly via actual osmotic adaptation, probably via the activity of the sensors, the cell combines effective feedback control with signalling competence, as required for a homeostatic process. However, there are clearly additional mechanisms that impinge on HOG pathway activity and feedback control, and understanding their contribution at a quantitative and time-resolved level will be future challenges. The crucial role of the protein phosphatases was already mentioned above. Quite clearly, their activity is required to inactivate Hog1 once the cell adapts and signalling from the sensors ceases. The protein phosphatases are also central in controlling signalling noise and thresholds; if they require upregulation to perform, their function in adaptation is less obvious. In addition, it has been reported that Ssk1 in the Sln1 pathway is destabilised in osmostressed cells and this mechanism may contribute to pathway downregulation, although it has to take into account maintenance of signalling competence (Sato *et al.*, 2003). Moreover, Ste11 in the Sho1 pathway is an unstable protein and degraded at least during pheromone response (Esch and Errede, 2002).

There are a number of exciting future directions where quantitative analyses and a combination of modelling and experimentation will provide new insights into the mechanisms of osmotic homeostasis. For instance, the contribution of the two HOG pathway branches to signalling have so far only been studied in mutants blocking one branch, which may give an incorrect impression of their roles in wild type cells. Monitoring the activity of different proteins in the two branches by more sophisticated experimental setups in combination with modelling may help in elucidating the precise roles of the Sln1 and Sho1 branches in the overall response. Another interesting question concerns the quantitative and temporal contributions of different control mechanisms in glycerol accumulation:

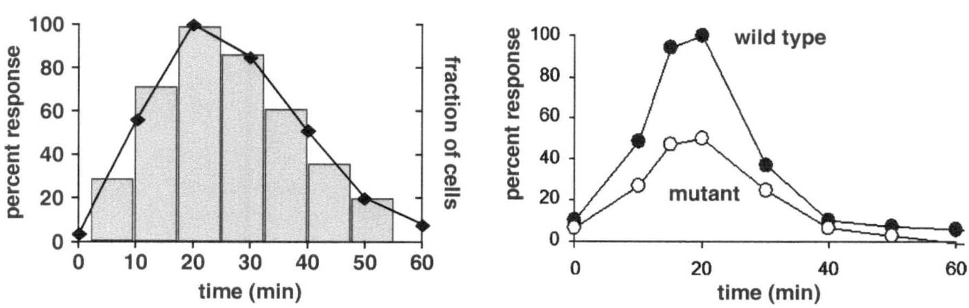

Figure 6 Response profiles. A bell-shaped response profile derived from cell populations could be interpreted in different ways: either all cells in the population show such behaviour or individual cells could show all-or-nothing responses at different time points. In the right panel the mutant phenotype could be interpreted such that all mutant cells respond to maximally 50% or that 50% of the cells show a full response. In both examples the interpretation for thresholds, noise control and response mechanisms were very different.

Fps1 closure, control of glycolytic flux as well as upregulation of enzymes involved in glycerol production and uptake. Finally, most measurements done so far monitor osmotic responses in cell populations. Such response profiles may give a misleading impression since they could be composed of single cell responses that follow a very different profile (Figure 6). Such single cell measurements are feasible using specific fluorescent reporters combined with flow cytometry, time-lapse microscopy as well as microfluidics combined with optical manipulations tools. Hence, systems level understanding of cell responses also requires acquisition of new types of data and, therefore, not only encompasses computational but also experimental challenges.

ACKNOWLEDGEMENTS

The author thanks Bodil Nordlander, Carl-Fredrik Tiger and Elzbieta Petelenz for critical reading of the manuscript. Work in the author's lab is supported by grants from the Swedish Research Council, Vinnova as well as the European Commission (QUASI: LSHG-CT2003-503230; SystemsBiology: MEST2-CT2004-514169).

REFERENCES

Albertyn, J., Hohmann, S., Thevelein, J.M. and Prior, B.A. (1994). *GPD1*, which encodes glycerol-3-phosphate dehydrogenase is essential for growth under osmotic stress in *Saccharomyces cerevisiae* and its expression is regulated by the high-osmolarity glycerol response pathway. *Molecular and Cellular Biology*, **14**, 4135–4144.

Alepuz, P.M., de Nadal, E., Zapater, M., Ammerer, G. and Posas, F. (2003). Osmostress-induced transcription by Hot1 depends on a Hog1-mediated recruitment of the RNA Pol II. *The EMBO Journal*, **22**, 2433–2442.

Alepuz, P.M., Jovanovic, A., Reiser, V. and Ammerer, G. (2001). Stress-induced MAP kinase Hog1 is part of transcription activation complexes. *Molecular Cell*, **7**, 767–777.

Ansell, R., Granath, K., Hohmann, S., Thevelein, J.M. and Adler, L. (1997). The two isoenzymes for yeast NAD^+-dependent glycerol 3-phosphate dehydrogenase encoded by *GPD1* and *GPD2* have distinct roles in osmoadaptation and redox regulation. *The EMBO Journal*, **16**, 2179–2187.

Bilsland, E., Molin, C., Swaminathan, S., Ramne, A. and Sunnerhagen, P. (2004). Rck1 and Rck2 MAPKAP kinases and the HOG pathway are required for oxidative stress resistance. *Molecular Microbiology*, **53**, 1743–1756.

Bilsland-Marchesan, E., Arino, J., Saito, H., Sunnerhagen, P. and Posas, F. (2000). Rck2 kinase is a substrate for the osmotic stress-activated mitogen-activated protein kinase Hog1. *Molecular and Cellular Biology*, **20**, 3887–3895.

Brewster, J.L., de Valoir, T., Dwyer, N.D., Winter, E. and Gustin, M.C. (1993). An osmosensing signal transduction pathway in yeast. *Science*, **259**, 1760–1763.

Catlett, N.L., Yoder, O.C. and Turgeon, B.G. (2003). Whole-genome analysis of two-component signal transduction genes in fungal pathogens. *Eukaryot Cell*, **2**, 1151–1161.

Causton, H.C., Ren, B., Koh, S.S., Harbison, C.T., Kanin, E., Jennings, E.G., Lee, T.I., True, H.L., Lander, E.S. and Young, R.A. (2001). Remodeling of yeast genome expression in response to environmental changes. *Molecular Biology of the Cell*, **12**, 323–337.

Clotet, J., Escote, X., Adrover, M.A., Yaakov, G., Gari, E., Aldea, M., de Nadal, E. and Posas, F. (2006). Phosphorylation of Hsl1 by Hog1 leads to a G2 arrest essential for cell survival at high osmolarity. *The EMBO Journal*, **25**, 2338–2346.

Davenport, K.R., Sohaskey, M., Kamada, Y., Levin, D.E. and Gustin, M.C. (1995). A second osmosensing signal transduction pathway in yeast. Hypotonic shock activates the PKC1 protein kinase-regulated cell integrity pathway. *The Journal of Biological Chemistry*, **270**, 30157–30161.

Davenport, K.D., Williams, K.E., Ullmann, B.D. and Gustin, M.C. (1999). Activation of the *Saccharomyces cerevisiae* filamentation/invasion pathway by osmotic stress in high-osmolarity glycogen pathway mutants. *Genetics*, **153**, 1091–1103.

De Nadal, E., Alepuz, P.M. and Posas, F. (2002). Dealing with osmostress through MAP kinase activation. *EMBO Reports*, **3**, 735–740.

De Nadal, E., Zapater, M., Alepuz, P.M., Sumoy, L., Mas, G. and Posas, F. (2004). The MAPK Hog1 recruits Rpd3 histone deacetylase to activate osmoresponsive genes. *Nature*, **427**, 370–374.

Dihazi, H., Kessler, R. and Eschrich, K. (2004). High osmolarity glycerol (HOG) pathway-induced phosphorylation and activation of 6-phosphofructo-2-kinase are essential for glycerol accumulation and yeast cell proliferation under hyperosmotic stress. *The Journal of Biological Chemistry*, **279**, 23961–23968.

Esch, R.K. and Errede, B. (2002). Pheromone induction promotes Ste11 degradation through a MAPK feedback and ubiquitin-dependent mechanism. *Proceedings of the National Academy of Sciences of the United States of America*, **99**, 9160–9165.

Escote, X., Zapater, M., Clotet, J. and Posas, F. (2004). Hog1 mediates cell-cycle arrest in G1 phase by the dual targeting of Sic1. *Nature Cell Biology*, **6**, 997–1002.

Ferrigno, P., Posas, F., Koepp, D., Saito, H. and Silver, P.A. (1998). Regulated nucleo/cytoplasmic exchange of HOG1 MAPK requires the importin beta homologs NMD5 and XPO1. *The EMBO Journal*, **17**, 5606–5614.

Furukawa, K., Hoshi, Y., Maeda, T., Nakajima, T. and Abe, K. (2005). *Aspergillus nidulans* HOG pathway is activated only by two-component signalling pathway in response to osmotic stress. *Molecular Microbiology*, **56** 1246–1261.

Gasch, A.P., Spellman, P.T., Kao, C.M., Carmel-Harel, O., Eisen, M.B., Storz, G., Botstein, D. and Brown, P.O. (2000). Genomic expression programs in the response of yeast cells to environmental changes. *Molecular Biology of the Cell*, **11**, 4241–4257.

Hedfalk, K., Bill, R.M., Mullins, J.G., Karlgren, S., Filipsson, C., Bergstrom, J., Tamas, M.J., Rydstrom, J. and Hohmann, S. (2004). A regulatory domain in the C-terminal extension of the yeast glycerol channel Fps1p. *The Journal of Biological Chemistry*, **279**, 14954–14960.

Hohmann, S. (2002). Osmotic stress signaling and osmoadaptation in yeasts. *Microbiology Molecular Biology Reviews*, **66**, 300–372.

Hohmann, S., Nielsen, S. and Agre, P. (2001). *Aquaporins*. Academic Press, San Diego, CA.

Jacoby, T., Flanagan, H., Faykin, A., Seto, A.G., Mattison, C. and Ota, I. (1997). Two protein-tyrosine phosphatases inactivate the osmotic stress response pathway in yeast by targeting the mitogen-activated protein kinase, Hog1. *The Journal of Biological Chemistry* **272**, 17749–17755.

Jung, U.S. and Levin, D.E. (1999). Genome-wide analysis of gene expression regulated by the yeast cell wall integrity signalling pathway. *Molecular Microbiology*, **34**, 1049–1057.

Karlgren, S., Filipsson, C., Mullins, J.G., Bill, R.M., Tamas, M.J. and Hohmann, S. (2004). Identification of residues controlling transport through the yeast aquaglyceroporin Fps1 using a genetic screen. *European Journal of Biochemistry*, **271**, 771–779.

Karlgren, S., Pettersson, N., Nordlander, B., Mathai, J.C., Brodsky, J.L., Zeidel, M.L., Bill, R.M. and Hohmann, S. (2005). Conditional osmotic stress in yeast: a system to study transport through aquaglyceroporins and osmostress signaling. *The Journal of Biological Chemistry*, **280**, 7186–7193.

Klipp, E., Nordlander, B., Kruger, R., Gennemark, P. and Hohmann, S. (2005). Integrative model of the response of yeast to osmotic shock. *Nature Biotechnology*, **23**, 975–982.

Krantz, M., Becit, E. and Hohmann, S. (2006a). Comparative analysis of HOG pathway proteins to generate hypotheses for functional analysis. *Current Genetics*, **49**, 152–165.

Krantz, M., Becit, E. and Hohmann, S. (2006b). Comparative genomics of the HOG-signalling system in fungi. *Current Genetics*, **49**, 137–151.

Krantz, M., Nordlander, B., Valadi, H., Johansson, M., Gustafsson, L. and Hohmann, S. (2004). Anaerobicity prepares *Saccharomyces cerevisiae* cells for faster adaptation to osmotic shock. *Eukaryotic Cell*, **3**, 1381–1390.

Luyten, K., Albertyn, J., Skibbe, W.F., Prior, B.A., Ramos, J., Thevelein, J.M. and Hohmann, S. (1995). Fps1, a yeast member of the MIP family of channel proteins, is a facilitator for glycerol uptake and efflux and is inactive under osmotic stress. *The EMBO Journal*, **14**, 1360–1371.

Maeda, T., Takekawa, M. and Saito, H. (1995). Activation of yeast PBS2 MAPKK by MAPKKKs or by binding of an SH3-containing osmosensor. *Science*, **269**, 554–558.

Maeda, T., Tsai, A.Y.M. and Saito, H. (1993). Mutations in a protein tyrosine phosphatase gene (*PTP2*) and a protein serine/threonine phosphatase gene (*PTC1*) cause a synthetic growth defect in *Saccharomyces cerevisiae. Molecular and Cellular Biology*, **13**, 5408–5417.

Maeda, T., Wurgler-Murphy, S.M. and Saito, H. (1994). A two-component system that regulates an osmosensing MAP kinase cascade in yeast. *Nature*, **369**, 242–245.

Mattison, C.P. and Ota, I.M. (2000). Two protein tyrosine phosphatases, Ptp2 and Ptp3, modulate the subcellular localization of the Hog1 MAP kinase in yeast. *Genes & Development*, **14**, 1229–1235.

Norbeck, J., Påhlman, A.K., Akhtar, N., Blomberg, A. and Adler, L. (1996). Purification and characterization of two isoenzymes of DL-glycerol-3-phosphatase from *Saccharomyces cerevisiae*. Identification of the corresponding *GPP1* and *GPP2* genes and evidence for osmotic regulation of Gpp2p expression by the osmosensing mitogen-activated protein kinase signal transduction pathway. *The Journal of Biological Chemistry*, **271**, 13875–13881.

Oliveira, R., Lages, F., Silva-Graca, M. and Lucas, C. (2003). Fps1p channel is the mediator of the major part of glycerol passive diffusion in *Saccharomyces cerevisiae*: Artefacts and re-definitions. *Biochimica Biophysica Acta*, **1613**, 57–71.

O'Rourke, S.M. and Herskowitz, I. (1998). The Hog1 MAPK prevents cross talk between the HOG and pheromone response MAPK pathways in *Saccharomyces cerevisiae. Genes & Development*, **12**, 2874–2886.

O'Rourke, S.M. and Herskowitz, I. (2004). Unique and redundant roles for HOG MAPK pathway components as revealed by whole-genome expression analysis. *Molecular Biology of the Cell*, **15**, 532–542.

O'Rourke, S.M., Herskowitz, I. and O'Shea, E.K. (2002). Yeast go the whole HOG for the hyperosmotic response. *Trends in Genetics*, **18**, 405–412.

Pettersson, N., Filipsson, C., Becit, E., Brive, L. and Hohmann, S. (2005). Aquaporins in yeasts and filamentous fungi. *Biological of the Cell*, **97**, 487–500.

Pokholok, D.K., Zeitlinger, J., Hannett, N.M., Reynolds, D.B. and Young, R.A. (2006). Activated signal transduction kinases frequently occupy target genes. *Science*, **313**, 533–536.

Posas, F., Chambers, J.R., Heyman, J.A., Hoeffler, J.P., de Nadal, E. and Arino, J. (2000). The transcriptional response of yeast to saline stress. *The Journal of Biological Chemistry*, **275**, 17249–17255.

Posas, F. and Saito, H. (1997). Osmotic activation of the HOG MAPK pathway via Ste11p MAPKKK: scaffold role of Pbs2p MAPKK. *Science*, **276**, 1702–1705.

Posas, F., Wurgler-Murphy, S.M., Maeda, T., Witten, E.A., Thai, T.C. and Saito, H. (1996). Yeast HOG1 MAP kinase cascade is regulated by a multistep phosphorelay mechanism in the SLN1-YPD1-SSK1 "two-component" osmosensor. *Cell*, **86**, 865–875.

Proft, M., Gibbons, F.D., Copeland, M., Roth, F.P. and Struhl, K. (2005). Genomewide identification of Sko1 target promoters reveals a regulatory network that operates in response to osmotic stress in *Saccharomyces cerevisiae. Eukaryotic Cell*, **4**, 1343–1352.

Proft, M., Mas, G., de Nadal, E., Vendrell, A., Noriega, N., Struhl, K. and Posas, F. (2006). The stress-activated Hog1 kinase is a selective transcriptional elongation factor for genes responding to osmotic stress. *Molecular Cell*, **23**, 241–250.

Proft, M., Pascual-Ahuir, A., de Nadal, E., Arino, J., Serrano, R. and Posas, F. (2001). Regulation of the Sko1 transcriptional repressor by the Hog1 MAP kinase in response to osmotic stress. *The EMBO Journal*, **20**, 1123–1133.

Proft, M. and Struhl, K. (2002). Hog1 kinase converts the Sko1-Cyc8-Tup1 repressor complex into an activator that recruits SAGA and SWI/SNF in response to osmotic stress. *Molecular Cell*, **9**, 1307–1317.

Proft, M. and Struhl, K. (2004). MAP kinase-mediated stress relief that precedes and regulates the timing of transcriptional induction. *Cell*, **118**, 351–361.

Raitt, D.C., Posas, F. and Saito, H. (2000). Yeast Cdc42 GTPase and Ste20 PAK-like kinase regulate Sho1-dependent activation of the Hog1 MAPK pathway. *The EMBO Journal*, **19**, 4623–4631.

Reiser, V., Raitt, D.C. and Saito, H. (2003). Yeast osmosensor Sln1 and plant cytokinin receptor Cre1 respond to changes in turgor pressure. *The Journal of Cell Biology*, **161**, 1035–1040.

Reiser, V., Salah, S.M. and Ammerer, G. (2000). Polarized localization of yeast Pbs2 depends on osmostress, the membrane protein Sho1 and Cdc42. *Nature Cell Biology*, **2**, 620–627.

Rep, M., Krantz, M., Thevelein, J.M. and Hohmann, S. (2000). The transcriptional response of *Saccharomyces cerevisiae* to osmotic shock. Hot1p and Msn2p/Msn4p are required for the induction of subsets of high osmolarity glycerol pathway-dependent genes. *The Journal of Biological Chemistry*, **275**, 8290–8300.

Rep, M., Proft, M., Remize, F., Tamas, M., Serrano, R., Thevelein, J.M. and Hohmann, S. (2001). The *Saccharomyces cerevisiae* Sko1p transcription factor mediates HOG pathway-dependent osmotic regulation of a set of genes encoding enzymes implicated in protection from oxidative damage. *Molecular Microbiology*, **40**, 1067–1083.

Rep, M., Reiser, V., Holzmüller, U., Thevelein, J.M., Hohmann, S., Ammerer, G. and Ruis, H. (1999b). Osmotic stress-induced gene expression in *Saccharomyces cerevisiae* requires Msn1p and the novel nuclear factor Hot1p. *Molecular and Cellular Biology*, **19**, 5474–5485.

Ruis, H. and Schüller, C. (1995). Stress signaling in yeast. *Bioessays*, **17**, 959–965.

Saito, H. and Tatebayashi, K. (2004). Regulation of the osmoregulatory HOG MAPK cascade in yeast. *Journal of Biochemistry (Tokyo)*, **136**, 267–272.

Sato, N., Kawahara, H., Toh-e, A. and Maeda, T. (2003). Phosphorelay-regulated degradation of the yeast Ssk1p response regulator by the ubiquitin-proteasome system. *Molecular and Cellular Biology*, **23**, 6662–6671.

Siderius, M., Van Wuytswinkel, O., Reijenga, K.A., Kelders, M. and Mager, W.H. (2000). The control of intracellular glycerol in *Saccharomyces cerevisiae* influences osmotic stress response and resistance to increased temperature. *Molecular and Microbiology*, **36**, 1381–1390.

Tamás, M.J., Karlgren, S., Bill, R.M., Hedfalk, K., Allegri, L., Ferreira, M., Thevelein, J.M., Rydstrom, J., Mullins, J.G. and Hohmann, S. (2003). A short regulatory domain restricts glycerol transport through yeast Fps1p. *The Journal of Biological Chemistry*, **278**, 6337–6345.

Tamás, M.J., Luyten, K., Sutherland, F.C.W., Hernandez, A., Albertyn, J., Valadi, H., Li, H., Prior, B.A., Kilian, S.G., Ramos, J., Gustafsson, L., Thevelein, J.M. and Hohmann, S. (1999). Fps1p controls the accumulation and release of the compatible solute glycerol in yeast osmoregulation. *Molecular and Microbiology*, **31**, 1087–1104.

Tao, W., Deschenes, R.J. and Fassler, J.S. (1999). Intracellular glycerol levels modulate the activity of Sln1p, a *Saccharomyces cerevisiae* two-component regulator. *The Journal of Biological Chemistry*, **274**, 360–367.

Tatebayashi, K., Yamamoto, K., Tanaka, K., Tomida, T., Maruoka, T., Kasukawa, E. and Saito, H. (2006). Adaptor functions of Cdc42, Ste50, and Sho1 in the yeast osmoregulatory HOG MAPK pathway. *The EMBO Journal*, **25**, 3033–3044.

Teige, M., Scheikl, E., Reiser, V., Ruis, H. and Ammerer, G. (2001). Rck2, a member of the calmodulin-protein kinase family, links protein synthesis to high osmolarity MAP kinase signaling in budding yeast. *Proceedings of the National Academy of Sciences of the United States of America*, **98**, 5625–5630.

Thorsen, M., Di, Y., Tangemo, C., Morillas, M., Ahmadpour, D., Van der Does, C., Wagner, A., Johansson, E., Boman, J., Posas, F., Wysocki, R. and Tamas, M.J. (2006). The MAPK Hog1p Modulates Fps1p-dependent Arsenite Uptake and Tolerance in Yeast. *Molecular Biology of the Cell*, **17**, 4400–4410.

Tong, A.H., Lesage, G., Bader, G.D., Ding, H., Xu, H., Xin, X., Young, J., Berriz, G.F., Brost, R.L., Chang, M., Chen, Y., Cheng, X., Chua, G., Friesen, H., Goldberg, D.S., Haynes, J., Humphries, C., He, G., Hussein, S., Ke, L., Krogan, N., Li, Z., Levinson, J.N., Lu, H., Menard, P., Munyana, C., Parsons, A.B., Ryan, O., Tonikian, R., Roberts, T., Sdicu, A.M., Shapiro, J., Sheikh, B., Suter, B., Wong, S.L., Zhang, L.V., Zhu, H., Burd, C.G., Munro, S., Sander, C., Rine, J., Greenblatt, J., Peter, M., Bretscher, A., Bell, G., Roth, F.P., Brown, G.W., Andrews, B., Bussey, H. and Boone, C. (2004). Global mapping of the yeast genetic interaction network. *Science*, **303**, 808–813.

Van Wuytswinkel, O., Reiser, V., Siderius, M., Kelders, M.C., Ammerer, G., Ruis, H. and Mager, W.H. (2000). Response of *Saccharomyces cerevisiae* to severe osmotic stress: Evidence for a novel activation mechanism of the HOG MAP kinase pathway. *Molecular Microbiology*, **37**, 382–397.

Warmka, J., Hanneman, J., Lee, J., Amin, D. and Ota, I. (2001). Ptc1, a type 2C Ser/Thr phosphatase, inactivates the HOG pathway by dephosphorylating the mitogen-activated protein kinase Hog1. *Molecular and Cellular Biology*, **21**, 51–60.

Warringer, J., Ericson, E., Fernandez, L., Nerman, O. and Blomberg, A. (2003). High-resolution yeast phenomics resolves different physiological features in the saline response. *Proceedings of the National Academy of Sciences of the United States of America*, **100**, 15724–15729.

Wu, C., Jansen, G., Zhang, J., Thomas, D.Y. and Whiteway, M. (2006). Adaptor protein Ste50p links the Ste11p MEKK to the HOG pathway through plasma membrane association. *Genes & Development*, **20**, 734–746.

Wurgler-Murphy, S.M., Maeda, T., Witten, E.A. and Saito, H. (1997). Regulation of the *Saccharomyces cerevisiae* Hog1 mitogen activated protein kinase by the Ptp2 and Ptp3 protein tyrosine phosphatases. *Molecular and Cellular Biology*, **17**, 1289–1297.

Wysocki, R., Chery, C.C., Wawrzycka, D., Van Hulle, M., Cornelis, R., Thevelein, J.M. and Tamás, M.J. (2001). The glycerol channel Fps1p mediates the uptake of arsenite and antimonite in *Saccharomyces cerevisiae*. *Molecular Microbiology*, **40**, 1391–1401.

Yale, J. and Bohnert, H.J. (2001). Transcript expression in *Saccharomyces cerevisiae* at high salinity. *The Journal of Biological Chemistry*, **276**, 15996–16007.

CHAPTER **9**

Oxidative Stress, Fragmentation and Cell Death During Bioreactor Growth of Filamentous Fungi

Geoffrey D. Robson, Jeff Puci, William Throndset and **Nigel Dunn-Coleman**

Contents

1. Introduction 130
2. Fragmentation of Fungal Hyphae in Submerged Culture 131
3. Oxidative Stress in Fungal Fermentations 131
4. Cell Death During Fed-Batch Cultivation of *A. niger* and *T. reesei* 132
5. Differences in the Patterns of Cell Death and Fragmentation Between *A. niger* and *T. reesei* During Fed-Batch Cultivation 132
6. Conclusions 138
Acknowledgements 139
References 139

Abstract

The filamentous fungi posses a highly developed secretory pathway enabling them to secrete a broad range of extracellular hydrolases into the environment so as to degrade and utilise an array of environmental polymers. This capacity has been exploited industrially for decades for the production of a range of commercially important enzymes. In particular, *Aspergillus niger* (glucoamylase), *Aspergillus oryzae* (α-amylase) and *Trichoderma reesei* (cellulase) are used commercially as production strains and are reported to produce between 20 g l^{-1} and 40 g l^{-1} of extracellular enzyme in fed-batch fermentations under optimised conditions. During the fed-batch phase, cultures become denser and growth rate decreases. During this phase, cultures come under increasing environmental stress leading to reactive oxygen accumulation, cell death and autolysis. In this study, we examined changes in morphology, cell death and fragmentation during the fed-batch fermentation of industrial production strains of *A. niger* and *T. reesei* and demonstrate that cell death and fragmentation proceeds differently in these two strains. Further, we demonstrate that a localised build up of reactive oxygen species (ROS) appears to precede cell death and fragmentation.

British Mycological Society Symposia Series
Published by Elsevier Ltd.

1. INTRODUCTION

The filamentous fungi posses a highly developed secretory pathway enabling them to secrete a broad range of extracellular hydrolases into the environment to degrade and utilize an array of environmental polymers (Annis and Goodwin, 1997; de Vries and Visser, 2001; Robson *et al.*, 2005). This capacity has been exploited industrially for decades for the production of a range of commercially important enzymes. In particular, *Aspergillus niger* (glucoamylase), *Aspergillus oryzae* (α-amylase) and *Trichoderma reesei* (cellulase) are used commercially as production strains and are reported to produce between 20 g l^{-1} and 40 g l^{-1} of extracellular enzyme in fed-batch fermentations under optimized conditions (Finkelstein, 1987; Durand *et al.*, 1988). In addition, the ability to secrete such high levels of a single protein along with GRAS (Generally Regarded As Safe, US Food and Drug Administration) status makes them attractive hosts for the production of heterologous proteins of fungal and non-fungal origin (Archer, 2000; Conesa *et al.*, 2001; Punt *et al.*, 2002).

In submerged cultivation, the morphology of filamentous fungi can vary between freely dispersed hyphal elements and pellets composed of agglomerates of many individual hyphae, either loosely tangled or tightly packed together (Metz *et al.*, 1979; Tucker and Thomas, 1992; Cox *et al.*, 1998; Znidarsic and Pavko, 2001). The morphology of filamentous fungi in submerged culture is influenced by many factors including the strain, medium composition, agitation and pH (van Suijdam and Metz, 1981; Papagianni, 2004). While some products such as citric acid are produced in higher levels in pelleted cultures, the majority of enzymes including cellulase and glucoamylase, are produced in significantly higher quantities when the culture is dispersed and filamentous (Mackenzie *et al.*, 1994; Domingues *et al.*, 2000). While dispersed growth leads to more homogenous cultures, it leads to higher culture viscosities and decreased oxygen transfer from the gaseous phase (Metz *et al.*, 1979; Wittler *et al.*, 1986; Gibbs *et al.*, 2000). Fragmentation of hyphal elements occurs continuously during growth in submerged cultures and is essential in the maintenance of dispersed filamentous growth as it generates new hyphal elements for growth functionally similar to cell division in unicellular microbes. While physical agitation is thought to be important in this process, particularly when hyphae exceed a certain length (Tanaka, 1976; van Suijdam and Metz, 1981; Nielsen and Krabben, 1995), other factors may also play an important role. For example, Rhigelato *et al.* (1968) found higher rates of fragmentation in chemostat cultures of *Penicillium chrysogenum* when the growth rate was at the maintenance energy level that was correlated with a high level of vacuolation. Paul *et al.* (1994) reported a progressive increase in both vacuolation and hyphal fragmentation during fed-batch fermentation of *P. chrysogenum* after the onset of carbon limitation leading to the suggestion that fragmentation occurred due to the effect of physical shear forces on physiologically older, weaker, vacuolated compartments. Fragmentation therefore appears to result from interplay between the physiological status of the organism and the physical effects of shear stress.

2. FRAGMENTATION OF FUNGAL HYPHAE IN SUBMERGED CULTURE

In industrial bioreactors, particularly when growing in a dispersed form, filamentous fungi are subjected to a variety of stresses. One of these that has an important impact on the growth and morphology of fungi is the agitation rate of the impellers of the bioreactor (van Suijdam and Metz, 1981; Shamlou *et al.*, 1994). New hyphal fragments are generated through a process of fragmentation whereby a mycelial element becomes physically broken into two or more elements. This mechanism of generating new shorter hyphal fragments is balanced by growth of the fragments; thus this mechanism plays a key role in maintaining dispersed mycelial growth during the growth of the mould (Smith *et al.*, 1990; Shamlou *et al.*, 1994; Nielsen and Krebben, 1995). Although an important process, there have been few attempts to study the underlying mechanism of fragmentation, which is also difficult to quantitate. The rate of hyphal fragmentation is known to be proportional to the energy input into the impellers with increased fragmentation at increased impeller speeds (Nielson and Krabben, 1995). In most models of submerged growth, hyphal fragmentation rate is often assumed to not change during the course of the fermentation although some evidence suggests the tensile strength of hyphae may change significantly during the fermentation (Li *et al.*, 2000, 2002). However, these studies only consider fragmentation as a purely physical random process and few studies have considered any possible biological or physiological input that may dictate or play a role in hyphal fragmentation although it is clear that cell-wall composition and septation are important (McIntyre *et al.*, 2001a).

3. OXIDATIVE STRESS IN FUNGAL FERMENTATIONS

Oxidative stress is the direct result of aerobic growth due to the release of reactive oxygen species (ROS, namely superoxide anion, hydrogen peroxide, hydroxyl radical and nitric oxide) due to the incomplete reduction of some oxygen radicals in the mitochondria (Bai *et al.*, 2003). A build up of ROS within cells is toxic causing general cellular and membrane damage that can ultimately lead to cell death. To combat these deleterious effects, cells have developed a number of strategies to prevent or remove ROS (Bai *et al.*, 2003). However in older cells of yeast and filamentous fungi during stationary phase, ROS accumulation has been reported and is associated with cell death and autolysis (Laun *et al.*, 2001; Sami *et al.*, 2001; Bai *et al.*, 2003; Emri *et al.*, 2004a). Moreover, addition of oxidising agents leads to loss of fungal viability whereas addition of antioxidants has been reported to both increase yeast cell longevity in stationary phase and to reduce autolysis in *A. nidulans* and *S. cerevisiae* (Emri *et al.*, 2004b; Skulachev and Longo, 2005).

4. CELL DEATH DURING FED-BATCH CULTIVATION OF *A. NIGER* AND *T. REESEI*

Most large-scale industrial fermentations employ fed-batch fermentation with either continuous or pulse feeding in order to extend the productivity of the fermentation. Due to the high biomass density during this phase of the fermentation, oxygen transfer rates are usually limiting (Metz *et al.*, 1979; Olsvik and Kristiansen, 1994) and the mycelium begins to die. To date, few studies have attempted to estimate biomass viability during fed-batch fermentation. Harvey *et al.* (1998) estimated the percentage of biomass undergoing autolysis during the batch fermentation of *Penicillium chrysogenum* by quantifiying the percentage of hyphal length lacking cytoplasm by image analysis. Percentage autolysis was seen to rise steadily from <5 to approximately 70% over a period of 200 h post-stationary phase whereas McKintyre *et al.* (2001b) reported that up to 15% of the biomass was empty of cytoplasm during the log phase rising rapidly to >50% in the stationary phase. To investigate the relationship between oxidative stress, cell death and fragmentation in fed-batch fermentations, we investigated these parameters in two industrial production strains used by Genencor International, *A. niger* (dgr246) and *T. reesei* (morph1) in 14 l fed-batch fermentations. To estimate the extent of cell death during industrial fed-batch fermentations, a dual-staining fluorescent image analysis approach was developed using calcofluor white (total mycelium, DAPI filter set) and propidium iodide (viability, rhodamine filter set). Pairs of images were collected randomly, converted into jpeg format and analysed using image tool (UTHSCSA, version 3). Images were thresholded to mask the fluorescent areas and converted into binary images and the number of white pixels recorded. The percentage of dead biomass was calculated by dividing the total white pixels counted for the propidium iodide image (dead hyphal area) by the total white pixels counted for the matching calcofluor white image (total hyphal area). Using this method, it was possible to estimate the percentage viability of the biomass over time (Figure 1). In *A. niger*, the percentage of dead biomass increased progressively throughout the fermentation reaching approximately 30% after 138 h. However, by contrast, the percentage of dead biomass in *T. reesei* increased to approximately 8% and remained almost constant throughout the fermentation.

5. DIFFERENCES IN THE PATTERNS OF CELL DEATH AND FRAGMENTATION BETWEEN *A. NIGER* AND *T. REESEI* DURING FED-BATCH CULTIVATION

In *T. reesei* during the feeding stage, the pre-existing biomass became increasingly fragmented. Fragmentation could be seen microscopically as both empty compartments (pre-fragmentation, Figures 4 and 6) and as sites where fragmentation had occurred as residual wall material adjacent to a septum (Figure 7). Moreover, the septa adjacent to a fragmented area often became the site of new hyphal growth (intrahyphal growth, Figure 5). Intrahyphal growth has previously been

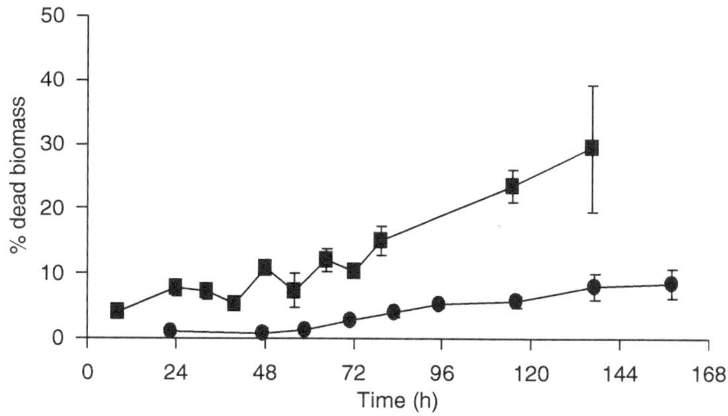

Figure 1 Percentage of dead biomass estimated from the percentage of biomass that stained with propidium Iodide during the fed-batch fermentation of industrial strains of *A. niger* (■) and *T. reesei* (●). Each point is the mean of at least 15 fields of view ± SEM.

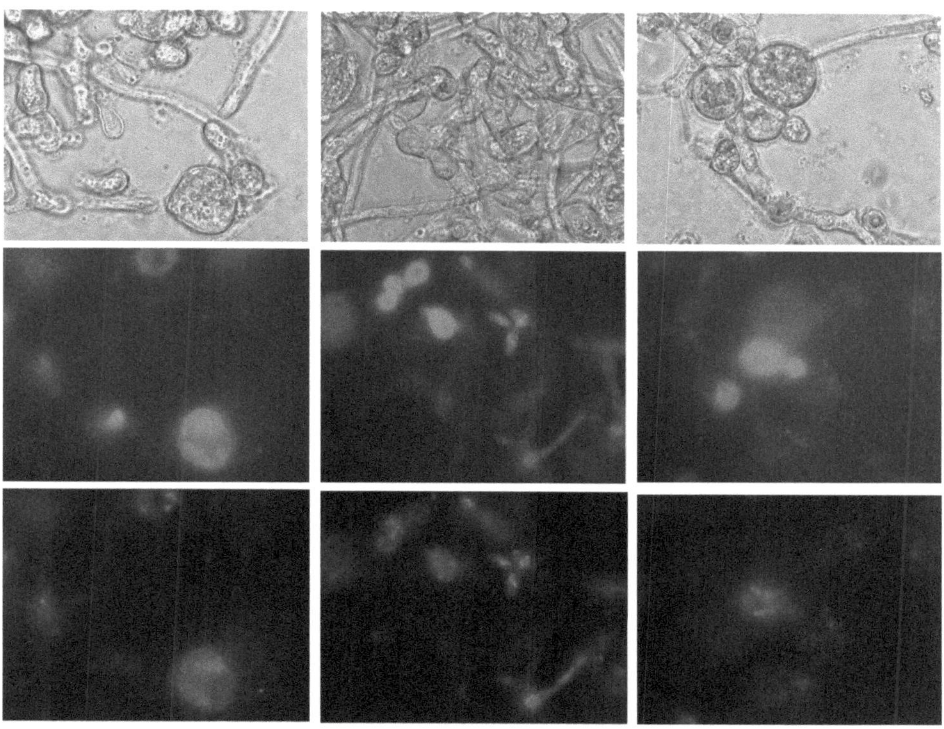

Figure 2 Accumulation of ROS and PI during the fed-batch fermentation of *A.niger*. Top panels, bright field, middle panels, ROS accumulation, Bottom panels, PI accumulation. **(See Colour Section)**

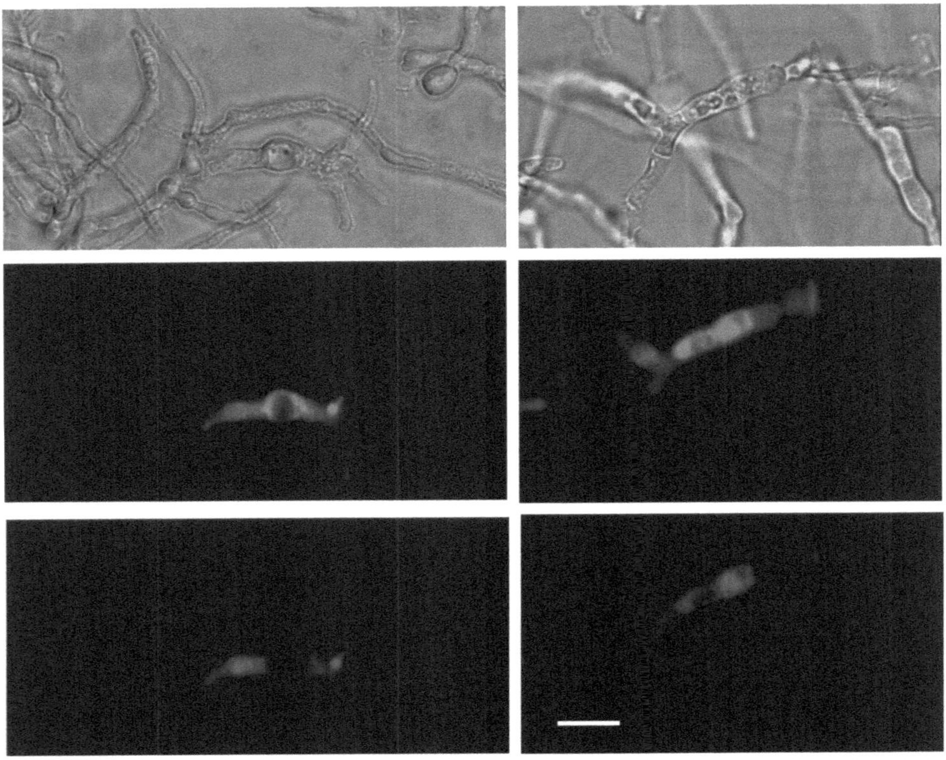

Figure 3 Compartmentalised accumulation of ROS and PI during the fed-batch fermentation of *T.reesei*. Top panels, bright field, middle panels, ROS accumulation, Bottom panels, PI accumulation. **(See Colour Section)**

reported to occur in the stationary phase of a number of fungi and to proliferate in some gene disruptants (Miller and Anderson, 1961; Trinci and Righelato, 1970; Schrickx *et al.*, 1995; Horiuchi *et al.*, 1999) but here it was extensive and present throughout the fed-batch phase. When cultures were dual stained with propidium iodide (PI) and the cell-permeant dye 5-(and-6)-carboxy-2′,7′ dichlorodihydrofluorescein diacetate (DHCR), which following deacetylation in the cell reacts with ROS to produce an oxidised fluorescent dye, compartments could be clearly identified that were non-viable (stained strongly with propidium iodide) and/or had accumulated ROS (Figure 3). All compartments that were stained with PI also counterstained with DHCR suggesting localised ROS were present in these compartments. In numerous instances, compartments could be seen stained with DHCR, but either did not or only faintly with PI, suggested ROS accumulation prior to death and fragmentation (Figure 3). In most instances, ROS and PI staining were confined to single compartments and dead compartments were visible that were clearly in the process of fragmenting (Figure 4). Toward the end of the fermentation, chlamydospores began to form both at hyphal tips and intrahyphally (Figure 9). Chlamydospores are resting structures

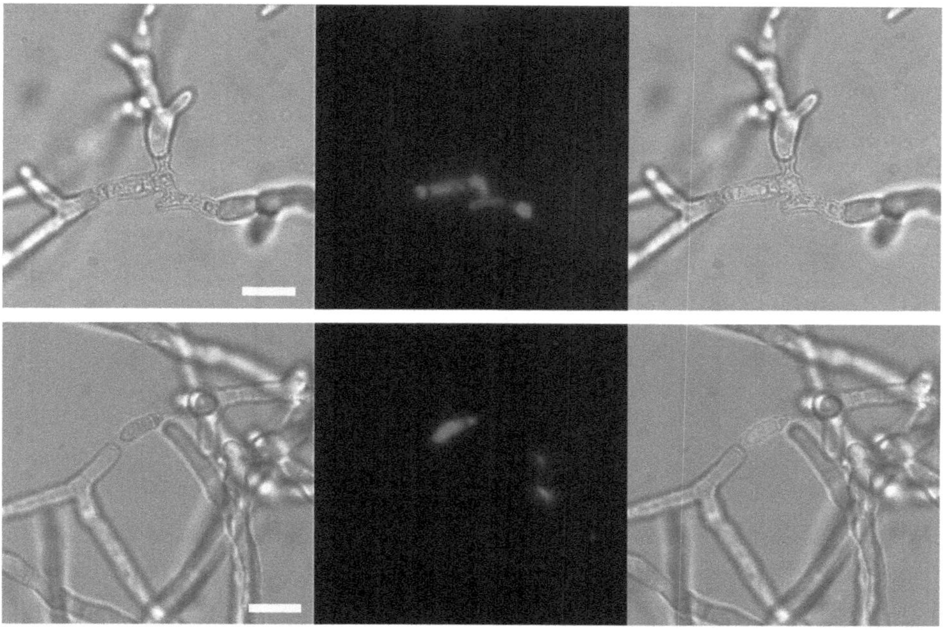

Figure 4 Dead hyphal compartments stained with PI during the fed-batch fermentation of *T.reesei*. Left panel, bright field, middle panel, PI accumulation, Right panel, merged image. **(See Colour Section)**

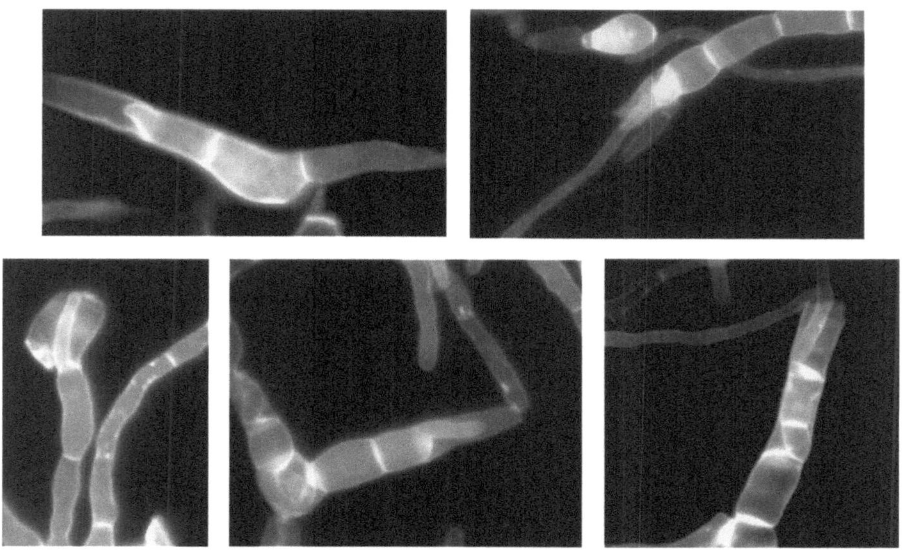

Figure 5 *T. reesei* during fed-batch fermentation showing frequent intrahyphal growth from septa adjacent to dead compartments. Samples were stained with calcofluor white.

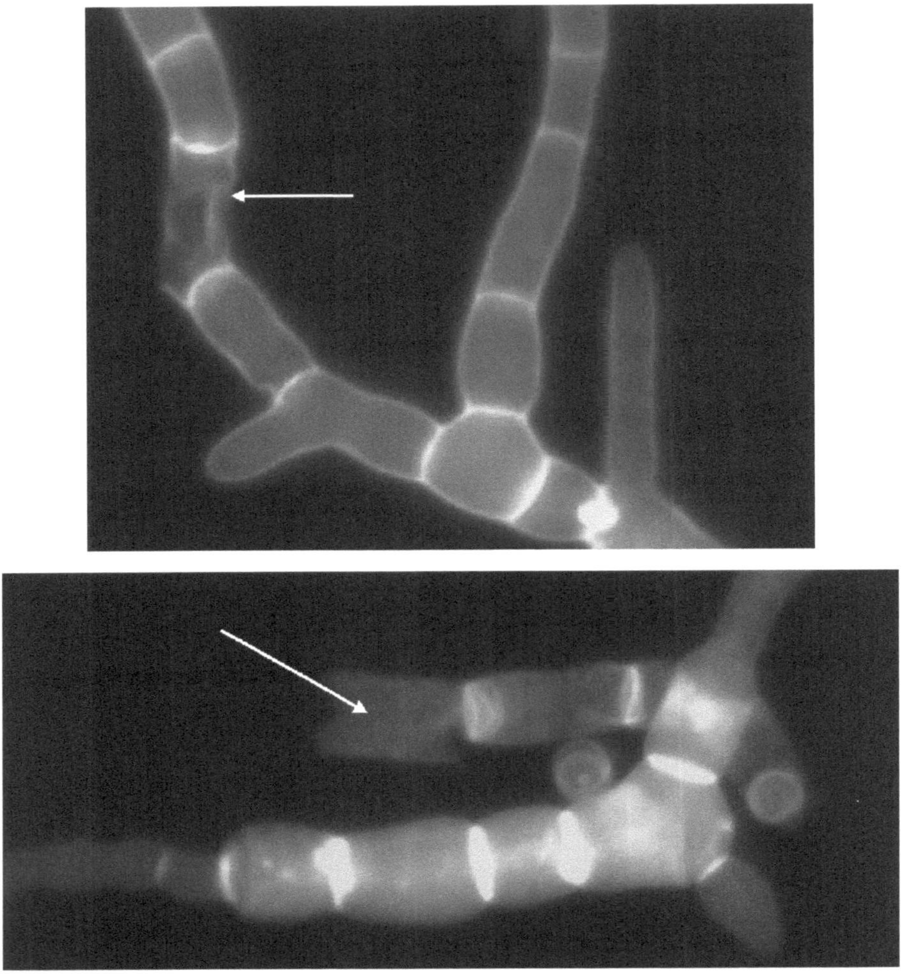

Figure 6 *T.reesei* during fed-batch fermentation showing a dead compartment about to fragment (upper panel) and a recently fragmented compartment (lower panel). Samples were stained with calcofluor white.

produced in the older regions of colonies or during the stationary phase and are commonly formed by many *Species of Trichoderma* and *Giocladium* sp. (Domsch *et al.*, 1980; Gams and Bisset, 1998; Samuels *et al.*, 2002).

In the *A. niger* production strain, dispersed hyphal elements were also the predominant morphology during the early stages of fermentation. In contrast to *T. reesei*, cell death as determined by PI staining was less localised, often with two, three or more adjacent compartments affected (Figure 2) Moreover, cell death and hyphal disintegration progressively increased throughout the fermentation. Like *T. reesei*, all compartments that stained with PI also had accumulated ROS, with some compartments accumulating ROS prior to PI staining, again suggesting cell

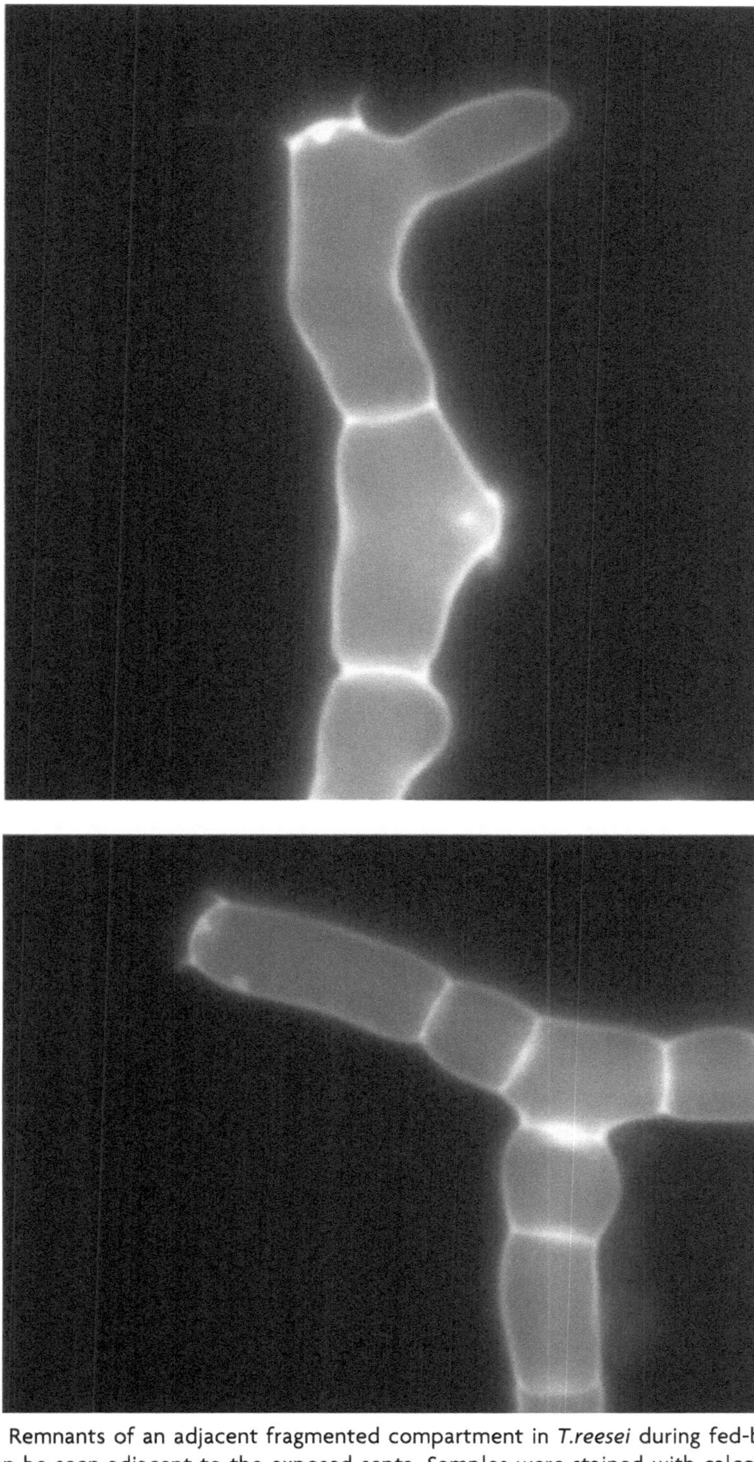

Figure 7 Remnants of an adjacent fragmented compartment in *T.reesei* during fed-batch growth can be seen adjacent to the exposed septa. Samples were stained with calcofluor white.

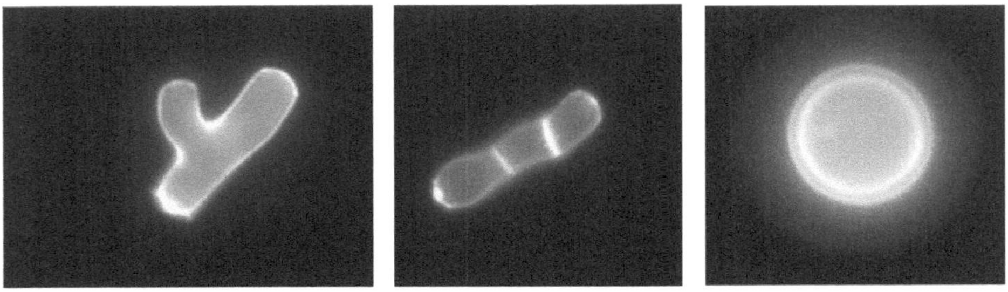

Figure 8 Survival structures seen toward the end of a fed-batch fermentation of *A.niger*. Samples were stained with calcofluor white.

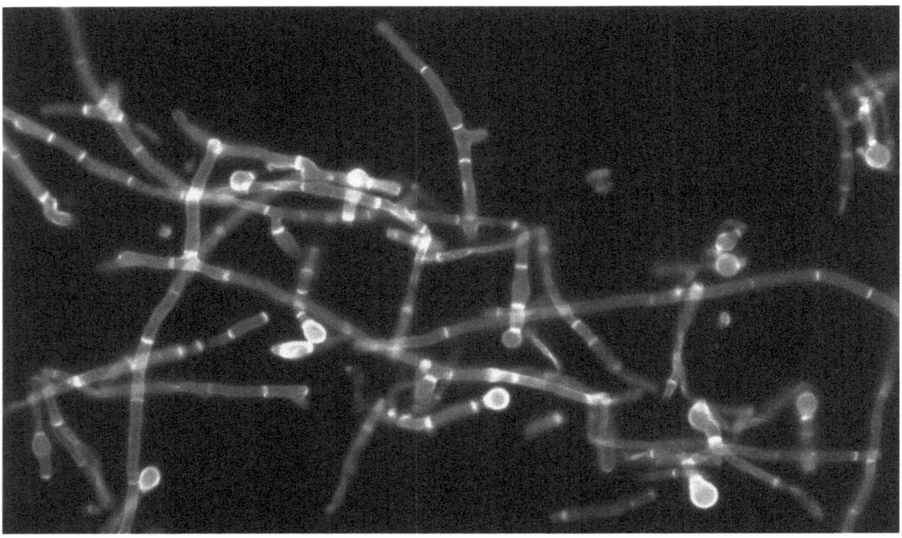

Figure 9 Accumulation of chlamydospores toward the end of a fed-batch fermentation of *T.reesei*. Samples were stained with calcofluor white.

death triggered by ROS accumulation (Figure 2). Unlike *T. reesei*, where resting chlamydospores were formed towards the end of the fermentation, no sporulating structures were observed although increasingly small viable elements consisting of one or two compartments or yeast-like structures were visible (Figure 8). Yeast-like structures have been previously reported in autolysing cultures of *A. nidulans* and *Penicillium chrsysogenum* and may be physiologically adapted to survive while the majority of the mycelium dies (Sami *et al.*, 2001; Emri *et al.*, 2004a).

6. CONCLUSIONS

Both *A. niger* and *T. reesei* are important industrial fungi used for the large-scale commercial production of both native and recombinant enzymes. During fed-batch fermentation, key differences in the behaviour of these strains became evident. While both showed evidence of ROS accumulation, cell death and fragmentation; in *A. niger* the proportion of dead biomass rose progressively and was not confined to single compartments leading to the progressive autolysis of the culture. By contrast, a low but constant proportion of dead biomass was maintained in *T. reesei* and was often confined to single compartments with the result that there was progressive fragmentation of the biomass that ultimately culminated in the production of chlamydospores. Accumulation of ROS in fungal mycelia in submerged culture has been reported previously in a number of fungal strains and is thought to trigger subsequent autolytic processes (Emri *et al.*, 2004a, 2004b; Radman *et al.*, 2004). The differing way in which ROS accumulation and subsequent cell death occurs in the two species may reflect their preferred substrate. *A. niger* is well known as a prodigious glucoamylase secretor which can degrade starch efficiently, enabling fast growth rates. Glucoamylase is also a growth-associated product that is manufactured and secreted during the log phase (Swift *et al.*, 1998). By contrast, *T. reesei* is well adapted to cellulose degradation that is far more difficult to degrade leading to much lower growth rates due to the relatively slow release of sugars. *T. reesei* cellulases have been shown to be growth dissociated and are only produced and secreted at low growth rates (Pakula *et al.*, 2005). Thus on cellulose, low growth rates and the length of time required to degrade the substrate would be a disadvantage if ROS accumulation was not controlled and compartmentalised, whereas significant ROS accumulation in *A. niger* occurs when most of the substrate has already been utilised. Further work is required to determine the impact that ROS accumulation may have on protein production and yield and to determine if compartmental cell death prior to fragmentation involves a programmed cell-death pathway recently uncovered in filamentous fungi (Mousavi and Robson, 2003, 2004; Emri *et al.*, 2005; Fedorova *et al.*, 2005).

ACKNOWLEDGEMENTS

The authors would like to acknowledge the support of the BBSRC (GAPSIA) and Genencor International.

REFERENCES

Annis, S.L. and Goodwin, P.H. (1997). Recent advances in the molecular genetics of plant cell wall-degrading enzymes produced by plant pathogenic fungi. *European Journal of Plant Pathology*, **103**, 1–14.

Archer, D.B. (2000). Filamentous fungi as microbial cell factories for food use. *Current Opinions in Biotechnology*, **11**, 478–483.

Bai, Z., Harvey, L.M. and McNeil, B. (2003). Oxidative stress in submerged cultures of fungi. *Critical Reviews in Biotechnology*, **23**, 267–302.

Conesa, A., Punt, P.J., van Luijk, N. and van den Hondel, C.A.M.J.J. (2001). The secretion pathway in filamentous fungi: A biotechnological view. *Fungal Genetics and Biology*, **33**, 155–171.

Cox, P.W., Paul, G.C. and Thomas, C.R. (1988). Image analysis of the morphology of filamentous micro-organisms. *Microbiology*, **144**, 817–827.

Domingues, F.C., Queiroz, J.A., Cabral, J.M. and Fonseca, L.P. (2000). The influence of culture conditions on mycelial structure and cellulase production by *Trichoderma reesei* Rut C-30. *Enzyme and Microbial Technology*, **26**, 394–401.

Domsch, K.H., Gams, W. and Anderson, T.H. (1980). *Compendium of Soil Fungi*, Vol. 1. Academic Press, London, UK.

Durand, H., Clanet, M., Tiraby, G. (1988). Genetic improvement of *Trichoderma reesei* for large scale cellulase production. *Enzyme and Microbial Technology*, **10**, 341–346.

Emri, T., Molnar, Z. and Pócsi, I. (2005). The appearances of autolytic and apoptotic markers are concomitant but differently regulated in carbon-starving *Aspergillus nidulans* cultures. *FEMS Microbiology Letters*, **251**, 297–303.

Emri, T., Molnar, Z., Pusztahelyi, T. and Pocsi, I. (2004a). Physiological and morphological changes in autolyzing *Aspergillus nidulans* cultures. *Folia Microbiologica*, **49**, 277–284.

Emri, T., Molnar, Z., Pusztahelyi, T., Rosen, S. and Pocsi, I. (2004b). Effect of vitamin E on autolysis and sporulation of *Aspergillus nidulans*. *Applied Biochemistry and Biotechnology*, **118**, 337–348.

Fedorova, N.D., Badger, J.H., Robson, G.D., Wortman, J.R. and Nierman, W.C. (2005). Comparative analysis of programmed cell death pathways in filamentous fungi. *BMC Genomics*, 6, 177.

Finkelstein, D.B. (1987). Improvement of enzyme production in *Aspergillus*. *Antonie van Leeuwenhoek*, **53**, 349–352.

Gams, W. and Bisset, J. (1998). Morphology and identification of *Trichoderma*. In: *Trichoderma & Gliocladium* (G.E. Harman and C.P. Kubicek, eds.), Vol. 1. Taylor and Francis, London.

Gibbs, P.A., Seviour, R.J. and Schmid, F. (2000). Growth of filamentous fungi in submerged culture: Problems and possible solutions. *Critical Reviews in Biotechnology*, **20**, 17–48.

Harvey, L.M., McNeil, B., Berry, D.R. and White, S. (1998). Autolysis in batch cultures of *Penicillium chrysogenum* at varying agitation rates — energy-dependent proteases and their targets. *Enzyme and Microbial Technology*, **22**, 446–458.

Horiuchi, H., Fujiwara, M., Yamashita, S., Ohta, A. and Takagi, M. (1999). Proliferation of Intrahyphal Hyphae Caused by Disruption of *csmA*, which encodes a Class V chitin synthase with a myosin motor-like domain in *Aspergillus nidulans*. *Journal of Bacteriology*, **181**, 3721–3729.

Laun, P., Pichova, A., Madeo, F., Fuchs, J., Ellinger, A., Kohlwein, S., Dawes, I., Frohlich, K.-U. and Breitenbach, M. (2001). Aged mother cells of *Saccharomyces cerevisiae* show markers of oxidative stress and apoptosis. *Molecular Microbiology*, **39**, 1166–1173.

Li, Z.J., Shukla, V., Fordyce, A.P., Pedersen, A.G., Wenger, K.S. and Marten, M.R. (2000). Fungal morphology and fragmentation behavior in a fed-batch *Aspergillus oryzae* fermentation at the production scale. *Biotechnology and Bioengineering*, **70**, 300–312.

Li, Z.J., Shukla, V., Wenger, K., Fordyce, A., Pedersen, A.G. and Marten, M. (2002). Estimation of hyphal tensile strength in production-scale *Aspergillus oryzae* fungal fermentations. *Biotechnology and Bioengineering*, 77, 601–613.

MacKenzie, D.A., Gendron, L.C.G., Jeenes, D.J. and Archer. D.B. (1994). Physiological optimisation of secreted protein production by *Aspergillus niger*. *Enzyme and Microbial Technology*, **16**, 276–280.

McIntyre, M., Dynesen, J. and Nielsen, J. (2001a). Morphological characterization of *Aspergillus nidulans*: Growth, septation and fragmentation. *Microbiology*, **147**, 239–246.

McIntyre, M., Eade, J.K., Cox, P.W., Thomas, C.R., White, S., Berry, D.R. and McNeil, B. (2001b). Quantification of autolysis in *Penicillium chrysogenum* by semiautomated image analysis. *Canadian Journal of Microbiology*, 47, 315–321.

Metz, B., Kossen, W.F. and van Suijdam, J.C. (1979). The rheology of mould suspensions. *Advances in Biochemical Engineering*, **11**, 103–156.

Miller, C.V. and Anderson, N.A. (1961). Proliferation of conidiophores and intrahyphal hyphae in *Aspergillus niger*. *Mycologia*, **53**, 433–436.

Mousavi, A. and Robson, G.D. (2003). Entry into the stationary phase is associated with a rapid loss of viability and an apoptotic-like phenotype in the opportunistic pathogen *Aspergillus fumigatus*. *Fungal Genetics & Biology*, **31**, 221–229.

Mousavi, S.A.A. and Robson, G.D. (2004). Oxidative and amphotericin-mediated cell death in the opportunistic pathogen *Aspergillus fumigatus* is associated with an apoptotic-like phenotype. *Microbiology,* **150**, 1937–1945.

Nielsen, J. and Krabben, P. (1995). Hyphal growth and fragmentation of *Penicillium chrysogenum* in submerged cultures. *Biotechnology and Bioengineering,* **46**, 588–598.

Olsvik, E.S. and Kristiansen, B. (1994). Rheology of filamentous fermentations. *Biotechnology Advances,* **12**, 1–39.

Pakula, T.M., Salonen, K., Uusitalo, J. and Penttilä, M. (2005). The effect of specific growth rate on protein synthesis and secretion in the filamentous fungus *Trichoderma reesei. Microbiology,* **151**, 135–143.

Papagianni, M. (2004). Fungal morphology and metabolite production in submerged mycelial processes. *Biotechnology Advances,* **22**, 189–255.

Paul, G.C., Kent, C.A. andThomas, C.R. (1994). Hyphal vacuolation and fragmentation in *Penicillium chrysogenum. Biotechnology and Bioengineering,* **44**, 655–660.

Punt, P.J., van Biezen, N., Conesa, A., Albers, A., Mangnus, J. and van den Hondel, C.A.M.J.J. (2002). Filamentous fungi as cell factories for heterologous protein production. *Trends in Biotechnology,* **20**, 200–206.

Radman, R., Bucke, C. and Keshavarz, T. (2004). Elicitor effects on reactive oxygen species in liquid cultures of *Penicillium chrysogenum. Biotechnology Letters,* **26**, 147–152.

Righelato, R., Trinci, A.P.J., Pirt, S.J. and Peat, A. (1968). The influence of maintenance energy and growth rate on the metabolic activity, morphology and conidiation of *Penicillium chrysogenum. Journal of General Microbiology,* **50**, 399–412.

Robson, G.D., Huang, J., Wortman, J. and Archer, D.B. (2005). A preliminary analysis of the process of protein secretion and the diversity of putative secreted hydrolases encoded in *Aspergillus fumigatus*: Insights from the genome. *Medical Mycology,* **43**, 41–47.

Sami, L., Emři, T. and Pócsi, I. (2001). Autolysis and ageing of *Penicillium chrysogenum* cultures undercarbon starvation: Glutathione metabolism and formation of reactive oxygen species. *Mycological Research,* **105**, 1246–1250.

Samuels, G.J., Dodd, S.L., Gams, W., Castlebury, L.A. and Petrini, O. (2002). *Trichoderma* species associated with the green mold epidemic of commercially grown *Agaricus bisporus. Mycologia,* **94**, 146–170.

Schrickx, J.M., Stouthamer, A.H. and van Verseveld, H.W. (1995). Growth behaviour and glucoamylase production by *Aspergillus niger* N402 and a glucoamylase overproducing transformant in recycling culture without a nitrogen source. *Applied Microbiology and Biotechnology,* **43**, 109–116.

Shamlou, P.A., Makagiansar, H.Y., Ison, A.P. and Lilly, M.D. (1994). Turbulent breakage of filamentous microorganisms in submerged culture in mechanically stirred bioreactors. *Chemical Engineering Science,* **49,** 2621–2631.

Skulachev, V.P. and Longo, V.D. (2005). Aging as a mitochondria-mediated atavistic program. Can aging be switched off? *Annals of the NewYork Acadamy of Sciences,* **1057**, 145–164.

Smith, J.J., Lilly, M.D. Fox, R.I. (1990). The effect of agitation on the morphology and penicillin production of *Penicillium chrysogenum. Biotechnology and Bioengineering,* **35**, 1011–1023.

van Suijdam, J.C. and Metz, B. (1981). Influence of engineering variables upon the morphology of filamentous molds. *Biotechnology and Bioengineering,* **23,** 111–148.

Swift, R.J., Wiebe, M.G., Robson, G.D. and Trinci, A.P.J. (1998). Recombinant glucoamylase production by *Aspergillus niger* B1 in chemostat and pH auxostat cultures. *Fungal Genetics and Biology,* **25**, 100–109.

Tanaka, H. (1976). Studies on the effect of agitation on mycelia in submerged culture. *Journal of Fermentation Technology,* **54**, 818–819.

Trinci, A.P.J. and Righelato, R.C. (1970). Changes in constituents and ultrastructure of hyphal compartments during autolysis of glucose-starved *Penicillium chrysogenum. Journal of General Microbiology,* **60**, 239–249.

Tucker, K.G. and Thomas, C.R. (1992). Mycelial morphology: The effect of spore inoculum level. *Biotechnology Letters,* **14**, 1071–1074.

de Vries, R.P. and Visser, J. (2001). *Aspergillus* enzymes involved in degradation of plant cell wall polysaccharides. *Microbiology and Molecular Biology Reviews,* **65**, 497–522.

Wittler, R., Baumgartl, H., Lübbers, D.W. and Schügerl, K. (1986). Investigations of oxygen transfer into *Penicillium chrysogenum* pellets by microprobe measurements. *Biotechnology and Bioengineering*, **28**, 1024–1031.

Znidarsic, P. and Pavko, A. (2001). The morphology of filamentous fungi in submerged cultivations as a bioprocess parameter. *Food Technology and Biotechnology*, **39**, 237–252.

CHAPTER 10

Weak Organic Acid Resistance of Spoilage Yeasts

Mehdi Mollapour and **Peter W. Piper**

Contents

1. Introduction 144
2. How Weak Organic Carboxylic Acids Act On *S. cerevisiae* 145
3. The Different Routes to Weak Acid Adaptation by *S. cerevisiae* 147
4. Hog1p-Dependent Degradation of Fps1p is the Key Determinant of Acetate Resistance 148
5. Overcoming the Constant weak Acid stress Imposed on *S. Cerevisiae* by the Presence of Moderately Lipophilic Organic Acids 150
   5.1 Weak Acid Resistance in the *Zygosaccharomyces* 152
Acknowledgements 153
References 153

Abstract

Reduced reliance on the use of chemical preservatives is desirable in the light of recent evidence that raises concerns over whether the large-scale consumption of these compounds is completely safe. Resistance of spoilage yeasts to the organic acid preservatives currently approved for use in foods and beverages is often a major factor preventing a lowering of preservative levels. Recent work has unravelled the stress responses whereby *Saccharomyces cerevisiae* becomes resistant to such acids. Resistance to high acetic acid is achieved by loss of the plasma membrane channel, the Fps1p aquaglyceroporin, which facilitates diffusional entry of this acid into cells. Acetic acid stress activates the Hog1p MAP kinase, whereupon this active Hog1p causes phosphorylation, ubiquitination and delivery of Fps1p via endocytic pathway to the vacuole for degradation. Other carboxylic acid preservatives (propionic, sorbic or benzoic acids) are too large to enter the yeast cell through the Fps1p channel, but being more lipophilic than acetic acid, can enter cells at appreciable rates by passive diffusion across the lipid bilayer. Resistance to these involves the induction of an activity for catalysing active efflux of the preservative anion from the cell. In *S. cerevisiae*, this is the plasma membrane Pdr12p ABC-transporter, regulated by a transcription factor (War1p). The major role of War1p appears to be the induction of *PDR12* gene in cells stressed by these moderately lipophilic acids. Still

British Mycological Society Symposia Series
Published by Elsevier Ltd.

other mechanisms of weak acid resistance are present in the *Zygosaccharomyces*, most notably a capacity for oxidative degradation of sorbic and benzoic acids that is absent in *S. cerevisiae*.

1. INTRODUCTION

Plant materials often contain compounds with antimicrobial properties, these being intrinsic to many foods and spices. Examples include eugenol in cloves, allicin in garlic, cinnamic aldehyde and eugenol in cinnamon, allyl isothiocyanate in mustard, eugenol and thymol in sage, and carvacrol (isothymol) and thymol in oregano. Other naturally occurring antimicrobials, compounds currently in large-scale use as food preservatives, are the weak organic acids (e.g. sorbic and benzoic acids, both to be found naturally at high level in many fruits and berries). It is possible to trace the practical use of weak acids as preservatives back many centuries, to the burning of sulphur to sterilise the wooden vessels used in cider making and the use of acetic acid in vinegars and pickles (Stratford and Anslow, 1996).

Figure. 1 shows the carboxylate weak acid preservatives currently approved for use in large-scale food and soft drink preservation (www.cfsan.fda.gov/~comm/ift4-3.html). This chapter addresses how they act on spoilage yeasts, describing recent discoveries that reveal how such yeasts acquire a substantial resistance to these compounds. It is the desire to achieve substantial reductions in the large-scale use of preservatives that has prompted much of the research in this area. Not only is there considerable consumer demand for less preserved foods (perceived as more natural), but recent evidence raises significant concerns over whether the widespread use of organic acid preservatives is indeed a completely safe practice. Benzoic acid can react with the vitamin C (ascorbic acid) in soft drinks to produce benzene (Gardner and Lawrence, 1993). Both sorbic and benzoic acids are mutagenic towards the mitochondrial genome (Piper, 1999).

The yeasts generally termed "spoilage yeasts" include a number of the *Zygosaccharomyces* (notably *Z. rouxii*, *Z. bailii* and *Z. lentus*), as well as *Saccharomyces cerevisiae* (Fleet, 1992; Steels *et al.*, 1999, 2000; Piper *et al.*, 2001). These pose a significant spoilage threat for many materials preserved by low pH, low water activity and/or the presence of the highest weak organic acid preservative levels

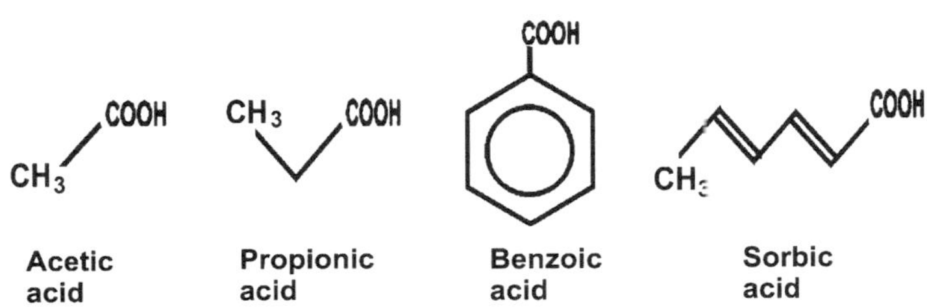

Figure 1 Carboxylic acids approved for large-scale use as food preservatives.

(acetic, propionic, sorbic or benzoic acids) allowed in food and beverage preservation. This resistance is a major obstacle preventing use of lower levels of these preservatives in foods and beverages. *Z. bailii* is also an important spoilage agent of wine, a property that accrues from its high-ethanol tolerance (Kalathenos *et al.*, 1995). As described below, distinct stress responses allow *S. cerevisiae* to acquire resistance to the organic acids in Fig. 1; one for acetic acid and another for the more lipophilic propionic, sorbic and benzoic acids. Though data on the resistance mechanisms of *Zygosaccharomyces* is still incomplete, these appear to share certain features of the resistance systems of *S. cerevisiae* but also to possess some unique features.

2. HOW WEAK ORGANIC CARBOXYLIC ACIDS ACT ON *S. cerevisiae*

The weak organic acids in Fig. 1 are much more potent inhibitors of microbial growth at low pH. Figure. 2 shows the general, textbook description of why this is so. At low pH, acetic acid (pK_a 4.75), propionic acid (pK_a 4.75), sorbic acid (pK_a 4.76) or benzoic acid (pK_a 4.19) are substantially in their undissociated state (XCOOH). At neutral pH, they are almost completely dissociated to the anion ($XCOO^-$). It is generally accepted that high extracellular levels of these acids pose a much greater threat at lower pH due to the uncharged form of the acid crossing the membrane, and therefore penetrating the cell, much more readily than the charged acid anion.

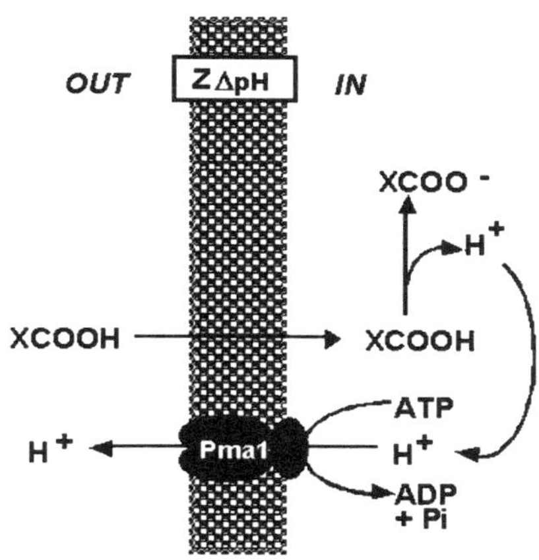

Figure 2 The general conception of how weak acids act on cells. If the undissociated acid (XCOOH) is freely permeable to the membrane, it's concentration inside and outside the cell should be the same. The ratio of acid (XCOOH) to anion ($XCOO^-$) however, is governed by the pH and the dissociation constant of the acid. A higher pH in the cytosol will cause a substantial fraction of acid to dissociate to the anion ($XCOO^-$), a form that is relatively membrane impermeant and will therefore accumulate inside the cell. An electrochemical potential difference is maintained across the yeast plasma membrane (ZΔpH), largely by means of plasma membrane H^+-ATPase (Pma1p)-catalysed proton extrusion.

In low pH cultures, undissociated acid that enters the cell will dissociate in the higher pH environment of the cytosol. Such dissociation generates protons and the acid anion (H^+, $XCOO^-$; Fig. 2). *S. cerevisiae* cannot metabolise sorbate or benzoate (Mollapour and Piper, 2001b). Also, in its glucose-repressed state, it is unable to assimilate large amounts of acetate or propionate. Cells exposed to high levels of these acids at low pH will therefore accumulate a high pool of the preservative anion (this, being charged, cannot very readily diffuse from the cell). This high anion pool may generate an abnormally high internal turgor pressure. For the more lipophilic preservatives, it also elevates free-radical production by the mitochondrial respiratory chain, severe oxidative stress being a major component of sorbic and benzoic acid stress in *S. cerevisiae* (see below). The proton release could potentially acidify the cytosol (reduce the intracellular pH, pHi), though it is probable that this is efficiently counteracted by the combined effects of intracellular buffering and the action of the plasma membrane H^+-ATPase, the latter an ATP-driven proton-efflux pump (Pma1p; Figs. 2 and 3). Proton efflux by the H^+-ATPase is a critical activity for the maintenance of homeostasis in all fungal and plant cells, as it constitutes the main activity for generating the electrochemical potential across the cell membrane (ZΔpH; Fig. 2), a potential that drives nutrient uptake and which regulates ion and pH balance (Serrano, 1991). Any pH_i reduction due to the intracellular dissociation of weak organic acid can be counteracted simply by increasing the H^+-ATPase-catalysed proton extrusion from the cell (Figs. 2 and 3). Indeed, H^+-ATPase does appear to be a key factor for counteracting weak acid stress, since such stress is associated with strong activation of the plasma membrane H^+-ATPase (Viegas and Sa Correia, 1991; Holyoak *et al.*, 1996; Piper *et al.*, 1997). Lowering the expression of the PMA1 gene makes yeast hyper-sensitive to sorbic acid (Holyoak *et al.*, 1996).

Until recently the inhibitory effects of weak organic acids on yeast were always attributed to this intracellular acidification and anion accumulation (Fig. 2). It is now clear that the story is much more complex; the acids in Fig. 1 are not all operating identically to inhibit growth and that *S. cerevisiae* uses different stress responses to acquire resistance to acetic acidate (Mollapour and Piper, 2006) or resistance to the more lipophilic preservatives (Piper *et al.*, 1998; Schüller *et al.*, 2004). This was initially indicated from physiological studies. Side-by-side comparisons of the effects of acetic acid and sorbic acid (acids of identical pK_a) on *S. cerevisiae* indicated that, while acetic acid might possibly be acting as in Figure. 2, the more hydrophobic sorbic acid inhibits cells mainly through a disordering of membrane structure (Stratford and Anslow, 1996, 1998; Bracey *et al.*, 1998). The decreases in pH_i seen with inhibitory acetic acid stress of *S. cerevisiae* (Arneborg *et al.*, 2000), are not apparent with inhibitory sorbic acid (Bracey *et al.*, 1998). Instead sorbic-stressed yeast may be suffering more from the effects of appreciable membrane disruption, and when oxygen is present, the oxidative stress caused by a dramatically increased endogenous production of oxygen-free radicals (Piper, 1999). A strong action of monocarboxylic acid preservatives on membranes is also apparent from the strong propensity of these acids to become much more inhibitory as they become more lipophilic (Stratford and Anslow, 1998; Holyoak *et al.*, 1999; Piper, 1999). For example, quite high acetic acid (80–150 mM) is needed to inhibit *S. cerevisiae* at pH 4.5, whereas only 1–3 mM of the more liposoluble sorbic acid achieves the same degree of growth inhibition (Stratford and Anslow, 1996; Bauer *et al.*, 2003; Mollapour and Piper, 2006).

3. THE DIFFERENT ROUTES TO WEAK ACID ADAPTATION BY *S. cerevisiae*

We found recently that *S. cerevisiae* acquires resistance to high acetic acid level by a mechanism different to the manner in which it becomes resistant to the more lipophilic propionic, sorbic, and benzoic acids (Figs. 1 and 3). Acetic acid is of sufficiently low molecular weight to enter the cell by passive diffusion through the glycerol channel (aquaglyceroporin) of the plasma membrane, Fps1p (Fig. 3). When pH 4.5 cultures of glucose-repressed *S. cerevisiae* are suddenly challenged by 100 mM acetate, the open Fps1p provides the route by which acetic acid floods into the cells as the acetate accumulation by these cells is almost entirely abolished by the *fps1* deletion mutation (our unpublished data). As described in greater detail below, acquisition of acetic acid resistance involves the loss of Fps1p from the membrane, thereby eliminating Fps1p-facilitated entry of toxic

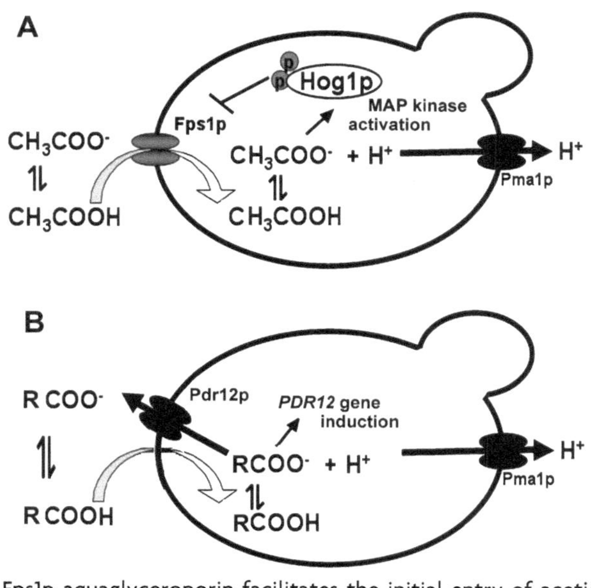

Figure 3 (A) The Fps1p aquaglyceroporin facilitates the initial entry of acetic acid to glucose-repressed *S. cerevisiae*, in this way triggering its own destruction due to the accumulation of an intracellular acetate pool activating the Hog1p that destabilises the open Fps1p channel. Fps1p loss removes the major route of acetic acid entry to the cell, thereby allowing survival in high acetic acid levels. (B) Induction of continuous preservative anion extrusion (Pdr12p) is essential for resistance to weak organic acids more lipophilic than acetic acid.

acetic acid into the cell (Fig. 3A). Remarkably, *S. cerevisiae* elicits a specific acetic acid stress response to do this, a response that we believe is not dependent on any *de novo* gene expression, but involves Hog1p MAP kinase phosphorylating directly Fps1p in order to target this channel for endocytosis to the vacuole.

The other carboxylic acids in Fig. 1 are too large to traverse the pore of the Fps1p channel. Instead, being more lipophilic than acetic acid, these enter cells mainly by dissolving in the phospholipid bilayer. Resistance to these larger acids requires, not the loss of diffusional entry of the acid as for acetic acid, but the induction of a system for active efflux of the preservative anion from the cell (Fig. 3B). This efflux is catalysed by a plasma membrane ABC-transporter, Pdr12p. Again inhibitory acid levels induce a stress response, but in the case of these more lipophilic acids, transcriptional changes are important for the adaptation (most importantly, induction of the *PDR12* gene directed by the transcription factor War1p (Schüller *et al.*, 2004)). So strong is this *PDR12* induction with sorbic acid stress that Pdr12p is one of the most abundant plasma membrane proteins in cells adapted to grow on sub-inhibitory levels of sorbic acid (Piper *et al.*, 1998, 2001). Note that *S. cerevisiae* cultures exposed to acids more lipophilic than acetic acid are subject to a constant weak acid stress; carboxylic acid that is entering them continuously by passive diffusion, must be continuously extruded by Pdr12p (Fig. 3B).

In summary, resistances to the weak acids in Fig. 1 centre around three key plasma membrane activities, the ATP-driven proton efflux pump H^+-ATPase (Pma1p), Fps1p and Pdr12p (Fig. 3). Acetic acid resistance is acquired by a *loss of function* (elimination of the channel that facilitates passive diffusional influx of acetic acid to the cell, Fps1p). Resistance to more lipophilic carboxylate preservatives is acquired by a *gain of function* (induction of Pdr12p). The presence or absence of Fps1p does not affect the resistances to the latter acids (our unpublished data). Equally, Pdr12p does not confer any resistance to acetic acid (Bauer *et al.*, 2003; Hatzixanthis *et al.*, 2003). Both Fps1p and Pdr12p, though, besides these roles in weak acid resistance, have other important functions. Pdr12p is also used to eliminate from the cell a number of organic acids produced in amino acid catabolism (Hazelwood *et al.*, 2006). Fps1p is the glycerol channel of the plasma membrane, its rapid interconversion between the open- and closed-channel states being of critical importance in osmoregulation. Fps1p closure allows cells to accumulate glycerol with hyperosmotic stress, but a rapid opening to allow the loss of this accumulated glycerol is essential if such cells that have accumulated a high osmolyte pool are to survive a subsequent hypoosmotic stress (Luyten *et al.*, 1995; Sutherland *et al.*, 1997; Tamas *et al.*, 1999, 2000).

4. Hog1p-DEPENDENT DEGRADATION OF FPS1P IS THE KEY DETERMINANT OF ACETATE RESISTANCE

High acetic acid exposure activates both the Hog1p and Slt2(Mpk1)p stress-activated MAP kinases of *S. cerevisiae*, but only the active Hog1p, not Slt2p, is needed for this yeast to become acetic acid resistant (Mollapour and Piper, 2006). Acetic acid enters the cells, and Hog1p becomes activated, considerably more rapidly in pH 4.5, as compared to pH 6.8 cultures. In contrast, the expression of a non-regulated, constitutively open Fps1p channel leads to an increased acetate accumulation, a more sustained Hog1p activation, and hypersensitivity to high acetic acid levels. Nonetheless, in an *fps1* deletant, there is no acetic acid uptake and no Hog1p activation in response to an acetic acid challenge. Furthermore such an *fps1Δ* mutant is even more acetic acid resistant than wild-type yeast (manuscript in preparation). Initial diffusional entry of acetic acid into yeast suddenly challenged by a high level of this acid is therefore largely through the Fps1p channel, creating an intracellular acetate pool that generates the signal for the transient activation of Hog1p (Fig. 3A).

This acetate activation of Hog1p is absent in the *ssk1Δ* and *pbs2Δ* mutants, but present in *sho1Δ* and *ste11Δ*, revealing that it involves the Sln1p branch of high-osmolarity glycerol (HOG) pathway signalling to Pbs2p (Mollapour and Piper, 2006). Despite this, the acetate-activated Hog1p, though it confers acetic acid resistance in *S. cerevisiae* growing at low pH, does not generate the *GPD1* gene induction or the intracellular glycerol accumulation that are hallmarks of Hog1p becoming activated by hyperosmotic stress (Mollapour and Piper, 2006). The response to acetic acid stress is therefore distinct from the response to osmostress in low pH cultures, even though both responses involve activation of Hog1p MAP kinase.

We have recently found that this Hog1p MAP kinase, when activated by acetic acid stress (or is activated artificially in unstressed cells by the expression of an overactive form of Pbs2p), directly phosphorylates residues T321 and S537 on Fps1p domains on the cytosolic face of the plasma membrane. These phosphorylations cause the Fps1p to become ubiquitinated and endocytosed to the vacuole for degradation (manuscript in preparation). As the Fps1p channel is the major route of acetic acid entry to cells, degradation of this channel abolishes such entry and allows the cells to survive high, potentially toxic acetic acid concentrations (Fig. 3A). Lethal levels of acetic acid are known to cause an apoptotic cell death in *Z. bailii* and *S. cerevisiae* (Ludovico *et al.*, 2001, 2003). Remarkably, this targeting of Fps1p for degradation appears to be the sole requirement for the active Hog1p in acetic acid resistance, since Hog1p is completely dispensible for acetic acid resistance in cells that lack Fps1p (whereas the *hog1Δ* mutant is acetic acid sensitive (Mollapour and Piper, 2006), *fps1Δ* single, and *fps1Δ hog1Δ* double mutant cells are both more acetic acid resistant than wild-type yeast (manuscript in preparation)).

Hog1p is also activated under conditions of hyperosmotic stress, but under these circumstances, the Fps1p channel is not degraded. Instead, osmotic stress causes an extremely rapid, turgor mediated, yet Hog1p-independent transition of Fps1p to the closed-channel state (Luyten *et al.*, 1995; Sutherland *et al.*, 1997; Tamas *et al.*, 1999, 2000). In this closed state, Fps1p appears to be protected from the Hog1-mediated degradation shown in Figure. 3A.

5. OVERCOMING THE CONSTANT WEAK ACID STRESS IMPOSED ON *S. cerevisiae* BY THE PRESENCE OF MODERATELY LIPOPHILIC ORGANIC ACIDS

Propionic, sorbic and benzoic acid (Fig. 1) resistance is acquired through induction of the Pdr12p ABC transporter (Fig. 3B). This induction is due to the activation of a zinc finger transcription factor, War1p (Kren *et al.*, 2003), and as with *pdr12Δ* cells, cells lacking this War1p are hypersensitive to propionic, sorbic and benzoic acids, but not to acetic acid (Bauer *et al.*, 2003; Hatzixanthis *et al.*, 2003). The War1p regulon comprises only a few genes, only one of which (*PDR12*) appears to be required for sorbic acid resistance (War1p-independent *PDR12* expression from the *GAL1-10* promoter fully restored sorbic acid resistance in a strain lacking War1p (Schüller *et al.*, 2004)). Promoter deletion analysis has identified a *cis*-acting weak acid-response element in the *PDR12* promoter, a sequence that mediates the War1p-directed induction of *PDR12*. War1p was found to occupy this element both in the absence and the presence of sorbic acid stress, revealing this transcription factor is bound constitutively to its DNA response element *in vivo* (Kren *et al.*, 2003).

Pdr12p gives resistance to water-soluble monocarboxylic acids of relatively short aliphatic carbon chain length (Piper *et al.*, 1998; Holyoak *et al.*, 1999; Hatzixanthis *et al.*, 2003). It also confers some resistance to short-chain aliphatic alcohols (Hatzixanthis *et al.*, 2003), compounds whose toxic effects are thought to be mainly due to their ability to dissolve in membranes (Weber and de Bont, 1996). However, *pdr12Δ* cells are not sensitive to dicarboxylic acids, highly lipophilic, long-chain fatty acids and alcohols. The toxic effects of alcohols are thought to be mainly due to a detergent disruption effect on membranes (Weber and de Bont, 1996; Holyoak *et al.*, 1999). It appears therefore that Pdr12p imparts resistance to organic acids or alcohols that can, to a reasonable extent, partition into both lipid and aqueous phases. This suggests that this ABC transporter might bind acid anions or alcohol molecules when they are incorporated within the inner leaflet of the plasma membrane. It may then transport them to the opposite (periplasmic) side of the membrane (i.e. act as a "flippase"), in order to release them into the aqueous phase of the periplasm. Such active efflux may lower the intracellular level of the acid anion or alcohol, on the basis that the polar groups on these carboxylate anion or alcohol substrates will slow their "flipping" across the cell membrane. Pdr12p activity can be readily visualised in living cells as the ability to catalyse an energy-dependent extrusion of fluorescein (Holyoak *et al.*, 1999).

There may be an additional benefit to the Pdr12p-catalysed extrusion of the preservative anion in Fig. 3B. The extent to which an increased H^+-ATPase activity *alone* can counteract the intracellular acidification due to intracellular dissociation of the acid may be limited, since there is a finite limit to the extent that H^+-ATPase action can enhance the charge component (Z) of the membrane electrochemical potential (ZΔpH; Fig. 2). One way to avoid this problem is to move a charge that compensates for the charge on an H^+-ATPase-extruded proton. Pdr12p-catalysed extrusion of an preservative anion could satisfy this requirement (Fig. 3B). In *S. cerevisiae* subjected to a constant weak acid stress (a continuous passive diffusional entry of a moderately lipophilic acid such as sorbic acid), the Pdr12p-catalysed anion extrusion might therefore have two beneficial effects. It may generate both lowered steady-state intracellular anion levels and, by movement of a charge that balances the charge on an H^+-ATPase-extruded proton, allow higher levels of *catalysed* proton extrusion. The *combined* actions of H^+-ATPase and Pdr12p may therefore be essential for *S. cerevisiae* to restore homeostasis to the point where growth can resume (Fig. 3B). The energy requirements for counteracting weak acid stress in this way are undoubtedly extremely high, as reflected in the dramatic reductions in biomass yield for cultures grown in the presence of this stress (Warth, 1988; Viegas and Sa Correia, 1991; Verduyn *et al.*, 1992; Stratford and Anslow, 1996; Piper *et al.*, 1997).

In aerobic *S. cerevisiae* cultures stressed with sorbic or benzoic acid, one of the major advantages of Pdr12p action is the reduction of oxidative stress. This yeast is considerably more sensitive to these acids when oxygen is present, due to the intracellular acids strongly enhancing the endogenous production of superoxide by the mitochondrial electron transport chain (Piper, 1999). Free-radical production is also the reason that sorbate and benzoate are strongly mutagenic towards the mitochondrial genome, as reflected in high rates of petite mutant segregation in aerobic *S. cerevisiae* cultures exposed to these acids (Piper, 1999). Excessive superoxide production is a major cause of the sensitivity of the *pdr12* mutant to sorbic and benzoic acids, as sensitivity is substantially rescued by the loss of superoxide dismutase activities (Piper, 1999). It is evidently less harmful for this superoxide ($pK_a \sim 5.7$) to diffuse from the weak acid-stressed cells in its uncharged, protonated state ($HO_2^{\bullet}$) (as probably occurs in the *pdr12,sod1* mutant), than that it should be dismutated to hydrogen peroxide, leading to the production of the extremely damaging hydroxyl radical, as would normally occur in *pdr12,SOD1+* cells (Piper, 1999).

Though the activities in Fig. 3 appear to be the major ones for acquired weak acid resistances, a wide spectrum of activities contribute to this resistance. In a recent screen of the collection of gene deletion mutants (http://www.sequence.stanford.edu/group/yeast_deletion_project/deletions3.html) for both an increased sensitivity and an increased resistance to sorbic acid at pH 4.5, 237 mutants were identified as unable to grow at pH 4.5 in presence of 2 mM sorbic acid, while 34 mutants exhibited higher sorbic acid resistance than the wild-type parental strain (Mollapour *et al.*, 2004). Several oxidative stress-sensitive mutants and mitochondrial mutants were found to be sorbic acid sensitive (Mollapour *et al.*, 2004). This reflects the importance of counteracting oxidative stress (Piper, 1999) and of sustaining a reducing intracellular environment (high-reduced glutathione levels and NADH/NAD and NADPH/NADP ratios) for the survival of sorbic acid stress. A subset of these sorbic acid sensitive mutants was also acetic acid sensitive. Thus resistances to both sorbic acid and acetic acid are very severely compromised in mutants lacking an acidified vacuole, in vacuolar protein sorting (*vps*) mutants, with defects in ergosterol biosynthesis (*erg* mutants) and with defects in actin and microtubule organisation (Mollapour *et al.*, 2004, 2006). In contrast, HOG pathway signalling was found to be required only for acetic acid resistance, not for sorbic acid resistance (Mollapour and Piper, 2006).

5.1 Weak Acid Resistance in the *Zygosaccharomyces*

Z. bailii differs from *S. cerevisiae* in possessing the ability to degrade the preservatives in Fig. 1 even when growing in the presence of high sugar concentrations. Unlike *S. cerevisiae,* it is able to catalyse oxidative degradation of sorbate and benzoate (Mollapour and Piper, 2001b). *Z. bailii,* unlike *S. cerevisiae,* can also assimilate acetate whilst growing on fermentable sugars (Sousa *et al.*, 1998).

The Fps1p aquaglyceroporin is probably a key factor for acetic acid resistance of *Zygosaccharomyces* sp., since the recently reported *fps1Δ* mutant of *Z. rouxii* (Tang *et al.*, 2005) displays an elevated acetic acid resistance (our unpublished data). Regarding the more lipophilic preservatives (Fig. 1), we were able to isolate a small *Z. bailii* gene (*ZbYME2*) which, when heterologously expressed in *S. cerevisiae,* confers to the latter yeast the ability to catabolise benzoate, sorbate and phenylalanine (Mollapour and Piper, 2001b). The product of *ZbYME2* has high sequence identity with the N-terminal, mitochondrial matrix domain of the *S. cerevisiae* protein Yme2p, and when expressed heterologously in *S. cerevisiae,* is mitochondrial and can confer Yme2p function (Mollapour and Piper, 2001b). It appears, therefore, that *ZbYME2* confers a broad-specificity monooxygenase function with benzoate-4-hydroxylase activity in the *Z. bailii* mitochondrion, a function that may have been lost by its *S. cerevisiae* homologue, *YME2*. Deletion of the *ZbYME2* gene in *Z. bailii* resulted in a mutant that lacks the ability to degrade benzoate or sorbate and that is, in addition, more sensitive to benzoic- and sorbic acid inhibition during pH 4.5 growth (Mollapour and Piper, 2001a, b). Therefore, when oxygen is present, *ZbYME2* elevates the sorbic- and benzoic acid resistances of *Z. bailii* by facilitating the degradation of these preservatives.

Using genomic fragments of *Z. bailii* DNA in an *S. cerevisiae* plasmid vector, we have searched for *Z. bailii* genes that would complement the sorbic acid sensitivity of the *S. cerevisiae war1Δ* and *pdr12Δ* mutants. No *Z. bailii* counterparts of Pdr12p and War1p were found by this approach, suggesting that *Z. bailii* may not possess an equivalent of the War1p-mediated *S. cerevisiae* response to sorbic acid stress. Also the plasma membrane of sorbic acid-stressed *Z. bailii,* unlike that of sorbic acid-stressed *S. cerevisiae,* does not display strong induction of a putative weak acid transporter (Piper *et al.*, 2001). Instead *Z. bailii* might also have much more efficient ways of changing its cell envelope so as to limit the diffusional entry of acids. This, in turn, will dramatically reduce any need for active anion extrusion. Indications that this may be the case come from observations that *Z. bailii* is much more resistant than *S. cerevisiae* to any short-term decrease in pH_i induced by acetic acid (Arneborg *et al.*, 2000); and trehalose induction by sorbic acid (Cheng *et al.*, 1999). When the benzoic acid sensitivities of several different yeast species were compared, these were, to a rough approximation, *inversely* proportional to rates of diffusional entry of propionate into the cells (Warth, 1989). Reducing diffusional entry of the moderately lipophilic acid preservatives into yeast is therefore probably a key, if still poorly understood, aspect of resistance.

ACKNOWLEDGEMENTS

We are indebted to our collaborators Peter Coote, Karl Kuchler and Frank Cooke for their valuable contributions to the understanding of sorbic acid resistance. This work has been supported by the BBSRC.

REFERENCES

Arneborg, N., Jesperson, L. and Jakobsen, M. (2000). Individual cells of *Saccharomyces cerevisiae* and *Zygosaccharomyces bailii* exhibit different short-term intracellular pH responses to acetic acid. *Archives of Microbiology*, **174**, 125–128.

Bauer, B.E., Rossington, D., Mollapour, M., Mamnun, Y., Kuchler, K. and Piper, P.W. (2003). Weak organic acid stress inhibits aromatic amino acid uptake by yeast, causing a strong influence of amino acid auxotrophies on the phenotypes of membrane transporter mutants. *European Journal of Biochemistry*, **270**, 3189–3195.

Bracey, D., Holyoak, C.D. and Coote, P.J. (1998). Comparison of the inhibitory effect of sorbic acid and amphotericin B on *Saccharomyces cerevisiae*: Is growth inhibition dependent on reduced intracellular pH? *Journal of Applied Microbiology*, **85**, 1056–1066.

Cheng, L., Moghraby, J. and Piper, P.W. (1999). Weak organic acid treatment causes a trehalose accumulation in low-pH cultures of *Saccharomyces cerevisiae*, not displayed by the more preservative-resistant *Zygosaccharomyces bailii*. *FEMS Microbiology Letters*, **170**, 89–95.

Fleet, G. (1992). Spoilage yeasts. *Critical Reviews in Biotechnology*, **12**, 1–44.

Gardner, L.K. and Lawrence, G.D. (1993). Benzene production from decarboxylation of benzoic acid in the presence of ascorbic acid and a transition metal catalyst. *Journal of Agricultural and Food Chemistry*, **40**, 693–695.

Hatzixanthis, K., Mollapour, M., Seymour, I., Bauer, E.B., Krapf, G., Schüller, C., Kuchler, K. and Piper, P.W. (2003). Moderately lipophilic carboxylate compounds are the selective inducers of the *Saccharomyces cerevisiae* Pdr12p ATP-binding cassette transporter. *Yeast*, **20**, 575–585.

Hazelwood, L.A., Tai, S.L., Boer, V.M., de Winde, J.H., Pronk, J.T. and Daran, J.M. (2006). A new physiological role for Pdr12p in *Saccharomyces cerevisiae*: Export of aromatic and branched-chain organic acids produced in amino acid catabolism. *FEMS Yeast Research*, **6**, 937–945.

Holyoak, C.D., Bracey, D., Piper, P.W., Kuchler, K. and Coote, P.J. (1999). The *Saccharomyces cerevisiae* weak acid-inducible ABC transporter Pdr12 transports fluorescein and preservative anions from the cytosol by an energy-dependent mechanism. *Journal of Bacteriology*, **181**, 4644–4652.

Holyoak, C.D., Stratford, M., McMullin, Z., Cole, M.B., Crimmins, K., Brown, A.J. and Coote, P.J. (1996). Activity of the plasma membrane H^+-ATPase and optimal glycolytic flux are required for rapid adaptation and growth of *Saccharomyces cerevisiae* in the presence of the weak-acid preservative sorbic acid. *Applied and Environment Microbiology*, **62**, 3158–3164.

Kalathenos, P., Sutherland, J.P. and Roberts, T.A. (1995). Resistance of some wine spoilage yeasts to combinations of ethanol and acids present in wine. *Journal of Applied Bacteriology*, **78**, 245–250.

Kren, A., Mamnun, Y.M., Bauer, B.E., Schüller, C., Wolfger, H., Hatzixanthis, K., Mollapour, M., Gregori, C., Piper, P. and Kuchler, K. (2003). War1p, a novel transcription regulator controlling weak acid stress response in yeast. *Molecular and Cellular Biology*, **23**, 1775–1785.

Ludovico, P., Sansonetty, F., Silva, M.T. and Corte-Real, M. (2003). Acetic acid induces a programmed cell death process in the food spoilage yeast *Zygosaccharomyces bailii*. *FEMS Yeast Research*, **3**, 91–96.

Ludovico, P., Sousa, M.J., Silva, M.T., Leao, C. and Corte-Real, M. (2001). *Saccharomyces cerevisiae* commits to a programmed cell death process in response to acetic acid. *Microbiology*, **147**, 2409–2415.

Luyten, K., Albertyn, J., Skibbe, W.F., Prior, B.A., Ramos, J., Thevelein, J.M. and Hohmann, S. (1995). Fps1, a yeast member of the MIP family of channel proteins, is a facilitator for glycerol uptake and efflux and is inactive under osmotic stress. *The EMBO Journal*, **14**, 1360–1371.

Mollapour, M., Fong, D., Balakrishnan, K., Harris, N., Thompson, S., Schüller, C., Kuchler, K. and Piper, P. (2004). Screening the yeast deletant mutant collection for hypersensitivity and hyperresistance to sorbate, a weak organic acid food preservative. *Yeast*, **21**, 927–946.

Mollapour, M., Phelan, J.P., Millson, S.H., Piper, P.W. and Cooke, F.T. (2006). Weak acid and alkali stress regulate phosphatidylinositol bisphosphate synthesis in *Saccharomyces cerevisiae*. *The Biochemical Journal*, **395**, 73–80.

Mollapour, M. and Piper, P.W. (2001a). Targeted gene deletion in *Zygosaccharomyces bailii*. *Yeast*, **18**, 173–186.

Mollapour, M. and Piper, P.W. (2001b). The *ZbYME2* gene from the food spoilage yeast *Zygosaccharomyces bailii* confers not just the functions of the native *YME2* in *Saccharomyces cerevisiae*, but also a capacity for catabolism of sorbate and benzoate, two major weak organic acid preservatives. *Molecular Microbiology*, **42**, 919–930.

Mollapour, M. and Piper, P.W. (2006). Hog1p MAP kinase determines acetic acid resistance in *Saccharomyces cerevisiae*. *FEMS Yeast Research*, **6**, 1274–1280.

Piper, P., Mahé, Y., Thompson, S., Pandjaitan, R., Holyoak, C., Egner, R., Mühlbauer, M., Coote, P. and Kuchler, K. (1998). The Pdr12 ABC transporter is required for the development of weak organic acid resistance in yeast. *The EMBO Journal*, **17**, 4257–4265.

Piper, P., Ortiz Calderon, C., Hatzixanthis, K. and Mollapour, M. (2001). Weak acid adaptation: The stress response that confers yeasts with resistance to organic acid food preservatives. *Microbiology*, **147**, 2635–2642.

Piper, P.W. (1999). Yeast superoxide dismutase mutants reveal a prooxidant action of weak organic acid food preservatives. *Free Radical Biology & Medicine*, **27**, 1219–1227.

Piper, P.W., Ortiz-Calderon, C., Holyoak, C., Coote, P. and Cole, M. (1997). Hsp30, the integral plasma membrane heat shock protein of *Saccharomyces cerevisiae*, is a stress-inducible regulator of plasma membrane ATPase. *Cell Stress & Chaperones*, **2**, 12–24.

Schüller, C., Mamnun, Y.M., Mollapour, M., Krapf, G., Schuster, M., Bauer, B., Piper, P.W. and Kuchler, K. (2004). Global phenotypic analysis and transcriptional profiling defines the weak acid stress response regulon in *Saccharomyces cerevisiae*. *Molecular Biology of the Cell*, **15**, 706–720.

Serrano, R. (1991). Transport across yeast vacuolar and plasma membranes. In: *The Molecular Biology of the Yeast Saccharomyces* (J.R. Broach, J. Pringle and E.W. Jones, eds.), pp. 523–586. Cold Spring Harbor Laboratory, New York.

Sousa, M.J., Rodrigues, F., CorteReal, M. and Leao, C. (1998). Mechanisms underlying the transport and intracellular metabolism of acetic acid in the presence of glucose in the yeast *Zygosaccharomyces bailii*. *Microbiology*, **144**, 665–670.

Steels, H., James, S.A., Roberts, I.N. and Stratford, M. (1999). *Zygosaccharomyces lentus*: A significant new osmophilic, preservative-resistant spoilage yeast, capable of growth at low temperature. *Journal of Applied Microbiology*, **87**, 520–527.

Steels, H., James, S.A., Roberts, I.N. and Stratford, M. (2000). Sorbic acid resistance: The inoculum effect. *Yeast*, **16**, 1173–1183.

Stratford, M. and Anslow, P.A. (1996). Comparison of the inhibitory action on *Saccharomyces cerevisiae* of weak-acid preservatives, uncouplers, and medium-chain fatty acids. *FEMS –Microbiology Letters*, **142**, 53–58.

Stratford, M. and Anslow, P.A. (1998). Evidence that sorbic acid does not inhibit yeast as a classic "weak acid preservative". *Letters in Applied Microbiology*, **27**, 203–206.

Sutherland, F.C., Lages, F., Lucas, C., Luyten, K., Albertyn, J., Hohmann, S., Prior, B.A. and Kilian, S.G. (1997). Characteristics of Fps1-dependent and -independent glycerol transport in *Saccharomyces cerevisiae*. *Journal of Bacteriology*, **179**, 7790–7795.

Tamas, M.J., Luyten, K., Sutherland, F.C., Hernandez, A., Albertyn, J., Valadi, H., Li, H., Prior, B.A., Kilian, S.G., Ramos, J., Gustafsson, L., Thevelein, J.M. and Hohmann, S. (1999). Fps1p controls the accumulation and release of the compatible solute glycerol in yeast osmoregulation. *Molecular Microbiology*, **31**, 1087–1104.

Tamas, M.J., Rep, M., Thevelein, J.M. and Hohmann, S. (2000). Stimulation of the yeast high osmolarity glycerol (HOG) pathway: Evidence for a signal generated by a change in turgor rather than by water stress. *FEBS Letters*, **472**, 159–165.

Tang, X.M., Kayingo, G. and Prior, B.A. (2005). Functional analysis of the *Zygosaccharomyces* rouxii Fps1p homologue. *Yeast*, **22**, 571–581.

Verduyn, C., Postma, E., Scheffers, W.A. and Van Dijken, J.P. (1992). Effect of benzoic acid on metabolic fluxes in yeasts: A continuous-culture study on the regulation of respiration and alcoholic fermentation. *Yeast*, **8**, 501–517.

Viegas, C.A. and Sa Correia, I. (1991). Activation of plasma membrane ATPase of *Saccharomyces cerevisiae* by octanoic acid. *Journal of General Microbiology*, **137**, 645–651.

Warth, A.D. (1988). Effect of benzoic acid on growth yield of yeasts differing in their resistance to preservatives. *Applied and Environmental Microbiology*, **54**, 2091–2095.

Warth, A.D. (1989). Transport of benzoic and propanoic acids by *Zygosaccharomyces bailii*. *Journal of General Microbiology*, **135**, 1383–1390.

Weber, F.J. and de Bont, J.A.M. (1996). Adaptation mechanisms of microorganisms to the toxic effects of organic solvents on membranes. *Biochimica et Biophysica Acta*, **1286**, 225–245.

CHAPTER 11

Heavy Metal Pollution and Genetic Adaptations in Ectomycorrhizal Fungi

Jan V. Colpaert

Contents

1. Metal Toxicity Triggers Evolutionary Processes in Plants and their Associated Microorganisms 157
2. Mycorrhizal Fungi in Metal-Polluted Soils 160
3. Metal-Tolerant Mycorrhizal Fungi 163
4. Mechanisms of Metal Tolerance 165
5. Does Adaptive Metal Tolerance in the Mycorrhizal Fungus Increase Host Plant Fitness? 168
6. Conclusions 170

References 170

Abstract

Heavy metal toxicity is a strong trigger for evolutionary adaptation in terrestrial biota that colonise metalliferous soils. Here, I will focus on the occurrence of metal tolerance in ectomycorrhizal (ECM) fungi, the predominant group of root symbionts of pioneer trees that try to colonise severely polluted sites. A considerable amount of literature exists on metal-tolerant plants, which is in sharp contrast to what we know about the tolerance in the fungal symbiotic partners that associate with these plants on metal-polluted soils. I will deal with the ecological and evolutionary processes that drive plant and fungal communities and populations on metal-contaminated sites. The few examples of true metal tolerance in ECM fungi are described and mechanisms possibly involved in this tolerance are briefly summarised. How true metal tolerance in ECM fungi can affect a host plant is discussed in the final Section.

1. METAL TOXICITY TRIGGERS EVOLUTIONARY PROCESSES IN PLANTS AND THEIR ASSOCIATED MICROORGANISMS

An unfortunate effect of many industrial activities has been the contamination of ecosystems with heavy metals. At its worst, heavy metal pollution can have

British Mycological Society Symposia Series
Published by Elsevier Ltd.

deleterious effects on communities and populations of all kinds of biota. It progressively wipes out the more sensitive species and finally can lead to an almost complete elimination of organisms. The most visible effect of metal toxicity in terrestrial ecosystems can be observed on vegetation. Along transects, towards metal smelters, a progressive reduction in floristic diversity is obvious and in extreme cases metal deserts may develop in the immediate vicinity of emission sources (Ernst, 1990). Such metal-contaminated soils strongly differ from unpolluted ones with respect not only to heavy metal concentrations but also to numerous additional physico-chemical factors. For most plant taxa, these constraints are too strong to be overcome. However, ecosystems and species exhibit considerable resilience and even extremely metal-toxic soils eventually become colonised by vegetation. Extreme toxicity induces a strong selection pressure for adaptation and a small number of organisms eventually manage to build up metal-tolerant populations. The phenomenon is well described for prokaryotes (Mergeay *et al.*, 2003) and plants in terrestrial environments (Bradshaw and McNeilly, 1981; Al-Hiyaly *et al.*, 1990). In aquatic ecosystems, metal adaptation has been observed in benthic organisms such as worms and algae and in prokaryotes surviving in contaminated sediments.

Once an ecosystem has been exposed to severe metal contamination, the initial decrease in species diversity may again slowly increase because of the evolution of metal resistance in a number of local species, most often grasses, and because of immigration of metal-tolerant organisms from other sites (Pauwels *et al.*, 2005). Soils that are naturally enriched with heavy metals typically harbour unique plant communities that are well adapted and more species-rich than recently polluted sites (Ernst, 1990). The potential for the development of metal tolerance—the ability to survive on metalliferous soils—is not widespread in the plant kingdom (Antonovics *et al.*, 1971) and it is remarkable that within wide geographic ranges the same plant species seem to develop pioneer metallophytes, even when different metals are responsible for the toxicity. This is best known for some monocotyledonous plants, in particular grasses (Schat *et al.*, 2000). Evolution for metal tolerance in grasses can be very rapid because normal, non-adapted populations often contain a low frequency (0.1–0.5%) of metal-tolerant individuals (Bradshaw and McNeilly, 1981; Al-Hiyaly *et al.*, 1993). Rapid evolution towards metal tolerance is probably more likely in *r*-strategists with a high reproductive rate and the production of a large number of offspring. A rapid establishment of metal-tolerant grass ecotypes has been observed on many polluted sites and against many different metals (Schat *et al.*, 2000). It is remarkable that the tolerance is usually quite specific for the particular metal that is present in toxic concentrations; suggesting that different genes must be involved and selected for. Relatively few dicots have been found to develop metal-tolerant ecotypes. Most metal-tolerant dicots are confined to metalliferous soils and only a few species, for example, *Silene vulgaris*, *Thlaspi caerulescens* and *Arabidopsis halleri*, have both metallicolous and non-metallicolous populations (Ernst, 1990; Assunção *et al.*, 2003; Pauwels *et al.*, 2005).

Many of the (pseudo)metallophytes can develop in the absence of suitable mycorrhizal fungi, which is probably an advantage during the primary succession of metal wastes or deserts. In general, there is little evidence that mycorrhizal fungi are absolutely essential for the initial stages of primary succession of plant communities on soils subjected to major environmental perturbations (Smith and Read, 1997; Wu *et al.*, 2004). This is not to say that these pioneer plants are never colonised by mycorrhizal fungi; many of these plants are facultative mycotrophs and most grasses and even pioneer species within the Brassicaceae and Polygonaceae family, taxa traditionally considered as being non-mycorrhizal, become colonised with arbuscular mycorrhizal (AM) fungi during the initial stages of primary succession (Wu *et al.*, 2004), smoothing the path for more obligate mycotrophs. This pattern is not different for the primary succession of vegetation on metal-contaminated soils (Vangronsveld *et al.*, 1996). There is even evidence that the AM symbiosis can increase the fitness of these host plants that usually exhibit 'low mycorrhizal response' (Hetrick *et al.*, 1994; Shetty *et al.*, 1995; Vogel-Mikus *et al.*, 2006).

Woody plants often are not considered as primary colonisers of metal-polluted soils (Schat *et al.*, 2000), but on a number of sites pioneer tree species, such as willows, poplars, birches and pines are able to build up small pioneer populations (Vrålstad *et al.*, 2002; Colpaert *et al.*, 2004; Adriaensen *et al.*, 2005). In trees with their long-reproductive cycles, the adaptive potential for metal tolerance seems to be low (Meharg and Cairney, 2000) and even if there is selection for individuals with a higher tolerance, it may take many decades before a reasonably tolerant population is build up. Apart from the lack of fast evolution for metal tolerance in woody plants, there is a second reason why colonisation of polluted soils by trees may be very slow. Woody pioneers rely much more on their ECM fungi than herbaceous pioneer plants rely on their AM mycobionts, irrespective of any soil pollution (Ashkannejhad and Horton, 2006). Recent investigations clearly illustrate the great importance of generalist ECM fungi for the establishment of woody species during primary succession on a volcanic desert (Nara, 2006a, 2006b). After severe disturbances, when mycorrhizal propagules are very scarce, ECM plants appear relatively late during primary succession. This is completely different in secondary successions where tree seedlings rapidly recruit ECM fungi, more often specialists, from dormant spore banks or other resistant propagules (Izzo *et al.*, 2006; Nara, 2006b). On severely metal-polluted soils, an additional constraint is toxicity. Suppose a relatively rare occasion when a germinating pioneer tree and a germinating ECM propagule comes close enough to form a compatible symbiotic association, at that time both partners must be able to withstand the metal toxicity during all stages of colonisation events. It is likely that the fungus is the most vulnerable partner in the pre-symbiotic phase. Despite this hostile environment, the ECM symbiosis persists on strongly metal-contaminated sites that are slowly colonised by mycotrophic tree species such as birches, pines and willows. Therefore, one can only conclude that trees resist extreme metal toxicity through a large phenotypic plasticity and through their association with a small guild of well-adapted ECM fungi (Wilkinson and Dickinson, 1995). Plant adaptation to selective pressures is often considered to be regulated by the plant genome, but it is absolutely clear that also mutualistic microorganisms can alleviate heavy metal stress in plants

(Hall, 2002; Schützendübel and Polle, 2002; Adriaensen *et al.*, 2004). Not only mycorrhizal fungi should be mentioned here. Also numerous metal-tolerant rhizosphere and endophytic bacteria that form loose or intimate associations with plants are known to improve plant fitness on metal-contaminated soils (Vivas *et al.*, 2006).

2. MYCORRHIZAL FUNGI IN METAL-POLLUTED SOILS

Plants of facultative metallophyte populations from metalliferous soil almost invariably exhibit higher levels of metal tolerance than those of normal soil populations (Schat *et al.*, 2000). Compared to the vast body of literature published on the impact of metal toxicity on plant communities and populations, relatively little information is available on the existence of metal-tolerant fungal species, populations or communities in natural habitats. Up to now, no highly adapted fungi with a distribution restricted to contaminated soils were detected. There is only some circumstantial evidence that mycorrhizal communities, both ECM and AM, are affected by heavy metal pollution. Species diversity seems to be lower on the most polluted areas (Staudenrausch *et al.*, 2005), but low ECM fungal diversity is rather typical for pioneer conditions. What we can be sure is that succession is slower on polluted soils because the invasion of new plants and microorganisms is much slower than in unpolluted conditions. *In situ* treatment of 3 ha Maatheide soil with a metal-immobilising substance decreased toxicity significantly and resulted in a quick increase in plant species diversity of metal-tolerant grassland (Vangronsveld *et al.*, 1996). It remains unclear to what extent the slow succession on metal-contaminated soils is really affected by a lack of suitable saprotrophic and biotrophic microorganisms. In this respect it is worth noting that at least plant species richness is higher in primary successions on disturbed sites close to metal outcrops (Ernst, 1990). A larger pool of metal-adapted plants (genes) is present in these areas and it is tempting to assume that this might be true as well for biota from the other trophic levels of the ecosystem.

In our field surveys along the Zn pollution gradient in northern Limburg (Colpaert *et al.*, 2004), we focused on the occurrence of ECM fungi in pioneer pine forests. Only on the most polluted area in Maatheide a very low number of ECM morphotypes (4) was found on roots of 25 year-old pine trees. Most remarkable was the frequent occurrence of a dark ascomycete from the *Hymenoscyphus ericae* aggregate, which was rare or absent from less polluted plots. Similar ascomycetes were also present on pioneer pine trees that colonise the Folldal Cu mine spoil in Norway (Vrålstad *et al.*, 2002) and dark ascomycetes (Sordariaceae) were also dominant mycorrhizas on roots of young *Salix caprea* trees thriving on very toxic soil close to the Pb-smelter of Zerjav in Slovenia (Personal communication M. Regvar). These observations are in the same line as those noted by Bradley *et al.* (1981) who demonstrated that heather (*Calluna vulgaris*) was entirely dependent on the ascomycete *Hymenoscyphus ericae* for survival on acidic Cu, Zn and (or) Pb polluted soils in the UK. Whether this ericoid mycorrhizal fungus can improve fitness of an ECM host under metal stress remains unclear. There are indications that particular ascomycetes are more stress-resistant than basidiomycetes and show up more frequently in mycorrhizal communities facing harsh environmental conditions or after severe disturbances (Baar *et al.*, 1999; Trowbridge and Jumpponen, 2004). Apart from the dark ascomycetes, the metalliferous sites in Maatheide and in Folldal had also a basidiomycete ECM fungus in common: *Suillus luteus*, a typical mycobiont of young pine trees; a fungus that is not confined to polluted sites and that is very common on young pines in primary successions (Fig. 1). On the Belgian sites this fungus is the most prolifically fruiting species, usually on the bare soil and most distantly from the stem base of its host trees.

In general, we have only a very incomplete view of the biodiversity of mycorrhizal fungi in metal-polluted environments. Waiting for below-ground community studies, I would like to refer to two community studies, based on above-ground sporocarp observations, along metal gradients in north and south Sweden (Rühling *et al.*, 1984; Rühling and Söderström, 1990). Although molecular studies regularly find a considerable lack of correspondence between the above- and below-ground communities of ECM colonisers (Gardes and Bruns, 1996), the field studies of Rühling suggest that particular mycorrhizal

Figure 1 *Suillus luteus*, a pioneer ectomycorrhizal fungus that evolved metal-tolerant ecotypes on metal-polluted soils. (**See Colour Section**)

species disappear with increasing metal stress. Sporocarp production in *Picea abies* forests along a Cu–Zn gradient was investigated (Rühling *et al.*, 1984). The average number of species of macrofungi per plot (1,000 m^2) decreased significantly along the gradient: 35 species at a Cu concentration of approximately 100 μg/g organic matter, 25 species at 1,000 μg/g and only 15 species at 10,000 μg Cu/g organic matter. Some ECM taxa decreasing along the gradient were: *Cantharellus cibarius*, *Cortinarius* sp., *Dermocybe* sp., *Gomphidius* sp., *Hydnum* sp., *Lactarius* sp., *Paxillus involutus*, *Russula* sp. Taxa which were not affected or that increased in frequency, include: *Albatrellus ovinus*, *Amanita* sp., *Cantharellus tubaeformis*, *Laccaria laccata*, *Leccinum* sp. In north Sweden, a similar study was performed along a more complex metal (As, Cu, Pb, Cd, Zn) gradient but very similar results were obtained (Rühling and Söderström, 1990). In the least polluted plots, 4.4 species of macrofungi were found per 100 m^2 whereas only 1.3 species were observed in the most heavily polluted plots. The number of observations of the genus *Amanita* increased in the most heavily polluted area, the genera *Cortinarius*, *Lactarius* and *Russula* showed decreasing numbers of observations. The reduced sporocarp production and the decreasing above-ground diversity do not necessarily mean that the percentage of root colonisation decreased over the same gradient. It is possible that a few fungi dominate the community or that species without above-ground sporocarps take over. It would be interesting to know whether there is an increase in mycorrhizal ascomycetes over the metal gradients studied by Rühling. Other basidiomycete taxa that have been frequently found on heavily polluted soils include *Hebeloma* sp., *Pisolithus tinctorius* (Turnau *et al.*, 1988), *Rhizopogon* sp. (Turnau *et al.*, 1996), *Scleroderma* sp. (Jones and Hutchinson, 1986), and the Cd-accumulating *Amanita muscaria* (Gast *et al.*, 1988; Kalač *et al.*, 1991).

Detailed studies on AM fungal communities from polluted sites are also scarce but from these surveys, a reduced number of species seem to be able to invade heavy metal-polluted sites (da Silva *et al.*, 2005), although a relatively high number of AM fungi may be present in old vegetation on metal outcrops (Pawlowska *et al.*, 1996; Tonin *et al.*, 2001).

Field studies on mycorrhizal community or population structure are much more difficult and laborious to obtain than studies on the vegetation itself. Because mycorrhizal fungi are cryptic organisms and hyphal growth is indeterminate, establishing even the most basic population features, such as number and size of individuals and net secondary production, can be difficult and can only be estimated via indirect molecular techniques. In addition, AM fungi are coenocytic clonal organisms and it is still not very well understood how genetic traits are transferred to descendants since we know little about the behaviour of nuclei during the life cycle of these fungi. This knowledge on reproductive biology is essential to better understand the selection process for genes that confer improved fitness in fungi from metal-polluted substrates. It is of practical importance to know whether 'metal-tolerant' genes can be maintained in a population of nuclei for long periods when the selection pressure is removed.

The evolution towards heavy metal tolerance is expected to cause a strong genetic bottleneck during colonisation of polluted areas, thus leading to a reduction of gene diversity in metal-tolerant populations as compared to populations from normal soils. Many studies of pseudometallophytes exhibiting populational tolerance have therefore attempted to detect reduced genetic variation in these metallicolous populations as compared to normal ones. Surprisingly, most of them failed to detect any such strong effects (Pauwels *et al.*, 2005). We investigated the genetic structure of the *S. luteus* populations along the Zn gradient in N-Limburg (Muller *et al.*, 2004). Both AFLP and microsatellite markers revealed high levels of genetic diversity within the geographic sub-populations, but genetic differentiation between sub-populations was limited, suggesting substantial gene flow and frequent sexual reproduction. Although *a priori* we expected that heavy metal pollution would have a strong effect on the genetic structure of *S. luteus* populations, no evidence was found for a consistent reduction of the genetic variation of the sub-populations in polluted habitats. Plant population geneticists found similar results for several pseudometallophytes (Vekemans and Lefèbvre, 1997; Van Rossum *et al.*, 2004; Pauwels *et al.*, 2005). Several non-exclusive processes may explain the surprisingly high level of variation in the metal-tolerant *S. luteus* populations. Successive colonisation events—the species is a common pioneer—or a relatively high frequency of metal tolerance in natural populations could have reversed the initial genetic bottleneck. In addition, recurrent migration of tolerant genotypes that originate due to admixture in non-polluted areas may attribute to the high level of genetic diversity of populations inhabiting contaminated soils. Genetic mixing between the tolerant *S. luteus* populations and surrounding populations inhabiting non-polluted soils is very likely, as indicated by the low level of population differentiation, and it would explain the relatively high frequency of tolerance observed in some of the populations in the transition zone from a high to low degree of pollution (Colpaert *et al.*, 2004).

3. METAL-TOLERANT MYCORRHIZAL FUNGI

From the above Section one can presume that mycorrhizal fungi that survive and reproduce in metal-contaminated substrates must have acquired a high resistance against metal toxicity. However, it proved to be quite difficult to demonstrate unequivocally that metal toxicity exhibited selection pressure on fungal communities and populations. Particular mycorrhizal guilds may have a high constitutive metal tolerance so that evolution to higher tolerance is simply not necessary (Meharg and Cairney, 2000). The occurrence of dark ascomycetes in the most contaminated soils gives some support to this hypothesis. The absence of metal adaptation would be in contrast to what is known for plants and bacteria.

Mycorrhizal fungi that colonise acidic soils can be exposed to high levels of toxic metals such as aluminium, iron and manganese. These fungi must have evolved mechanisms for coping with these elements and it is possible that some of these mechanisms confer some resistance to other metals as well including Cu, Cd, Pb and Zn. However, this reasoning should also apply to the plants themselves and there is no evidence that Al- or Fe-tolerant species have a high constitutive tolerance for Cu or Zn or that plants from acidic soils are better colonisers of contaminated areas. Excess metals may cause partly similar problems in cells (e.g. oxidative stress) and this damage may be healed with similar remedies, but on the other hand each metal has different characteristics, different kinetics and pathways for entering and leaving cells, causes some specific molecular interactions and needs specific molecules for chelation and sequestration. Uptake and homeostasis of heavy metals, at least of the essential ones, is strictly controlled which is only possible through specific fine tuning of the activity of highly metal-specific transporters located in the plasma membrane, tonoplast and membranes of other organelles. For example in yeast the activity of the Zn transporters is regulated at both transcriptional and post-transcriptional levels in response to Zn (Eide, 2003). A fungus that copes well with excess Cu is not necessarily well equipped to withstand excess Fe or Zn and *vice versa*. Metal tolerance mechanisms are usually quite specific in all kinds of organisms although low levels of co-tolerance sometimes have been found.

It is remarkable that some organisms seem to be predestined to evolve specific metal-tolerance mechanisms. Such particular species are well known among prokaryotes (Mergeay *et al.*, 2003) and in the plant kingdom (Ernst, 1990; Schat *et al.*, 2000). These organisms can acquire tolerances against many different metals. In bacteria plasmids play a major role in this multiple heavy metal tolerance. In most organisms, true adaptation seems to be governed by a small number of genes. Nevertheless, such genetic modifications occur slowly, in particular in eukaryotic species with long reproductive cycles.

Organisms trying to survive in heavy metal-contaminated ecosystems have a whole battery of responses to their disposal to cope with heavy metal stress. Species with the least efficient detoxification systems will disappear from the ecosystem. In conditions where the heavy metal pollution is so severe that there are consistent detrimental effects on metabolism, organisms are subjected to selective pressures for increased resistance to toxic metals. These adaptations involve modifications in the genetic patrimony of an organism. There has been a long debate whether mycorrhizal fungi have evolved adaptive tolerance against particular heavy metals (Hartley *et al.*, 1997). A major reason was the lack of sufficient data from different populations from sites with high and low levels of pollution. Measuring and comparing metal tolerance can only be achieved by screening a significant number of individuals from one or more species. It is necessary to analyse the intra- and inter-population variation in metal tolerances. This is far easier for plants than for their fungal partners. The isolation and axenic cultivation of large numbers of mycorrhizal fungi can be quite troublesome, especially for AM fungi. Luckily some pioneer ECM fungi are easier to cultivate so that inter- and intraspecific comparisons for metal tolerance become possible. Nevertheless, interspecies comparisons *in vitro* can still be misleading because some fungi might be much more sensitive to stress *in vitro* than in symbiosis, in contrast to species that grow equally well *in vitro* as in nature. Intraspecific comparisons are probably less susceptible to such confounding factors. Ideally once tolerance has been identified *in vitro*, it should also be verified in a plant experiment.

Selection for adaptive metal tolerance in mycorrhizal fungi has been discovered in only a few higher fungi. *Pisolithus tinctorius* isolates from an old coal-mining site had higher Al tolerance than isolates from rehabilitated and forest sites (Egerton-Warburton and Griffin, 1995). In the pioneer forests around several Zn-smelters in Belgium adaptive Zn tolerance was found in *Suillus luteus*, *S. bovinus* and *Rhizopogon luteolus*, but not in *Paxillus involutus* (Colpaert *et al.*, 2004). Other ECM fungi could not be tested because of a lack of sufficient isolates. Monokaryons from Zn-tolerant *S. luteus*, obtained through basidiospore germination, were also more Zn-tolerant than monokaryons from sensitive genotypes (unpublished), confirming the heritability of the tolerance trait. Screening the dikaryotic *Suillus* isolates against elevated Cd revealed that isolates from the most polluted soil in Maatheide, had acquired adaptive Cd tolerance (Colpaert *et al.*, 2000). However, the Cd tolerance was less pronounced than the Zn tolerance and it was far less widespread, which might be due to a lower selection pressure exerted by Cd. Grasses in the same area also developed a high Zn tolerance and some additional Cd tolerance.

The occurrence of *S. luteus* associated with poorly growing pine seedlings invading a Cu-mine spoil prompted us to check this mine population for Cu tolerance. Indeed this particular population showed a significantly higher Cu tolerance than populations from uncontaminated or Zn-contaminated sites. The Cu-tolerant isolates were not tolerant to elevated Zn. This indicates that in fungi also the evolution of metal-specific tolerance mechanisms is strongly triggered by the pollution of the local environment (Adriaensen *et al.*, 2005). Cu-tolerance was so far not detected in the *S. luteus* isolates. The evolution towards adaptive Cu tolerance is a rare phenomenon; even in the plant kingdom relatively few plants seem to be able to evolve Cu-tolerant ecotypes (Wu and Lin, 1990; Schat and Vooijs, 1997). Cu is a well-known fungicide but there are few papers that report on the evolution towards additional Cu resistance in fungi that are frequently treated on Cu-containing fungicides.

AM fungi can also develop metal tolerance, although the trait seems to be less stable than in higher fungi and plants. *Glomus* cultures with a high tolerance to Mn lose the specific tolerance when cultivated in metal-free substrates in contrast to cultivation in original contaminated soil (Malcová *et al.*, 2003). Metal tolerance in AM fungi has not yet been assessed directly in vitro, it is usually based on tests for spore germination, germ tube growth and root colonisation. Relatively few genotypes were investigated. Shetty *et al.* (1995) also showed that the AM fungi isolated from Zn-contaminated soil surrounding mine tailing were more effective in increasing plant biomass at high levels of Zn, whereas AM fungi from a non-contaminated site promoted plant growth only in the soil containing lower Zn concentrations.

4. MECHANISMS OF METAL TOLERANCE

The mechanisms that are involved in metal homeostasis and detoxification of essential and non-essential metals in ECM fungi were recently reviewed by Bellion *et al.* (2006). Extracellular mechanisms such as chelation and cell-wall binding as well as cellular mechanisms such as binding to (non)-protein thiols and transport into intracellular compartments play a role in the metal housekeeping of ECM fungi. These mechanisms are not different from those that are present in other eukaryotes. Some of these mechanisms are constitutively present, whereas others come into full operation when excess metals show up in the cytoplasm. Additional antioxidative detoxification systems, which allow the fungus to counteract the accumulation of reactive-oxygen species directly or indirectly, are part of the detoxification response. Whether one or more of these mechanisms are modified in the evolution towards adaptive true tolerance in ECM fungi remains unclear.

Reduced uptake of metals into the cytosol might be achieved by extracellular chelation or precipitation of metals with organic compounds, mostly acids such as citrate and oxalate. Nevertheless, it is unlikely that metal detoxification is the primary function of organic acid excretion. It is probably more involved in nutrient (cation and phosphate) mobilisation from minerals and in maintaining charge balances over membranes. In many fungi organic acid production largely depends on the N source in the medium, on the phosphate limitation and on the carbon status of the culture. It remains to be demonstrated that there is sufficient organic acid excretion in the myco(rhizo)sphere of mycorrhizal fungi that colonise metal-polluted soils in the absence of a readily available C source. Organic acids also have a high turnover in a soil as a consequence of microbial degradation. Eventually organic acid exudation should reduce metal uptake into the fungal cells. This is not obvious. Martino *et al.* (2003) studied ZnO solubilisation in isolates from the ericoid mycorrhizal fungus, *Oidiodendron maius*. Strains from polluted sites had a lower ZnO solubilisation activity and a lower organic acid production than isolates from control sites. Such a reduction in acid exudation could be an adaptation to Zn pollution in this ericoid mycorrhizal fungus and may reflect specific strategies to maintain homeostasis of essential metals under different soil conditions (Martino *et al.*, 2003). This pattern of reduced Zn solubilisation was not observed in Zn-tolerant *S. luteus* and *S. bovinus* isolates exposed to poorly soluble $Zn_3(PO_4)_2{\cdot}2H_2O$ (hopeite), neither a precipitation of an organic Zn compound was found in the culture medium (Fomina *et al.*, 2004, 2005). The Zn-tolerant *Suillus* strains yielded more biomass, acidified the MMN medium more and dissolved more of the Zn phosphate than less tolerant strains.

Fungal cell walls have high capacities for metal binding (biosorption), but the trait is dependant on fungal species and metal. Both AM and ECM fungi can bind substantial amounts of metals (Blaudez *et al.*, 2000; Joner *et al.*, 2000), although there is no strong evidence that this binding is an efficient strategy to keep excess metals out of cells. It is likely that binding sites of growing hyphal tips are almost immediately saturated with metals in strongly contaminated soils. Once the equilibrium with the surrounding soil solution is reached, the role of the cell wall in metal immobilisation is over. Biosorption is a mechanism not depending on the metabolic activity of the fungus whereas excretion of organic substances such as acids or metal-binding polyphenolics relies on the activity of the hyphae. Hydrophobines covering hyphae may affect the transfer rate of nutrients and cations from soil solution towards the plasma membrane but it is also evident that the deposition of a hydrophobic layer around hyphae is not a metal-specific barrier and has more general functions such as maintaining hydraulic balances in fungi. It cannot be excluded that metal toxicity in fungi will lead to changes in cell-wall composition and hyphal growth in an effort to avoid or reduce metal exposure. Lanfranco *et al.* (2002) showed that changes in hyphal morphology occur when an ericoid mycorrhiza-forming ascomycete was treated with millimolar concentrations of Zn. This led to apical swellings and increased branching in the subapical parts as well as a significant increase in the amount of chitin in metal-treated hyphae.

Despite extracellular chelation and cell-wall binding, large amounts of metal eventually enter into hyphae that are exposed to excess metals. At low external concentrations, essential metals enter cells through metal-specific high affinity transporters. It is less clear how non-essential metals penetrate cells. Blaudez *et al.* (2000) have quantified the proportion of Cd in the cytosol and the vacuole of *Paxillus involutus* and estimated it to be 20 and 30%, respectively. At high external Zn concentrations, Zn influx in *S. luteus* is high irrespective of its Zn tolerance or metabolic activity (25 or 0°C). Although down-regulation of metal transporters plays a role in metal homeostasis in fungi (Eide, 2003) and plants (Clemens, 2001; Hall, 2002), it cannot explain the adaptive Zn tolerance in *S. luteus*. Evolutionary adaptation to Zn-enriched soils can be achieved in several ways. In some plants, Zn tolerance seems to be functionally related to transport processes that permit compartmentation of Zn ions and prevent the accumulation of toxic levels in the cytoplasm (Chardonnens *et al.*, 1999; Clemens, 2001; Hall, 2002). Improved accumulation of Zn in vacuoles, reduced uptake or increased extracellular efflux of Zn are all possible mechanisms that may reduce Zn toxicity in adapted plant ecotypes (Chardonnens *et al.*, 1999; Hall, 2002; Dräger *et al.*, 2004). In mycorrhizal fungi, similar mechanisms may operate, but little detailed information is available.

To avoid cellular damage, metals that are in excess in the cytosol must be efficiently removed. Efflux back into the apoplast or compartmentalisation into organelles is also an option in fungi. The vacuole is probably the best candidate for mass storage of excess metals such as Cd and Zn (Blaudez *et al.*, 2000; Eide, 2003). For more details on transporters possibly involved in metal homeostasis in mycorrhizal fungi, I refer to Eide (2003) and Bellion *et al.* (2006). In *S. luteus* there is evidence that Zn efflux over the plasma membrane may explain the adaptive Zn tolerance (Adriaensen, unpublished). A Zn exclusion mechanism explains the lower Zn uptake in Zn-tolerant Suilloid fungi (Colpaert *et al.*, 2005). In the ECM fungus *Pisolithus tinctorius*, differential Al tolerance could also be attributed to an Al exclusion mechanism in the mycelia of Al-tolerant isolates from a mine site (Egerton-Warburton and Griffin, 1995).

The reactivity and limited solubility of most metal ions require constant chelation once they are taken up into the cell. Free-metal concentrations are very low so that most metals are strongly complexed by protein and non-protein ligands, often compounds with a relatively high specificity for only one or few metal species. Synthesis and activity of these compounds is probably strictly controlled. Possible metal chelators in mycorrhizal fungi include phytochelatins, metallothioneins and organic acids. Courbot *et al.* (2004) found that glutathione increased under Cd exposure in *Paxillus involutus*, as well as γ-glutamylcysteine and a Cd-binding metallothionein (Courbot *et al.*, 2004). Phytochelatins were not detected in *P. involutus*. The expression of metallothionein was studied at the transcriptional level in *P. involutus* exposed to different metal stress, and the results indicated a correlation between metal exposure and expression level (Bellion *et al.*, 2006). Whether intracellular complexation of metals plays a role in adaptive metal tolerance is not known. In plants, these metal chelators are important for basic tolerance and homeostasis but there is little unequivocal evidence that they are involved in the generation of naturally selected tolerance (Clemens, 2001).

In general, our knowledge on metal fluxes, compartmentation and intracellular complexation in mycorrhizal fungi is still very limited. Different detoxification mechanisms may also have an indirect impact on host plants. Fungi that compartmentalise metals into vacuoles, may increase transfer of these metals to their host. Motile tubular vacuoles are an important vector in the transport chain of mineral nutrients from the site of uptake at hyphal tips to the exchange region in the mycorrhizal root (Ashford and Allaway, 2002). The Zn exclusion observed in Zn-tolerant Suilloid fungi will not only help to prevent the onset of metal stress in these fungi, but it will also contribute to the protection of the host plant. The assumption that adapted isolates do not store extra Zn in their vacuoles is an attractive prospect for a host plant. Some ECM fungi are known to accumulate specific metals in their sporocarps, even on soils that are non-polluted; a well-known example being the Cd accumulation in *Amanita muscaria* (Lepp *et al.*, 1987; Lepp, 1992). In this respect it could be interesting to investigate whether this accumulation also leads to an increased Cd transfer to a host plant. Accumulation of Zn in vacuoles could greatly increase the transfer of zinc to the host plant and this exactly was shown not happening in pines colonised with a Zn-tolerant *S. bovinus* isolate (Adriaensen *et al.*, 2004).

5. DOES ADAPTIVE METAL TOLERANCE IN THE MYCORRHIZAL FUNGUS INCREASE HOST PLANT FITNESS?

Alleviation of metal toxicity in plants through ectomycorrhizas has been demonstrated in a number of experiments (for review see Jentschke and Godbold, 2000). In most of these experiments, it was shown that mycorrhizal plants were better off with a suitable ECM fungus than without; and it was also shown that some fungi were more efficient in plant protection than others. Only a few researchers made intraspecific comparisons with fungi from different origins. For protection of a host plant that has the potential to grow on a particular contaminated soil, three major aspects are particularly important for a successful mycorrhizal association: (1) survival of the fungus in the toxic substrate, (2) efficient nutrient transfer to the host, (3) low transfer of the toxic metal to the host.

The survival of the fungus is essential to establish the symbiosis. The same is of course true for the roots of the germinating trees. There is good evidence that a number of fungi cannot withstand metal concentrations in soil solutions of polluted soils. The intraspecific comparisons with Zn- and Cu-tolerant *Suillus* species demonstrate that normal (sensitive) isolates suffer severely from Zn and Cu concentrations that do not affect growth of tolerant isolates (Adriaensen *et al.*, 2004, 2005, 2006). In particular, the external mycelium that colonises the soil was most sensitive to excess metals. However, close to active roots even sensitive isolates can survive for several weeks.

Improved plant nutrition is a major function of the mycorrhizal symbioses. This is not different for ECM plants on metal-polluted soils. Most of these soils are poor in essential nutrients and it is clear that pioneer ECM fungi can alleviate this low nutrient stress. Their external mycelia contribute largely to the nutrient uptake and transfer to the host. This can only be achieved when the fungus can maintain the growth of its mycelium. Eventually this improved nutrition should lead to a better health and growth of trees associated with the most tolerant isolates. In comparisons between mycorrhizal and non-mycorrhizal plants, this was demonstrated in a number of experiments (Jentschke and Godbold, 2000). Also the metal-tolerant *Suillus* isolates can outperform their sensitive counterparts at least in a long-term experiment (Adriaensen *et al.*, 2006).

Fungal metal uptake, translocation and transfer to the host are a complex process. It is usually analysed indirectly by measuring metal contents in plants. Mycorrhizal fungi can improve uptake of essential elements in their host when present at low concentrations. However, there is probably a lot of inter- and intraspecific variation among fungi in nutrient uptake and transfer (Munkvold *et al.*, 2004). In a recent experiment we detected a differential Zn transfer to pines colonised with *S. luteus* exposed to a low (not deficient) Zn concentration (Table 1). Non-mycorrhizal plants had the lowest Zn concentration, plants colonised with a tolerant isolate had a slightly higher concentration, but the highest Zn concentration was observed in plants colonised by a normal *S. luteus* isolate. Pine growth rate was similar in all plants. The differential transfer of Zn in *S. luteus* can probably be explained by the adaptation that occurred in the Zn-tolerant Suilloids. A reduced transfer of metals at high external concentrations was also demonstrated in the dose-response experiments by Adriaensen *et al.* (2004, 2005), although interpretation of these data becomes more difficult as a result of the decreasing transpiration stream with increasing toxicity (Jentschke and Godbold, 2000).

Table 1 Differential zinc uptake in shoots and roots of *Pinus sylvestris* seedlings inoculated wth, or without *Suillus luteus*

Fungal Treatment	Shoot ($\mu g/g^1$ dry weight)	Root ($\mu g/g^1$ dry weight)
Non-mycorrhizal	20.7 ± 1.9	49.0 ± 2.7
Zn-sensitive *S. luteus*	54.3 ± 2.9	79.9 ± 7.0
Zn-tolerant *S. luteus*	30.5 ± 3.0	57.8 ± 3.5

Note: Plants were exposed to a low Zn concentration. Values are means from 10 plants ± SE.

6. CONCLUSIONS

Despite the scepticism about the need to evolve true tolerance in mycorrhizal fungi, there is more and more evidence that heavy metal adaptation in some ECM fungi has occurred and that it is a prerequisite for the survival of both trees and fungi on the most toxic soils. However, compared to what is known about metal tolerance in plants, we still have a long way to go. There is a definite need for studies on mycorrhizal population and community dynamics in polluted soils and the search for more species with true tolerance should continue. Among metal-tolerant ECM fungi, there are metal excluders and accumulators; we have no idea whether this behaviour does affect the metal transfer to a host plant. Furthermore, we are still at the dawn of the elucidation of the molecular mechanisms involved in metal homeostasis, detoxification and tolerance in filamentous fungi in general. A better understanding of metal transport mechanisms including their regulation and in the underlying biochemical and physiological mechanisms of tolerance are of key interest.

REFERENCES

Adriaensen, K., van der Lelie, D., Van Laere, A., Vangronsveld, J. and Colpaert, J.V. (2004). A zinc-adapted fungus protects pines from zinc stress. *New Phytologist*, **161**, 549–555.

Adriaensen, K., Vangronsveld, J. and Colpaert, J.V. (2006). Zinc tolerant *Suillus bovinus* improves growth of Zn-exposed *Pinus sylvestris* seedlings. *Mycorrhiza*. **16**, 553–558.

Adriaensen, K., Vrålstad, T., Noben, J.P., Vangronsveld, J. and Colpaert, J.V. (2005). Copper-adapted *Suillus luteus*, a symbiotic solution for pines colonizing Cu mine spoils. *Applied and Environmental Microbiology*, **71**, 7279–7284.

Al-Hiyaly, S.A.K., McNeilly, T. and Bradshaw, A.D. (1990). The effect of zinc contamination from electricity pylons — Contrasting patterns of evolution in 5 grass species. *New Phytologist*, **114**, 183–190.

Al-Hiyaly, S.A.K., McNeilly, T., Bradshaw, A.D. and Mortimer, A.M. (1993). The effect of zinc contamination from electricity pylons. Genetic constraints on selection for zinc tolerance. *Heredity*, **70**, 22–32.

Antonovics, J., Bradshaw, A.D. and Turner, R.G. (1971). Heavy metal tolerance in plants. *Advances in Ecological Research*, **7**, 1–85.

Ashford, A.E. and Allaway, W.G. (2002). The role of the motile tubular vacuole system in mycorrhizal fungi. *Plant and Soil*, **244**, 177–187.

Ashkannejhad, S. and Horton, T.R. (2006). Ectomycorrhizal ecology under primary succession on coastal sand dunes: Interactions involving *Pinus contorta*, suilloid fungi and deer. *New Phytologist*, **169**, 345–354.

Assunção, A.G.L., Bookum, W.M., Nelissen, H.J.M., Vooijs, R., Schat, H. and Ernst, W.H.O. (2003). Differential metal-specific tolerance and accumulation patterns among *Thlaspi caerulescens* populations originating from different soil types. *New Phytologist*, **159**, 411–419.

Baar, J., Horton, T.R., Kretzer, A.M. and Bruns, T.D. (1999). Mycorrhizal colonization of *Pinus muricata* from resistant propagules after a stand-replacing wildfire. *New Phytologist*, **143**, 409–418.

Bellion, M., Courbot, M., Jacob, C., Blaudez, D. and Chalot, M. (2006). Extracellular and cellular mechanisms sustaining metal tolerance in ectomycorrhizal fungi. *FEMS Microbiology Letters*, **254**, 173–181.

Blaudez, D., Botton, B. and Chalot, M. (2000). Cadmium uptake and subcellular compartmentation in the ectomycorrhizal fungus *Paxillus involutus*. *Microbiology*, **146**, 1109–1117.

Bradley, R., Burt, A.J. and Read, D.J. (1981). Mycorrhizal infection and resistance to heavy metal toxicity in *Calluna vulgaris*. *Nature*, **292**, 335–337.

Bradshaw, A.D. and McNeilly, T. (1981). Evolution and pollution. *Studies in Biology*, **130**, 1–76.

Chardonnens, A.N., Koevoets, P.L.M., van Zanten, A., Schat, H. and Verkleij, J.A.C. (1999). Properties of enhanced tonoplast zinc transport in naturally selected zinc-tolerant *Silene vulgaris*. *Plant Physiology*, **120**, 779–785.

Clemens, S. (2001). Molecular mechanisms of plant metal tolerance and homeostasis. *Planta*, **212**, 475–486.

Colpaert, J.V., Adriaensen, K., Muller, L.A.H., Lambaerts, M., Faes, C., Carleer, R. and Vangronsveld, J. (2005). Element profiles and growth in Zn-sensitive and Zn-resistant Suilloid fungi. *Mycorrhiza*, **15**, 628–634.

Colpaert, J.V., Muller, L.A.H., Lambaerts, M., Adriaensen, K. and Vangronsveld, J. (2004). Evolutionary adaptation to zinc toxicity in populations of Suilloid fungi. *New Phytologist*, **162**, 549–559.

Colpaert, J.V., Vandenkoornhuyse, P., Adriaensen, K. and Vangronsveld, J. (2000). Genetic variation and heavy metal tolerance in the ectomycorrhizal basidiomycete *Suillus luteus*. *New Phytologist*, **147**, 367–379.

Courbot, M., Diez, L., Ruotolo, R., Chalot, M. and Leroy, P. (2004). Cadmium-responsive thiols in the ectomycorrhizal fungus *Paxillus involutus*. *Applied and Environmental Microbiology*, **70**, 7413–7417.

Dräger, D.B., Desbrosses-Fonrouge, A.G., Krach, C., Chardonnens, A.N., Meyer, R.C., Saumitou-Laprade, P. and Krämer, U. (2004). Two genes encoding *Arabidopsis halleri* MTP1 metal transport proteins co-segregate with zinc tolerance and account for high *MTP1* transcript levels. *Plant Journal*, **39**, 425–439.

Egerton-Warburton, L. and Griffin, B. (1995). Differential responses of *Pisolithus tinctorius* isolates to aluminium *in vitro*. *Canadian Journal of Botany*, **73**, 1229–1233.

Eide, D.J. (2003). Multiple regulatory mechanisms maintain zinc homeostasis in *Saccharomyces cerevisiae*. *Journal of Nutrition*, **133**, 1532s–1535s.

Ernst, W.H.O. (1990). Mine vegetations in Europe. In: *Heavy Metal Tolerance in Plants: Evolutionary Aspects* (A.J. Shaw, ed.), pp. 21–37. CRC Press, Boca Raton.

Fomina, M., Alexander, I.J., Colpaert, J.V. and Gadd, G.M. (2005). Solubilization of toxic metal minerals and metal tolerance of mycorrhizal fungi. *Soil Biology & Biochemistry*, **37**, 851–866.

Fomina, M., Alexander, I.J., Hillier, S. and Gadd, G.M. (2004). Zinc phosphate and pyromorphite solubilization by soil plant-symbiotic fungi. *Geomicrobiology Journal*, **21**, 351–366.

Gardes, M. and Bruns, T.D. (1996). Community structure of ectomycorrhizal fungi in a *Pinus muricata* forest: Above- and below-ground views. *Canadian Journal of Botany*, **74**, 1572–1583.

Gast, C.H., Jansen, E., Bierling, J. and Haanstra, L. (1988). Heavy metals in mushrooms and their relationship with soil characteristics. *Chemosphere*, **17**, 789–799.

Hall, J.L. (2002). Cellular mechanisms for heavy metal detoxification and tolerance. *Journal of Experimental Botany*, **53**, 1–11.

Hartley, J., Cairney, J.W.G. and Meharg, A.A. (1997). Do ectomycorrhizal fungi exhibit adaptive tolerance to potentially toxic metals in the environment? *Plant and Soil*, **189**, 303–319.

Hetrick, B.A.D., Wilson, G.W.T. and Figge, D.A.H. (1994). The influence of mycorrhizal symbiosis and fertilizer amendments on establishment of vegetation in heavy metal mine spoil. *Environmental Pollution*, **86**, 171–179.

Izzo, A., Canright, M. and Bruns, T.D. (2006). The effects of heat treatments on ectomycorrhizal resistant propagules and their ability to colonize bioassay seedlings. *Mycological Research*, **110**, 196–202.

Jentschke, G. and Godbold, D.L. (2000). Metal toxicity and ectomycorrhizas. *Physiologia Plantarum*, **109**, 107–116.

Joner, E.J., Briones, R. and Leyval, C. (2000). Metal-binding capacity of arbuscular mycorrhizal mycelium. *Plant and Soil*, **226**, 227–234.

Jones, M.D. and Hutchinson, T.C. (1986). The effect of mycorrhizal infection on the response of *Betula papyrifera* to nickel and copper. *New Phytologist*, **102**, 429–442.

Kalač, P., Burda, J. and Staskova, I. (1991). Concentrations of lead, cadmium, mercury and copper in mushrooms in the vicinity of a lead smelter. *Science of the Total Environment*, **105**, 109–119.

Lanfranco, L., Balsamo, R., Martino, E., Perotto, S. and Bonfante, P. (2002). Zinc ions alter morphology and chitin deposition in an ericoid fungus. *European Journal of Histochemistry*, **46**, 341–350.

Lepp, N.W. (1992). Uptake and accumulation of metals in bacteria and fungi. In: *Biogeochemistry of Trace Metals* (D.C. Adriano, ed.), pp. 277–298. Lewis Publishers, Boca Raton.
Lepp, N.W., Harrison, S.C.S. and Morrell, B.G. (1987). A Role for *Amanita muscaria* L. in the circulation of cadmium and vanadium in a non-polluted woodland. *Environmental Geochemistry and Health*, **9**, 61–64.
Malcová, R., Rydlova, J. and Vosatka, M. (2003). Metal-free cultivation of *Glomus* sp BEG 140 isolated from Mn-contaminated soil reduces tolerance to Mn. *Mycorrhiza*, **13**, 151–157.
Martino, E., Perotto, S., Parsons, R. and Gadd, G.M. (2003). Solubilization of insoluble inorganic zinc compounds by ericoid mycorrhizal fungi derived from heavy metal polluted sites. *Soil Biology & Biochemistry*, **35**, 133–141.
Meharg, A.A. and Cairney, J.W.G. (2000). Co-evolution of mycorrhizal symbionts and their hosts to metal-contaminated environments. *Advances in Ecological Research*, **30**, 69–112.
Mergeay, M., Monchy, S., Vallaeys, T., Auquier, V., Benotmane, A., Bertin, P., Taghavi, S., Dunn, J., van der Lelie, D. and Wattiez, R. (2003). *Ralstonia metallidurans*, a bacterium specifically adapted to toxic metals: Towards a catalogue of metal-responsive genes. *FEMS Microbiology Reviews*, **27**, 385–410.
Muller, L.A.H., Lambaerts, M., Vangronsveld, J. and Colpaert, J.V. (2004). AFLP-based assessment of the effects of environmental heavy metal pollution on the genetic structure of pioneer populations of *Suillus luteus*. *New Phytologist*, **164**, 297–303.
Munkvold, L., Kjøller, R., Vestberg, M., Rosendahl, S. and Jakobsen, I. (2004). High functional diversity within species of arbuscular mycorrhizal fungi. *New Phytologist*, **164**, 357–364.
Nara, K. (2006a). Ectomycorrhizal networks and seedling establishment during early primary succession. *New Phytologist*, **169**, 169–178.
Nara, K. (2006b). Pioneer dwarf willow may facilitate tree succession by providing late colonizers with compatible ectomycorrhizal fungi in a primary successional volcanic desert. *New Phytologist*, **171**, 187–198.
Pauwels, M., Saumitou-Laprade, P., Holl, A.C., Petit, D. and Bonnin, I. (2005). Multiple origin of metallicolous populations of the pseudometallophyte *Arabidopsis halleri* (Brassicaceae) in central Europe: The cpDNA testimony. *Molecular Ecology*, **14**, 4403–4414.
Pawlowska, T.E., Blaszkowski, J. and Rühling, A. (1996). The mycorrhizal status of plants colonizing a calamine spoil mound in southern Poland. *Mycorrhiza*, **6**, 499–505.
Rühling, A., Bååth, E., Nordgren, A. and Söderström, B. (1984). Fungi in metal-contaminated soil near the Gusum brass mill, Sweden. *Ambio*, **13**, 34–36.
Rühling, A. and Söderström, B. (1990). Changes in fruitbody production of mycorrhizal and litter decomposing macromycetes in heavy metal polluted coniferous forests in north Sweden. *Water Air and Soil Pollution*, **49**, 375–387.
Schat, H., Llugany, M. and Bernhard, R. (2000). Metal-specific patterns of tolerance, uptake and transport of heavy metals in hyperaccumulating and nonhyperaccumulating metallophytes. In: *Phytoremediation of Contaminated Soil and Water* (N. Terry, G. Banuelos, eds.), pp. 171–188. Lewis, Boca Raton.
Schat, H. and Vooijs, R. (1997). Multiple tolerance and co-tolerance to heavy metals in *Silene vulgaris*: A co-segregation analysis. *New Phytologist*, **136**, 489–496.
Schützendübel, A. and Polle, A. (2002). Plant responses to abiotic stresses: Heavy metal-induced oxidative stress and protection by mycorrhization. *Journal of Experimental Botany*, **53**, 1351–1365.
Shetty, K.G., Hetrick, B.A.D. and Schwab, A.P. (1995). Effects of mycorrhizae and fertilizer amendments on zinc tolerance of plants. *Environmental Pollution*, **88**, 307–314.
da Silva, G.A., Trufem, S.F.B., Junior, O.J.S. and Maia, L.C. (2005). Arbuscular mycorrhizal fungi in a semiarid copper mining area in Brazil. *Mycorrhiza*, **15**, 47–53.
Smith, S.E. and Read, D.J. (1997). *Mycorrhizal Symbiosis*. 2nd Ed. Academic Press, London, UK.
Staudenrausch, S., Kaldorf, M., Renker, C., Luis, P. and Buscot, F. (2005). Diversity of the ectomycorrhiza community at a uranium mining heap. *Biology and Fertility of Soils*, **41**, 439–446.
Tonin, C., Vandenkoornhuyse, P., Joner, E.J., Straczek, J. and Leyval, C. (2001). Assessment of arbuscular mycorrhizal fungi diversity in the rhizosphere of *Viola calaminaria* and effect of these fungi on heavy metal uptake by clover. *Mycorrhiza*, **10**, 161–168.

Trowbridge, J. and Jumpponen, A. (2004). Fungal colonization of shrub willow roots at the forefront of a receding glacier. *Mycorrhiza*, **14**, 283–293.
Turnau, K., Gucwa, E., Mleczko, P. and Godzik, B. (1988). Metal content in fruit-bodies and mycorrhizas of *Pisolithus arrhizus* from zinc wastes in Poland. *Acta Mycologica*, **33**, 59–67.
Turnau, K., Kottke, I. and Dexheimer, J. (1996). Toxic element filtering in *Rhizopogon roseolus Pinus sylvestris* mycorrhizas collected from calamine dumps. *Mycological Research*, **100**, 16–22.
Vangronsveld, J., Colpaert, J.V. and van Tichelen, K.K. (1996). Reclamation of a bare industrial area contaminated by non-ferrous metals: Physicochemical and biological evaluation of the durability of soil treatment and revegetation. *Environmental Pollution*, **94**, 131–140.
Van Rossum, F., Bonnin, I., Fénart, S., Pauwels, M., Petit, D. and Saumitou-Laprade, P. (2004). Spatial genetic structure within a metallicolous population of *Arabidopsis halleri*, a clonal, self-incompatible and heavy-metal-tolerant species. *Molecular Ecology*, **13**, 2959–2967.
Vekemans, X. and Lefèbvre, C. (1997). On the evolution of heavy-metal tolerant populations in *Armeria maritima*: Evidence from allozyme variation and reproductive barriers. *Journal of Evolutionary Biology*, **10**, 175–191.
Vivas, A., Barea, J.M., Biro, B. and Azcon, R. (2006). Effectiveness of autochthonous bacterium and mycorrhizal fungus on *Trifolium* growth, symbiotic development and soil enzymatic activities in Zn contaminated soil. *Journal of Applied Microbiology*, **100**, 587–598.
Vogel-Mikus, K., Pongrac, P., Kump, P., Necemer, M. and Regvar, M. (2006). Colonisation of a Zn, Cd and Pb hyperaccumulator *Thlaspi praecox* Wulfen with indigenous arbuscular mycorrhizal fungal mixture induces changes in heavy metal and nutrient uptake. *Environmental Pollution*, **139**, 362–371.
Vrålstad, T., Myhre, E. and Schumacher, T. (2002). Molecular diversity and phylogenetic affinities of symbiotic root-associated ascomycetes of the Helotiales in burnt and metal polluted habitats. *New Phytologist*, **155**, 131–148.
Wilkinson, D.M. and Dickinson, N.M. (1995). Metal resistance in trees—the role of mycorrhizae. *Oikos*, **72**, 298–300.
Wu, B.Y., Isobe, K. and Ishii, R. (2004). Arbuscular mycorrhizal colonization of the dominant plant species in primary successional volcanic deserts on the Southeast slope of Mount Fuji. *Mycorrhiza*, **14**, 391–395.
Wu, L. and Lin, S.L. (1990). Copper tolerance and copper uptake of *Lotus purshianus* (Benth.) Clem. & Clem. and its symbiotic *Rhizobium loti* derived from a copper mine waste population. *New Phytologist*, **116**, 531–539.

CHAPTER 12

Lichens and Metals

O. William Purvis and **Barbara Pawlik-Skowrońska**

Contents

1.	Introduction	176
2.	Metal Influences on the Composition of Lichen Communities Growing on Rocks and Soils	179
3.	Metal Influences on the Composition of Lichen Communities Growing on Trees	182
4.	Metal Uptake and Extracellular Localisation	185
5.	Effects of Trace Metals at the Cellular Level	188
6.	Conclusions	190
	Acknowledgements	191
	References	191

Abstract

The ability of macrolichens to accumulate high levels of potentially toxic metals has led to their widespread use as biomonitors of metal deposition and consideration in biogeochemical prospecting. Metal bioavailability and pH influence the composition of lichen assemblages and play a role in metal fixation and retention. Areas traditionally regarded as highly toxic where few, if any, higher plants survive, support abundant lichens, including rare species restricted to such sites. Lichens grow directly on minerals including uranium, copper, lead and arsenic phosphates. Metal oxalates and lichen-acid metal complexes, melanin pigments and organic phosphates may help lichens to tolerate high metal concentrations. Phytochelatin (PC) production by lichen photobionts and thiol peptide status in both symbionts are two possible detoxification mechanisms operating at the cellular level in epiphytic macrolichens and a crustose lichen accumulating over 1.3% dry wt. Cu. Other tolerance mechanisms, including metal binding by metallothioneins, are yet to be considered. Lichens dominate 6% of the land's surface, are pioneers of fresh rock outcrops and contribute to soil formation. Understanding metal uptake and retention by lichens is important for environmental monitoring, understanding global biogeochemical cycles, learning how organisms tolerate potentially toxic elements and for conservation.

British Mycological Society Symposia Series
Published by Elsevier Ltd.

1. INTRODUCTION

Lichenised fungi play a role in the evolution of terrestrial life, ecosystem function and the maintenance of biodiversity (Hawksworth, 1988, 1991; Nash, 1996a; Schwartzman, 2002; Gorbushina, 2006). Composite organisms, lichens, include at least one fungus (mycobiont) and an alga or cyanobacterium (photobiont) living in symbiosis. Lichenised fungi are ecologically obligate biotrophs acquiring carbon from their photobionts (Honegger, 1997). The lichen symbiosis may involve multiple and different bionts at various stages of its life history (Hawksworth, 1988; Jahns, 1988). Lichens colonise bark, rocks and soil and occur in all terrestrial ecosystems, covering about 6% of the earth's land surface. They are dominant in arctic and antarctic tundra regions where they form an important part of the food chain. Arctic and sub-arctic lichen heaths are readily visible from space using remote sensing techniques (Tømmervik *et al.*, 1995, 1998, 2003). Lichens play a major role in plant ecology and the cycling of elements, such as C, N and P and radionuclides (Knops *et al.*, 1991; Schwartzman, 1993; Nash, 1996b). They contribute to soil formation and stabilisation (Jones, 1988). Metalliferous habitats that are so toxic that they support few, if any higher plants often have abundant crustose 'metallophyte' lichens, including rare species. Some grow directly on minerals (Gilbert, 1980, 2000; Purvis and Halls, 1996; Haas and Purvis, 2006). Lichens colonise contaminated sites near metal smelters (LeBlanc *et al.*, 1972; Nash, 1972, 1989, 1996b; Purvis *et al.*, 2000; Cuny *et al.*, 2004a). Relatively slow growing (up to a maximum of a few centimeters per annum), some are rapid colonizers of bare ground, including metal-contaminated sites (Gilbert, 1980, 1990). Lichens contain over 700, in many cases unique, secondary products 'stress metabolites', which are believed to aid survival in inhospitable regions (Lange, 1992; Elix, 1996; Huneck and Yoshimura, 1996; Huneck, 2004). These may be present up to 20% dry wt. (Elix, 1996; Huneck and Yoshimura, 1996). The ecological role of 'stress metabolites' is largely conjectural and complex (Lange, 1992) and their role in avoiding toxic effects of copper was also suggested (Purvis, 1985; Chisholm *et al.*, 1987; Purvis *et al.*, 1987, 1990).

Lacking a protective cuticle and roots, lichens absorb nutrients and trace elements, including metals from dry and wet atmospheric deposition. Lichens derive cations from the substrate and water running over their surfaces. Inputs of cations are further enhanced by foliar leaching from the canopy and the concentrating effects of stemflow (De Bruin and Hackenitz, 1986; Bargagli, 1998, 1990; Ceburnis and Steinnes, 2000; Loppi and Pirintsos, 2003; Hauck and Paul, 2005; Rusu *et al.*, 2006a). Trace element accumulation in saxicolous crustose lichens was considered 50 years ago by Lounamaa (1956) who established links between element contents in saxicolous lichens and rocks. At that time it was difficult to distinguish between incorporation of mineral fragments from biological fixation. Saxicolous lichens were subsequently considered as tools in biogeochemical prospecting (Le Roy and Koksoy, 1962). Exploration geologist Steve Czehura was one of the first to recognise the potential for lichens as indicators of ore deposits when he discovered dark green specimens of *Lecanora cascadensis* H. Magn. growing on rocks containing blebs of bornite (Cu_5FeS_4) and other copper minerals (Czehura, 1977). An interesting example of unconscious biogeochemical prospecting by lichens was by Tomassini *et al.* (1976) who discovered high copper levels around Conwoyto Lake in Northwest Territories, Canada. At a later date, a mining company independently discovered a copper-ore deposit in the area (Brooks, 1983).

Oliver Gilbert, over 40 years ago established links between sulphur and fluoride tissue concentrations and thallus deformities (Gilbert, 1969, 1971, 1973; Otnyukova, 2006). A relationship between element contents in epiphytic *Hypogymnia physodes* and macrolichen diversity near industrial plants in Finland was established (Laaksovirta and Olkkonen, 1977; Garty, 1993). These early studies, supported by laboratory and field experimental studies, paved the way for the wide-scale application of lichens as biomonitors of air pollution from local to national scales worldwide (Nimis *et al.*, 2002). A few countries routinely map metal deposition using lichens (e.g. Nimis *et al.*, 2000; Jeran *et al.*, 2006; USDA, 2006) to inform policy. A strict monitoring protocol is necessary with respect to sampling and preparation to determine trace metals in lichens (Bargagli and Nimis, 2002). In polluted regions, where native epiphytic species are largely absent, macrolichen transplants are frequently used to monitor metal contamination (Jeran *et al.*, 1995; Sloof, 1995; Mikhailova, 2002; Purvis *et al.*, 2004b; Adamo *et al.*, 2006; Rusu *et al.*, 2006a, 2006b). Priority areas for biomonitoring include polluted sites close to human habitations and areas of conservation interest (European Habitats Directive 79/409/EEC).

Dramatic changes have occurred in the distribution and abundance of some of the commonest lichens in our cities and the countryside over recent years. (Rose and Hawksworth, 1981; Hawksworth and McManus, 1989; Seaward and Letrouit-Galinou, 1991; Gilbert, 1992; Bates *et al.*, 2001; Nimis *et al.*, 2002; Purvis *et al.*, 2003; van Herk *et al.*, 2003; Hauck and Paul, 2005; Davies *et al.*, 2007; Larsen *et al.*, 2007; Saipunkaew *et al.*, 2007; Wolseley *et al.*, 2007; Isocrono *et al.*, 2007). The nature and impact of environmental pollutants is constantly changing. If lichen data are to be used to monitor or formulate regulatory decisions regarding air pollution levels, we need to know what levels are damaging to lichens and which gaseous pollutants (or other substances) are the primary or contributing cause of the observed damage or distribution change (Richardson, 1988; Nimis and Purvis, 2002). Understanding how lichen floras are influenced by changing atmospheric conditions is important for biomonitoring, conservation and policy (Richardson, 1988; Nimis *et al.*, 2002). The importance of cations and pH in governing the physiology and distribution of vascular plants is well established, although considerably less studied in lichens (Farmer *et al.*, 1991; Gauslaa and Holien, 1998; Hauck and Paul, 2005). Early research considered the atmosphere as the major, if not exclusive, source of cations for lichens supported by data showing a poor correlation between lichen and substratum element content (Brown, 1976).

Metals are either directly or indirectly involved in all aspects of fungal and plant growth (Gadd, 1993; Farago, 1994). Elements, including trace metals, have traditionally been considered as being 'essential' or 'non-essential'. It has been suggested that some plants require small amounts of almost every naturally occurring element in the periodic table though uncertainty remains over biological role(s) (Mehra and Farago, 1994; Bargagli, 1998). Approximately 30% of enzymes are metalloenzymes or require metal ions for activity (Hay, 1994). Most metals, including essential elements such as Ca, are toxic to plants and fungi when present in excess and or a particular chemical form (Gadd, 1993; Macnair and Baker, 1994). Many attempts have been made to define metals in relation to their physical, chemical and biological properties. pH and oxidation–reduction potential influence metal availability and toxicity. Ligand preferences and polarizing power (charge–radius ratio) are especially important in a biological context since these properties determine biological activity (Nieboer and Richardson, 1980; Gadd, 1993). Different metal species have distinctive binding affinities and co-ordination chemistries (Kosman, 2004). Metal plant and –fungus interactions are further complicated by synergistic interactions between metal species and other environmental variables including climate, nutrients and pollutants. Climate (heat and cold temperature stress) causes cellular damage to higher plants and algae (Davison and Barnes, 2002; Sanita di Toppi and Pawlik-Skowrońska, 2003) and influences the susceptibility of lichens to other pollutants. Dry lichens have survived the extreme conditions of outer space (L. Sanchez pers comm. to OWP). Interactions between metals may be both positive and negative according to the type and species of plant or fungus and prevailing environmental conditions (Gadd, 1993; Mehra and Farago, 1994). Geological factors are very important. For instance clay minerals adsorb metal ions in soils reducing their potential toxicity to fungi (Gadd, 1993).

The earth's aerosol burden comes partly from the effects of vapour pressure and resulting volatilization of bedrock components, and partly from anthropogenic sources (Goldberg, 1976; Takala *et al.*, 1998). The importance of anthropogenically derived emissions to atmospheric aerosol emissions has increased over the last few centuries to greatly surpass natural inputs (Nriagu, 1989; Shotbolt *et al.*, 2004). The range of metals used in manufacturing, our ability to disperse pollutants over a wide area and the intensity of emissions increased with population and energy consumption to peak in the UK in 1960s and 1970s (Shotbolt *et al.*, 2004). Metals are highly enriched near ore bodies and mining regions leading to naturally increased metal concentrations in vegetation (Brooks, 1983). Many lichen species occur in these habitats, which have a unique biodiversity (Purvis and Halls, 1996; Whiting *et al.*, 2004).

Particulate chemistry determines the potential effects that they may have on vegetation (Gilbert, 1976; Farmer, 1993, 2002). Damaged lichens are found in remote regions, which may not be due to pollution, but to entirely natural causes (Otnyukova, 2006). Toxic metal effects, including discoloured or pinkish necrotic patches and an absence of growth are readily observed in the field as for example beneath galvanised wire and metal structures (Seaward, 1974) or beneath leaded stained glass church windows. Closer examination often reveals tolerant species. Roadsides provide important habitats for lichens, some normally associated with mine sites. *Stereocaulon* species colonised polluted road-site verges during the period of high Pb emissions and *Vezdaea leprosa* occurs alongside motorway crash barriers in Germany and the UK (Ernst, 1995; Gilbert, 2000). *Psilolechia leprosa* is found on churches beneath copper lightening conductors throughout UK (Coppins and Purvis, 1987). Several books discuss the complex interactions between fungi, lichens, plants and metals (e.g. Shaw, 1989; Markert, 1993; Farago, 1994; Winkelmann and Winge, 1994; Nash, 1996b; Bargagli, 1998; Bell and Treshow, 2002; Gadd, 2006).

Review articles focus on biomonitoring metal deposition using macrolichens and experimental metal-uptake studies (Brown, 1991; Garty, 1993; Easton, 1994; Richardson, 1995; Wilson, 1995; Purvis, 1996; Branquinho *et al.*, 1999; Garty, 2001; Bargagli and Mikhailova, 2002; Haas and Purvis, 2006). Biogeochemical interest in crustose lichens has focused on their ability to weather rocks as a direct action of physical and chemical processes. These include metal chelation by lichen substances and organic acids (Lee, 2000; Fomina *et al.*, 2006; Haas and Purvis, 2006). Metallophyte lichens and their habitats are of special conservation interest (Purvis and James, 1985; Purvis, 1993, 2001; Church *et al.*, 1996; Purvis and Halls, 1996; Whiting *et al.*, 2004, 2005). A notable example is the rare red data book species 'Copper Lichen' (*Lecidea inops*) (Church *et al.*, 1996) included on Schedule 8 of the UK Wildlife and Countryside Act, 1981.

A diagrammatic representation of possible metal uptake mechanisms and metal localisation is shown (Figure 1). Metal uptake can result in distinctive colours, normally reflecting secondary metabolites and considered as important species characters. Unusually green-coloured lichen species containing copper were previously considered to represent distinct species (Purvis, 1985, 1997; Purvis *et al.*, 1985). This paper considers the influence of metals on the composition of lichen communities growing on rocks and trees, metal uptake and extracellular localization and the effects of metals at the cellular level.

2. METAL INFLUENCES ON THE COMPOSITION OF LICHEN COMMUNITIES GROWING ON ROCKS AND SOILS

Lichens colonizing rocks are intimately associated with 12 major elements (Si, Al, Fe, Ca, Mg, Na, K, Ti, P and Mn) including 10 metals that account for over 99.9% of crustal rocks (Taylor and McLennan, 1985; Davies, 1994). Striking evidence that cations and pH influence the composition of lichen assemblages is based on studies of naturally mineralised environments and on abandoned mine spoil heaps. The *Acarosporion sinopicae* Wirth ex James *et al.* community is dominated by *Acarospora sinopica* and other rust-coloured species characteristic of rocks containing high levels of Fe and Cu sulphides which through oxidative weathering, assisted by microbial activity, create a low pH (Wirth, 1972; Purvis, 1985; Purvis and Halls, 1996; Jenkins *et al.*, 2000) (Figure 2). Species found in this specialised habitat contain a single photobiont species (Beck, 1999). By contrast, the *Lecideion inopis* Purvis, grows on rocks containing copper minerals usually accompanied by significant calcite ($CaCO_3$), creating a higher pH environment (Purvis, 1985; Purvis and James, 1985; Purvis and Halls, 1996). At Coniston Copper Mines in the English Lake District, it occurs on malachite ($Cu_2CO_3(OH)_2$), azurite ($Cu_3(CO_3)2(OH)_2$) and with smaller amounts of

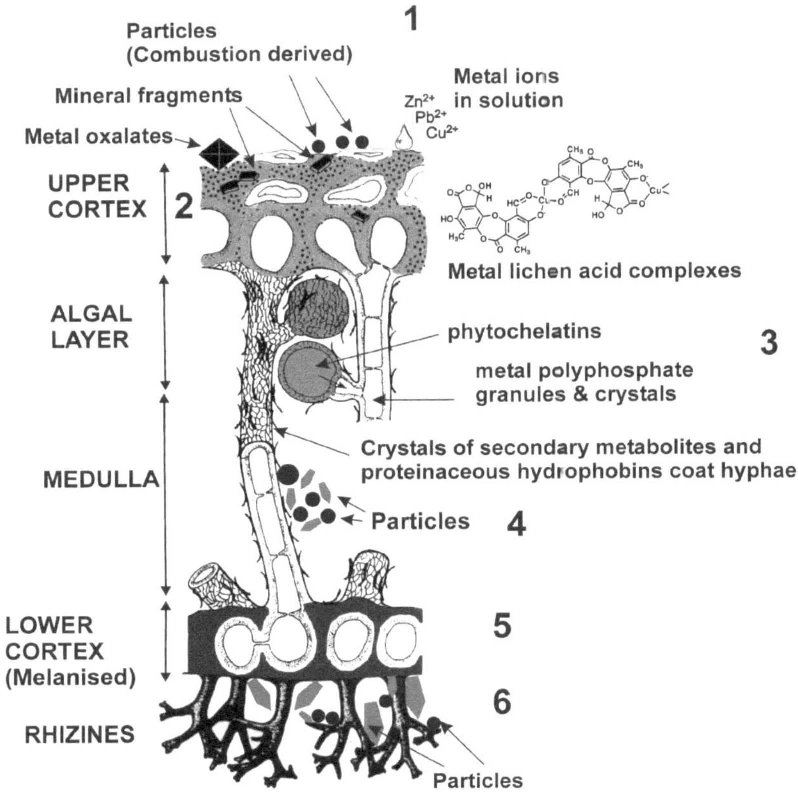

Figure 1 Diagrammatic representation of possible metal uptake mechanisms and metal localization in a foliose macrolichen (*Source:* Honegger (1997), Sarret *et al.* (1998), Purvis (2000), Haas and Purvis (2006)) (not to scale). (1) Accumulation of particles (mineral and combustion-derived) in dry and wet deposition, metal ion sorption via soluble phases, particulate entrapment in intercellular spaces. (2) Particulates are not necessarily inert and may be solubilized by acid precipitation and/or lichen-derived organic acids leading to metal sorption to e.g. extracellular hydrophilic β-glucans secreted by mycobiont cortical cells with negatively charged anionic sites (e.g. Sarret *et al.*, 1998); extracellular metal oxalate and metal lichen acid-complex formation (Cu-Norstictic acid shown) (Purvis *et al.*, 1987, 1990; Takani *et al.*, 2002). (3) Intracellular phytochelatin, thiol peptides containing metal-chelating sulphydryl groups of cysteine help protect photobionts from metal toxicity (Pawlik-Skowrońska *et al.*, 2002, 2006). (4) Particles (mineral, combustion—derived and biogenic e.g. oxalates) trapped in intercellular spaces of fungal hyphae and/or attached to medullary hyphae coated with hydrophobic mycobiont-derived secondary metabolites which may further act as sites for metal complexation (Purvis *et al.*, 1987, 1990). Also coated by water-repellant proteins, 'hydrophobins', which protect thalli from intracellular penetration of heavy metal-containing solutions. (5) Lower cortex containing extracellular hydrophilic β-glucans secreted by cortical cells, which may bind metals as in (2). Melanins are often present which may sorb metals (U, Cu and Fe) in melanized tissues in fruiting bodies (not shown) (McLean *et al.*, 1998; Purvis *et al.*, 2004a). (6) Particulates (mineral or combustion-derived) may become trapped by rhizines through surface flow; sorbed to cell walls (Goyal and Seaward, 1982) and in extracellular melanized regions (as in the outer wall of fruiting bodies, McLean *et al.*, 1998; Purvis *et al.*, 2004a). Rhizines occupy by far the bulk of the thallus in section and may extend to several millimetres, hyphae to several centimetres in the substrate. Particles and metals may also be removed from thalli by a variety of processes.

Figure 2 (a) *Cladonia rei* (foreground), the dominant lichen in Birch woodland (background) in Mid Urals where the vascular Flora was almost completely destroyed by smelter emissions (Udachin *et al.*, 2003; Purvis *et al.*, 2004b). The same species is known from other contaminated soils (Cuny *et al.*, 2004a); (b) *Trapelia involuta* growing on metatorbernite ($Cu(UO_2)_2(PO_4)_2.8H_2O$) (McLean *et al.*, 1998; Purvis *et al.*, 2004a). Other samples were growing in direct contact with metazeunerite ($Cu(UO_2)_2(AsO_4)_2$. $8H_2O$, autunite $Ca(UO_2)_2(PO_4)_2.10H_2O$, beudantite ($PbFe_3(AsO_4, PO_4)(SO_4)(OH)_6$, Pyrite ($FeS_2$), chalcopyrite ($CuFeS_2$), galena (PbS) and löllingite ($FeAs_2$) (Purvis *et al.*, 2004a); (c) *Acarospora sinopica* (deep rusty red) and *A. smaragdula* v. *lesdainii* f. *subochracea* Magn. (yellow-orange) characteristic of iron sulphide-rich rocks (Sweden); (d) Copper-rich *Lecanora polytropa* (Hoffm.) Rabenh. (pale turquoise-green) (Pawlik-Skowrońska *et al.*, 2002, 2006) and *Lecidea inops* Th. Fr. (grey-black) characteristic of Cu-rich, higher pH habitats. **(See Colour Section)**

chrysocolla ($Cu_2H_2Si_2O_5(OH)_4$), covelline (CuS) and bornite (Cu_5FeS_4) cementing fractured quartz (Stanley, 1979; Purvis and James, 1985). Field and herbarium evidence confirm that substrate mineralisation beneath the lichen is not essential to support these distinctive assemblages. *Lecidea inops* was observed growing on quartz near Røros in Norway where it was subject to irrigation from Cu-rich run-off from contaminated soils (Purvis, 1985; Purvis and Halls, 1996). The same species was collected on wood at a copper mine in the Czech Republic by Antonin Vězda (NHM herbarium). Another species belonging to this community, *Psilolechia leprosa*, is frequent on church walls beneath copper lightening conductors on mortar (Coppins and Purvis, 1987). Characteristic lichen assemblages occur on metalliferous soils polluted by industrial emissions and on abandoned mine wastes (Gilbert, 1980; Aptroot and van den Boom, 1998; Gilbert, 2000; Cuny *et al.*, 2004a). These sites are prone to disturbance (e.g. by mineral particle turnover and 'off-road biking') and are colonised by a range of minute crustose 'stress-tolerant ruderals' reproducing rapidly before their habitat is destroyed (Grime, 1979; Gilbert, 1990). One of the basic conditions for the colonization of metalliferous soils is the ability of plant species to evolve metal resistance (Antonovics *et al.*, 1971; Ernst, 1989). This aspect has not yet been investigated in lichens growing in metalliferous habitats (Figure 3).

3. METAL INFLUENCES ON THE COMPOSITION OF LICHEN COMMUNITIES GROWING ON TREES

Bark cation content and pH play a major influence in determining the composition of epiphytic lichen floras (Barkman, 1958; James *et al.*, 1977; Gauslaa, 1985; Farmer *et al.*, 1991; Gauslaa and Holien, 1998). Bark chemistry is influenced by several factors, including tree species, atmospheric chemistry, soil-nutrient status, vegetation cover, proximity to particulate sources (e.g. quarrying activities and roads), topographic and meteorological factors. Trees derive cations from the soil through root uptake, impaction of particulates on trunks and branches and by foliar interception of wet and dry deposition by the leaf canopy, which flows down trunks (stem flow) (Farago, 1994; Bargagli, 1998). Crown canopy architecture, surface microtopography and the balance between the process of deposition and the movement and loss of particles play a major role in particle interception and pollutant delivery to lichens (Otnyukova, 2006). Epiphytic lichens act as filters, trapping particulates washed down the tree trunk in stemflow (Brown *et al.*, 1994; Bargagli and Mikhailova, 2002). Particles are not necessarily inert and may be solubilised by rainfall (especially in polluted regions) and/or by lichen acid secretions in the thallus (Nieboer *et al.*, 1978; Nash, 1989; Haas and Purvis, 2006). Lichen species sensitive to air pollution were found to be restricted to trees having higher bark Ca, K and Mg contents and pH than individuals lacking these lichens (Farmer *et al.*, 1991). The mineral status of different components, like epiphytes, trees and soils are strongly coupled within a mycorrhiza-dependent forest ecosystem (Gauslaa and Holien, 1998). Bark cation element ratios influence lichen colonisation and growth (Hauck *et al.*, 2002a).

Particulate deposition can lead to damage, growth reductions and changes in lichen communities, either directly or indirectly through changes in substrate chemistry or pH (Gilbert, 1976; Farmer, 1993). Some particulates are chemically relatively inert, others highly reactive. Dust from limestone quarry, cement and from many roads is highly alkaline as well as contains high Ca levels, and hence reactive. Perhaps the most dramatic evidence for the effects of particulate deposition on plant communities has come from Oliver Gilbert's studies carried out near a lime-dust source in Derbyshire where he identified zones according to lichen diversity (Gilbert, 1976). Heavily dusted trees had few lichens, but this was followed by a zone containing lichens that are normally saxicolous together with species typical of nutrient-enriched bark. These species eventually declined and

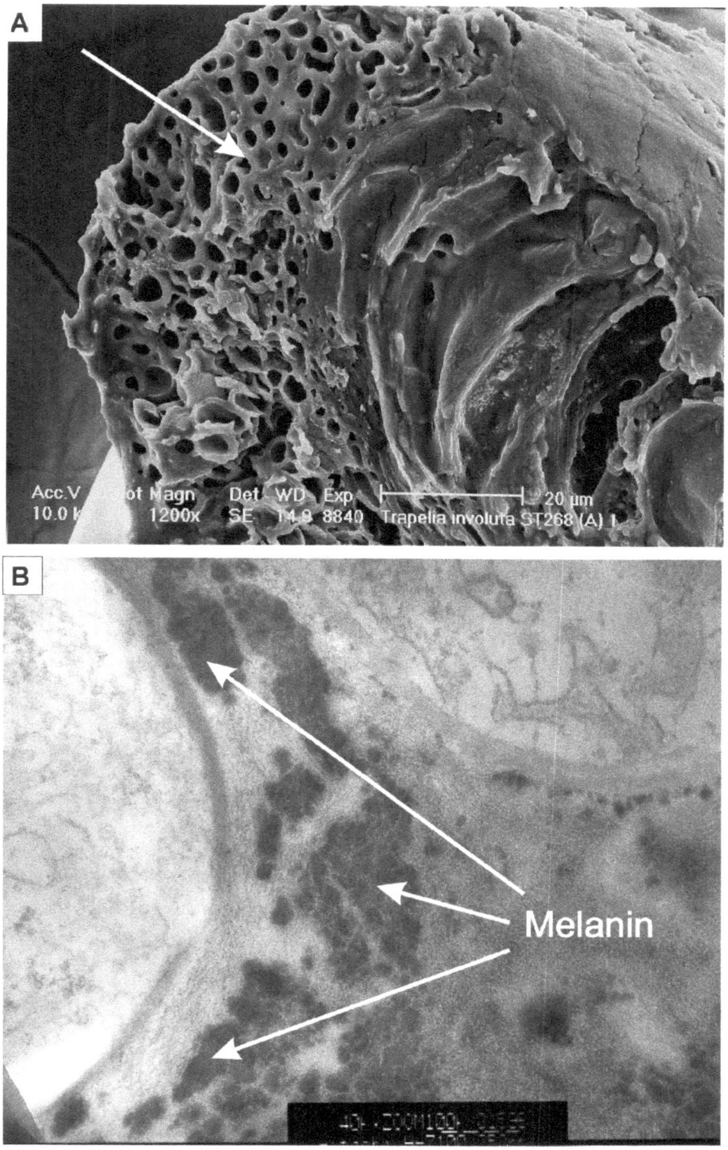

Figure 3 Secondary electron image using Hitachi S4500 FESEM showing strongly conglutinate, paraplectenchymatous, melanised exciple tissues with hyphal lumena in *Trapelia involuta* (Taylor) hertel growing on metatorbernite (see Figure 1b). (b) TEM image (Hitachi H-7100 TEM operating at 100 KV) of exciple showing extracellular melanin pigments. *T. involuta* samples were fixed with glutaraldehyde and OsO_4, embedded in an epoxy resin according to Kasama *et al.* (2001). After sectioning by ultramicrotomy, thin films were collected on Cu and Ni grids coated with collodion and C, and stained with uranyl acetate and lead citrate solutions to enhance contrast of the lichen's structures. Scale = 750 nm.

were replaced by those characteristics of acidified bark in a moderately polluted environment, including *Hypogymnia physodes*. Bark pH increased from 3.5 to 6.5 due to $CaCO_3$ near the source. Gilbert suggested that stimulation of the *Xanthorion* can be explained as an indirect pH effect, the range 6.0–7.7 being optimal for nitrification (which is non-existent below 3.7) and also enhanced phosphorus availability. He also drew attention to the availability of trace elements. Dust influences need not be negative. Alkaline dusts from urban demolition encourage lichen reinvasion of acidified substrates where formerly few or no species existed (Gilbert, 1992). In warmer drier Mediterranean areas dusts have a major influence on lichen community composition (Loppi and DeDominicis, 1996; Loppi and Pirintsos, 2000).

An elegant study concerned a copper mine in South Portugal, processing sulphide ore to produce Cu concentrates, where fugitive dust emissions from stockpiles and waste heaps were the major source of atmospheric pollution (Branquinho *et al.*, 1999; Branquinho, 2001). Field evidence suggested that fruticose lichens were most sensitive to the influence of Cu dusts, followed by foliose and crustose forms, supported by laboratory investigations (see below). Even amongst fruticose species, Branquinho established differences in sensitivity. She found that the capacity of *Ramalina fastigiata* to intercept Cu was approximately twice that of *Usnea* species due to a greater efficiency in intercepting particles, yet had a wider distribution, growing closer to the mine than did *Usnea* (Branquinho, 2001). Few other studies have reported changes in lichen community composition in response to increased levels of Cu and other metals (Folkeson and Anderssonbringmark, 1988; Loppi *et al.*, 1998; Branquinho, 2001).

Evidence for metal influences on epiphytic lichen abundance and diversity in regions remote from pollution sources comes from a series of studies carried out by Markus Hauck and colleagues investigating the relationship between lichen diversity on coniferous trees and bark and stemflow chemistry (Hauck *et al.*, 2002b; Hauck and Runge, 2002; Hauck, 2003; Hauck, 2005). Mn in tree bark is primarily soil derived, readily taken up by plant roots as Mn^{2+} via the transpiration flux in the stem and transported to all living tissues, including young foliage and to bark where it is stored to avoid toxic concentrations of free Mn^{2+} (Hauck and Paul, 2005). Solubility strongly increases with increasing acidity and water logging as insoluble Mn^{3+} and Mn^{4+} ions are converted into soluble Mn^{2+} in a reducing environment. Correlations between epiphytic lichen abundance and Mn supply were made in several coniferous forests in Germany and western and eastern North America (Hauck, 2003; Hauck and Paul, 2005). Decreasing cover values of the foliose lichen *Hypgogymnia physodes* with increasing Mn concentrations in bark or stemflow were repeatedly found in stands of *Picea abies* in the Harz Mountains, Germany. Bark concentrations were generally more important than stemflow, though in some cases Mn–Ca ratios were also considered important (Hauck *et al.*, 2002a). Leaching of Mn is strongly dependent on weather conditions and especially intense during snowmelt when the viscosity of stemflow and surface tension are high. On account of their mobility, Mn^{2+} ions are readily absorbed by tree canopies and much of the absorbed Mn remains on the canopy surface, especially in epicuticular waxes. Abundant evidence exist in several studies from different countries that foliar leaching of metals (Mn, Cs, Rb, Ba and Zn) strongly absorbed from the soil further enhance metal levels in epiphytic lichens and mosses (De Bruin and Hackenitz, 1986; Bargagli, 1990; Ceburnis and Steinnes, 2000; Loppi and Pirintsos, 2003; Hauck and Paul, 2005; Rusu *et al.*, 2006a). Microorganisms, especially bacteria, but also fungi, are known to catalyse Mn(II) oxidation. In the presence of oxygen and at high pH, Mn(III) and Mn(IV) phases predominate (Tebo *et al.*, 2004). Amongst the stronger oxidants found in the environment, Mn(III) and Mn(IV) phases react with the reduced forms of other metals (e.g. Se, As and Cr) thereby influencing toxic metal availability by oxidative precipitation or solubilisation (Tebo *et al.*, 2004). 40 km west of London, higher Mn concentrations were recorded in *Parmelia sulcata* than previously reported in this species (Purvis *et al.*, 2003, 2005, 2006). Recorded Pb concentrations were amongst the lowest recorded in specimens sampled over a 200-year period (Purvis *et al.*, 2007). A consequence of the success of measures to reduce emissions of heavy metals may be that labile elements like Mn more readily accumulate in lichens and mosses in the absence of competing protons and other cations (like Pb) with higher binding affinities (Purvis *et al.*, 2007). The potential for Mn species to influence lichen vitality in deciduous woodlands is yet to be established.

4. METAL UPTAKE AND EXTRACELLULAR LOCALISATION

Lichens accumulate metals principally by (i) metal sorption through ion-exchange processes according to the binding affinities of metals and the nature of organic ligands in the lichen thallus, (ii) intracellular accumulation through energy-dependent carrier-mediated systems, (iii) adsorption of particles and (iv) 'biomineral' or organo-metal complex formation (Brown, 1976, 1991; Nieboer *et al.*, 1978; Brown *et al.*, 1994; Richardson, 1995; Branquinho, 2001; Bargagli and Mikhailova, 2002; Haas and Purvis, 2006). The concept of avoidance as an adaptive biological mechanism allowing plants to resist or tolerate higher concentrations of metals than do non-adapted races/ecotypes of species is well established (Baker, 1981; Shaw, 1989; Macnair and Baker, 1994), though not yet addressed in lichens. How particles enter thalli is conjectural. Variations in surface morphology and higher surface to volume ratios may not always be the main factors explaining the ability for different species to capture and retain particles (Branquinho, 2001). Experimental research has been undertaken and widely reviewed confirming the dynamic nature of uptake and loss (Brown, 1976, 1991; Nieboer *et al.*, 1978; Nash, 1989; Richardson, 1995; Purvis, 1996; Purvis *et al.*, 2004b; Haas and Purvis, 2006).

Hyperaccumulating plants have the ability to take up extraordinary quantities of certain metal ions without succumbing to toxic effects (Brooks, 1983; Callahan *et al.*, 2006). Lange and Ziegler (1963) proposed that tolerance might involve transport of ions to regions external to the plasmalemma and the cell wall. Extracellular localisation of metal contaminants does appear to be a major reason explaining why lichens tolerate such high contents (Purvis, 1984; Wilson and

Jones, 1984; Sarret *et al.*, 1998; Nash, 1989), though there is a lack of experimental research. Reports of high metal contents, 'hyperaccumulation' (in excess of 1000 ppm) in lichens are normally due to the trapping of particulate matter and formation of biominerals. Possibly the highest copper concentrations (ca. 16% dry wt.) ever recorded were from a sample of *Acarospora rugulosa* Körber found growing on the mineral brochantite $Cu_4SO_4(OH)_6$ in Sweden corresponding to ca. 40% copper oxalate in an extracellular form (Chisholm *et al.*, 1987). Metal oxalate formation, widely reported in lichens (Jones *et al.*, 1981; Purvis, 1984; Wilson and Jones, 1984; Modenesi *et al.*, 1998; Giordani *et al.*, 2003; Pawlik-Skowrońska *et al.*, 2006) are potential extracellular metal detoxification mechanisms in lichens and non-lichenised fungi (Purvis, 1984; Wilson and Jones, 1984; Fomina *et al.*, 2006). Dihydrated oxalates of Mg, Fe^{2+}, Mn, Ni, Zn and Co are completely isomorphous with respect to each other and are likely to occur where lichens colonise suitable substrates (Wilson and Jones, 1984). Manganese oxalate was first reported in the saxicolous lichen *Pertusaria pseudocorallina* growing on cryptomelane (KMn_8O_{16}) containing discrete lamellar growths of hollandite ($BaMn_8O_{16}$) and nearly 1% Zn at a mine in Scotland (Wilson and Jones, 1984). Ferric oxalate was found in *Caloplaca callopisma* (Ascaso *et al.*, 1982) and magnesium oxalate (the mineral glushinskite) in *Lecanora* (*Tephromela*) *atra* (Wilson *et al.*, 1980). Zn and Pb were immobilised as extracellular oxalates on mycobiont hyphae in the hyperaccumulator *Diploschistes muscorum* (Sarret *et al.*, 1998). Studies by Hauck and co-workers confirmed that Mn^{2+} ions efficiently replace Ca^{2+} from extracellular binding sites under experimental conditions. Short-term exposure to Mn is therefore sufficient for the extracellular absorption of significant amounts of ions (Hauck and Paul, 2005). In the medulla, Mn was incorporated in Ca-oxalate crystals, especially on the surface of young growing hyphae. On a long-term basis, this is suspected to affect the integrity of the crystals, which fulfil important structural and physiological functions (Paul *et al.*, 2003). Manganese oxalate formation was suggested as a possible tolerance mechanism in *Hypogymnia physodes* transplanted into a forested site near Krakow, in Poland (Budka *et al.*, 2002). The biological basis for oxalate production remains poorly understood. Copper oxalate was not found in *Aspicilia mashiginenesis* in spite of growing on secondary copper minerals (Purvis, 1984). Biological control of the monohydrated whewellite (COM) and dihydrated weddellite (COD) in the upper cortex of lichens was attributed to the occurrence of polyuronic acid substances that strongly affects calcium oxalate mineralization (Giordani *et al.*, 2003). The occurrence of oxalates is, in most cases, very good evidence for biochemical weathering of a rock substrate, but cations may be introduced by runoff or even occult precipitation (Lee, 2000). As suggested by Lee, the location of oxalates within thalli may be a good guide to the origin of divalent cations (atmosphere *vs.* hydrosphere *vs.* substrate).

The first evidence for complexation of lichen acids with trace elements was for Cu. Atypically green-coloured (Cu-rich) morphs of crustose *Acarospora* A. Massal., *Buellia* De Not., *Bellemerea* Hafellner & Cl. Roux, *Lecidea* s.l. and *Lecanora* s.l. species, were found in C. Scandinavia, Germany, Greenland and UK containing the lichen substances norstictic and psoromic acids (Purvis, 1985; Purvis *et al.*, 1987, 1990).

Evidence for copper complexation was proposed on the basis of IR spectra of lichen samples and synthetic copper (II) complexes of lichen substances (Purvis, 1985; Purvis *et al.*, 1990; Takani *et al.*, 2002). *Lecanora cascadensis* H. Magn. and *L. polytropa*, although dark-green coloured in cupriferous habitats, lack norstictic and psoromic acids (Alstrup and Hansen, 1977) and remain to be investigated. A stable complex of Zn with diploschistesic and lecanoric acids was suggested in the lichen *Diploschistes muscorum* growing on calamine-rich soils in Belgium (Cuny *et al.*, 2004b). Pb complexation with the yellow anthraquinone, parietinic acid, was considered but rejected and complexation with carboxylic acid groups confirmed in the cell walls of *Xanthoria parietina* using EXAFS (extended X-ray absorption fine structure) spectroscopy (Sarret *et al.*, 1998). Complexation with lichen acids is a potential tolerance mechanism, though this remains to be tested experimentally.

Polysaccharides are widely considered important binding sites for metal sorption, yet less than 100 lichen species have been investigated for their polysaccharide content (Olafsdottir and Ingolfsdottir, 2001). Metal biosorption by melanin pigments (Figure 2b) was proposed as being responsible for high U, Cu and Fe concentrations in the outer wall of *Trapelia involuta* (McLean *et al.*, 1998; Purvis *et al.*, 2004a). This hypothesis is supported by field and laboratory research on non-lichenised fungi. Melanin from *Aureobasidium pullulans* can bind significant amounts of Cu^{2+} and Fe^{2+} (Gadd, 1984, 1993). Melanin formation is not restricted to metal stress. For instance, high radiation selection pressure following the Chernobyl accident led to genetic adaptation of fungal strains with a high frequency of melanised species (Zhdanova *et al.*, 2000; Fomina *et al.*, 2006).

Many particulates are stored within intercellular spaces of lax medullary tissues (Garty *et al.*, 1979), but how they reach these areas is not clear (Bargagli and Mikhailova, 2002). Metals are not necessarily retained by thalli, cations are replaced by those with higher binding affinities, elements like Pb by those of a different isotopic composition (Spiro *et al.*, 2004) and mineral particles by 'smelter-related particles' (Williamson *et al.*, 2004). Labile elements e.g. Rb, K, Na, Mn and Al tend to be lost from lichen thalli near strong chemical sources (Purvis *et al.*, 2004b, 2006; Williamson *et al.*, 2004). In many cases it is impossible to determine by wet analysis alone if these are from the lichen or from mineral particles associated with the thallus.

A range of other mycogenic secondary minerals can be associated with fungal hyphae and lichen thalli, including birnessite, desert varnish (MnO and FeO) (Fomina *et al.*, 2006). Precipitation, including crystallization will immobilize metals and limit bioavailability, as well as perhaps lead to release of nutrients like sulphate and phosphate (Fomina *et al.*, 2006). Secondary clay minerals formed within and beneath lichens reflect excretion of organic acids and acidic mucopolysaccharides by lichen hyphae and by microorganisms inhabiting their thalli (Barker and Banfield, 1996, 1998; Barker *et al.*, 1997). Discovery of amorphous silica (opal) and siliceous relics of phyllosilicates reported in several studies by Wilson and Jones at the Macaulay Institute, Scotland sparked controversy over their formation. Independent studies confirmed that these are biogenic in origin and not artefacts (Purvis, 1985, 2000; Lee, 2000; Haas and Purvis, 2006). Several crustose lichens are rust coloured, widely attributed to hydrated iron oxides i.e. 'rust' though few investigations have been carried out. Aluminium-containing goethite was determined in the rust-coloured species, *Tremolecia atrata* (Jones *et al.*, 1981). Schwertman and Taylor suggested that iron oxides and hydroxides associated with lichens are formed by oxidation of organic molecules that have complexed substrate-derived iron (Lee, 2000).

Lichen biomonitoring and multi-element analysis provide a practical method to detect metals and other elements present at low concentrations in the environment, especially those difficult to detect in samples of filtered air. For instance, increased Pt recorded in the saxicolous *Xanthoparmelia* analysed in Maricopa County, Arizona coincided with a dramatically increased traffic volumes and urbanisation (Nash *et al.*, 2003). REE monitoring was carried out using lichens with reference to different geological and environmental samples at Zlatna, Romania (Rusu *et al.*, 2006a) and the Orlovka-Spokoinoe Ta–Nb–Sn–W mining site and ore-processing complex in Eastern Transbaikalia (Russia) (Dolgopolova *et al.*, 2006). Results confirm efficient REE enrichment by epiphytic lichens relative to soils and preferential fixation (Rusu *et al.*, 2006a). REE patterns and elemental ratios may provide a reliable technique to trace dust and metals sources, dispersal and a new tool for environmental assessment studies (Dolgopolova *et al.*, 2006).

5. EFFECTS OF TRACE METALS AT THE CELLULAR LEVEL

Most studies on the effects of metals on lichens have been performed under controlled laboratory conditions, and only a few confirmed under field conditions (Branquinho, 2001). Few aspects of heavy metal resistance/tolerance have been studied (Sarrett *et al.*, 1998; Sanita di Toppi, 2005b; Pawlik-Skowrońska *et al.*, 2006). Evidence for intracellular storage of metals based on microscopy is much more limited in lichens compared with other organisms, however, heavy metals penetrate intracellularly both in mycobiont and photobiont cells. Cd and Cr in *Xanthoria parietina* were immobilised mainly by cell walls, but were detected also in concentric bodies and vacuoles of mycobiont and chloroplasts of photobiont (Sanita di Toppi *et al.*, 2004, 2005a). Mn accumulated as intracellular polyphosphate granules in the lower cortex of *Hypogymnia physodes* (Paul *et al.*, 2003) and U and P—bearing acicular nanocrystals within intracellular concentric bodies (proteinaceous organelles) and along cell walls within the extracellular gelatinous mucopolysaccharide matrix (Haas *et al.*, 1998; Suzuki and Banfield, 1999). Due to the higher mycobiont biomass in the lichen thallus (Collins and Farrar, 1978), this biont binds most metal supplied. A range of metal toxicity symptoms including loss of cell membrane integrity, potassium leakage, disruption to ultrastructure, chlorophyll degradation and oxidative stress have been reported in lichens (Puckett, 1976; Branquinho, 2001; Cuny *et al.*, 2004b; Paul *et al.*, 2004; Sanita di Toppi *et al.*, 2005b). However, their expression depends on the metal and lichen species concerned (Branquinho *et al.*, 1997; Hauck and Paul, 2005). Symbiotic photobionts are considered more sensitive to metals than their mycobionts, photobiont chlorophyll degradation being an obvious sign of damage (Branquinho, 2001; Garty, 2001; Bačkor and Vaczi, 2002; Sanita di Toppi *et al.*, 2005a; Garty *et al.*,

2006; Pawlik-Skowrońska *et al.*, 2006). Copper is widely known to cause cell membrane damage and adversely influence the photosynthetic apparatus of lichens (Puckett, 1976; Garty *et al.*, 2006) and also affects fungal and algal ultrastructure (Tarhanen, 1998). Using an elegant sequential elution technique to determine Cu compartmentalisation in different regions of the lichen thallus, Branquinho confirmed that *Ramalina fastigiata* was absent near a Portuguese copper mine when intracellular Cu concentrations exceeded a threshold of ca. 2.0 μmol/g (Branquinho *et al.*, 1999). *Usnea* species were determined to be more sensitive to Cu uptake than *Ramalina* in spite of the greater accumulation capacity of the latter. The study suggests that the chemical nature of the surface plays a major role in influencing particle uptake. Cellular responses to copper stress were investigated in *Lecanora polytropa* (Hoffm.) Rabenh. (Pawlik-Skowrońska *et al.*, 2006). Bright blue-green apothecia accumulated up to 1.3% Cu dry wt. (205 μmol Cu/g dry wt.), ca. 50% in an exchangeable form (see Figure 1.). Phytochelatin (PC) production and thiol peptide status (GSH/GSSG; reduced glutathione/ glutathione disulphide) were reported for the first time in a crustose lichen, representing two possible detoxification mechanisms in this Cu-tolerant species. PCs are cysteine-rich peptides derived from glutathione (GSH) that are synthesised in response to heavy metals to chelate them intracellularly, important constituents of the complex metal detoxification systems of higher plants, algae and some fungi (Zenk, 1996). Pawlik-Skowrońska *et al.* (2002) found that Cd, Pb and Zn in the laboratory induced biosynthesis of PCs (PC2, PC3 and PC4) and their des-Gly derivatives in the widespread epiphytic lichens *Xanthoria parietina*, *Physconia grisea* and *Physcia adscendens*. Only the photobiont partners (*Trebouxia*) were capable of PC production, aposymbiotically-grown mycobionts only producing GSH. The lack of PC production was also recently reported in apo-mycobiont *Cladonia cristatella* exposed to Cu (Bačkor *et al.*, 2006). In the Cu-tolerant *L. polytropa* short-chained PCs (PC2-PC3) occurred in an oxidised form (Pawlik-Skowrońska *et al.*, 2006), unlike the previous, where they were produced by photobionts in lichens subjected to Cd, Zn and Pb. It strongly suggests that –SH PC groups in photobiont were involved in Cu detoxification by reduction of Cu^{2+} to less toxic Cu^{1+}.

GSH, the precursor of PCs, is the principal low-molecular thiol and non-enzymatic antioxidant in both lichen symbionts (Kranner *et al.*, 2005). It plays a critical role in cellular defence against oxidative damage caused also by heavy metals. In the Cu-rich native *L. polytropa*, GSH occurred in a reduced form suggesting that no serious oxidative stress had occurred (Pawlik-Skowrońska *et al.*, 2006). Only at very high Cu^{2+} concentrations, destructive for photobiont pigments did partial oxidation of GSH to GSSG occur. As reported by Cuny *et al.* (2004b), GSH in the terricolous lichen *D. muscorum* living in metalliferous sites also occurred mostly in the reduced form and in higher concentrations than in the same lichen species living in unpolluted sites. In contrast, in *X. parietina* collected from an unpolluted area, high concentrations of Cd and Cr caused a decrease in GSH level (Sanita di Toppi *et al.*, 2004, 2005b). The ability to keep high level of reduced GSH seems to be one of the essential mechanisms of metal resistance in lichens.

Copper complexation by low molecular mass organic acids and non-protein thiols do not entirely account for the observed Cu-tolerance in *L. polytropa*

(Pawlik-Skowrońska *et al.*, 2006). Other tolerance mechanisms, such as Cu binding by metallothioneins reported in non-lichenised fungi (Kameo *et al.*, 2000), must also be considered. Lichenised fungi are rarely found in nature without their photobionts. This suggests that Trebouxioid lichens may gain an ecological advantage through avoidance of heavy metal stress by PC synthesis. However, total PC and desGly-PC concentrations were much lower than similar derivatives synthesised in other free-living chlorophycean microalgae in response to the same Cd and Pb concentrations, suggesting that the availability of metals to photobiont cells is limited as compared to free-living algal cells exposed to similar metal concentrations (Pawlik-Skowrońska, 2000, Pawlik-Skowrońska *et al.*, 2002). Photobionts are to a great extent protected from toxic metal invasion by the mycobiont partner, which takes up considerably more metals than do algal cells (Paul *et al.*, 2003). Hydrophobins, water-repellent proteins produced only by the mycobiont (Wessels, 2000) may further protect mycobiont from intracellular metal uptake.

Antioxidant mechanisms involving GSH and superoxide dismutase (SOD) reduced the effects of very high Cd, Pb and Zn soil concentrations in the terricolous lichen *Diploschistes muscorum* (Scop.) R. Sant. (Cuny *et al.*, 2004b). Increased levels of ascorbic acid (AA) and activity of guaiacol peroxidase were observed in the Cd- and Cr-exposed *X. parietina* (Sanita di Toppi *et al.*, 2004, 2005b). AA, a non-enzymatic antioxidant, produced under oxidative stress in lichens (Caviglia and Modenesi, 1999) acts probably also as a precursor of oxalate. GSH and AA present in aqueous cell compartments create a reducing power used in radical and non-radical redox reactions (Cuny *et al.*, 2002). Secondary metabolites of lichens seem to play essential role in heavy metal detoxification by means of toxic metal complexation (Purvis *et al.*, 1990; Takani *et al.*, 2002) and antioxidative activity. Increased level of physodalic acid was found in *H. physodes* transplanted to highly metal-polluted sites (Bialonska and Dayan, 2005). Secondary metabolites (lichenic acids, orcinol-type depsides and tridepsides) produced by lichen mycobionts may provide protection against lipid peroxidation by blocking toxic metals that might otherwise initiate free radical reactions and by direct free-radical scavenging (Stepanenko *et al.*, 2002; Behera *et al.*, 2005). Both mycobionts and photobionts of lichens possess defence mechanisms to avoid HM toxicity. However, it is still unclear why some lichens are better adapted to metal polluted environments than others. Lower metal (Cu) and higher proline contents were recently found in the Cu-resistant lichen apophotobiont *Trebouxia erici* (Bačkor *et al.*, 2003, 2004).

6. CONCLUSIONS

Lichen–metal interactions continue to stimulate research across diverse disciplines since the pioneering studies of Lounamaa in Finland 50 years ago (Lounamaa, 1956). Many 'new' elements are entering our environment through technological innovations and from natural sources. These include the rare earth elements (REE)—yttrium, lanthanum and lanthanides from cerium to lutetium, the platanoids (Pt, Pd, Rh, etc.) and other rare elements. The environmental fate and consequences of these elements are virtually unknown (Rusu *et al.*, 2006a). Acid deposition throughout the 20th century has resulted in severe acidification of woodland soils, leading to base-cation depletion and mobilisation of Mn, Fe and Al through clay weathering, enhancing bioavailability for these elements and higher plant-tissue concentrations (Blake *et al.*, 1999). The consequences for different lichen species remain to be determined. Major advances in microscopical, analytical and sample preparation techniques greatly facilitate our understanding of lichen–metal interactions (Wierzchos and Ascaso, 1994; Richardson, 1995; Sarret *et al.*, 1998; Williamson *et al.*, 1998; Lee, 1999; Adamo and Violante, 2000). Study of lichens in novel mineralogical and geochemical environments will yield new data concerning adaptation, tolerance and the fate of potential environmental contaminants. Modern analytical systems for multi-element determination at trace and ultratrace levels in parallel with SEM/electron probe microanalysis and biochemical methods provide valuable new opportunities to understand accumulation and their influence on lichens.

ACKNOWLEDGEMENTS

BPS gratefully acknowledges a Royal Society study visit to UK and two Royal Society Joint Collaborative Joint Project Grants for Sweden (Is/ESEP/JP) with Mats Wedin and Anna Crewe and for Russia (gf/fSU/JP) with Irina Mikhailova. We are grateful to Professor G.M. Gadd for a British Mycological Society travel award for OWP to attend the 'BMS Stress 2006 Meeting'. We thank H. Taylor for photographing *Trapelia involuta*, T. Ohnuki for arranging embedding for TEM in Japan and A. Ball for assistance with TEM operation at NHM. We are grateful to Rosmarie Honegger for allowing us to adapt Figure 15.1 in Honegger (1997), to Alan Orange for permission to include his drawing of *Physconia* rhizines, Takani Masako for the proposed Cu-norstitic acid diagram in Figure1. The authors dedicate this paper to the late Dr O.L. Gilbert for his inspiration and kindness.

REFERENCES

Adamo, P. and Violante, P. (1991). Weathering of volcanic rocks from Mt Vesuviusassociated with the lichen *Stereocaulon vesuvianum*. *Pedobiologia*, **35**, 209–217.

Adamo, P. and Violante, P. (2000). Weathering of rocks and neogenesis of minerals associated with lichen activity. *Applied Clay Science*, **16**, 229–256.

Adamo, P., Crisafulli, S., Giordano, S., Minganti, V., Modenesi, P., Monaci, F., Pittao, E., Tretiach, M. and Bargagli, R. (2007). Lichen and moss bags as monitoring devices in urban areas. Part II: Trace element content in living and dead biomonitors and comparison with synthetic materials. *Environmental Pollution*. **146**, 392–399.

Alstrup, V. and Hansen, E.S. (1977). Three species of lichens tolerant of high concentrations of copper. *Oikos*, **29**, 290–293.

Antonovics, J., Bradshaw, A.D. and Turner, R.G. (1971). Heavy metal tolerance in plants. *Advanced Ecological Research*, **7**, 1–85.

Aptroot, A. and van den Boom, P.P.G. (1998). *Pyrenocollema chlorococcum*, a new species with a chlorococcoid photobiont from zinc-contaminated soils and wood. *Cryptogamie Bryologie Lichenologie*, **19**, 193–196.

Ascaso, C., Galvan, J. and Ortega, C. (1982). The weathering of calcareous rocks by lichens. *Pedobiologia*, **24**, 219.

Bačkor, M., Fahselt, D., Davidson, R.D. and Wu, C.T. (2003). Effects of copper on wild and tolerant strains of the lichen photobiont *Trebouxia erici* (Chlorophyta) and possible tolerance mechanisms. *Archives of Environmental Contamination and Toxicology*, **45**, 159–167.

Bačkor, M., Fahselt, D. and Wu, C.T. (2004). Free proline content is positively correlated with copper tolerance of the lichen photobiont *Trebouxia erici* (Chlorophyta). *Plant Science*, **167**, 151–157.

Bačkor, M., Pawlik-Skowrońska, B., Tomko, J., Budova, J. and Sanita di Toppi, L. (2006). Response to copper stress in aposymbiotically grown lichen mycobiont *Cladonia cristatella* Tuck.: Uptake, viability, ergosterol and production of non-protein thiols. *Mycological Research*, **110**, 994–999.

Bačkor, M. and Vaczi, P. (2002). Copper tolerance in the lichen photobiont *Trebouxia erici* (Chlorophyta). *Environmental and Experimental Botany*, **48**, 11–20.

Baker, A.J.M. (1981). Accumulators and excluders—Strategies in the response of plants to heavy-metals. *Journal of Plant Nutrition*, **3**, 643–654.

Bargagli, R. (1990). Assessment of metal air pollution by epiphytic lichens: The incidence of crustal materials and of the possible uptake from substrate barks. *Studia geobotanica*, **10**, 97–103.

Bargagli, R. (1998). *Trace Elements in Terrestrial Plants: An Ecophysiological Approach to Biomonitoring and Biorecovery.* Springer, Berlin.

Bargagli, R. and Mikhailova, I. (2002). Accumulation of inorganic contaminants. In: *Monitoring with Lichens—Monitoring Lichens* (P.L. Nimis, C. Scheidegger and P.A. Wolseley, eds.), pp. 65–84. Kluwer Academic, Dordrecht.

Bargagli, R. and Nimis, P.L. (2002). Guidelines for the use of epiphytic lichens as biomonitors of atmospheric deposition of trace elements. In: *Monitoring with Lichens—Monitoring Lichens* (P.L. Nimis, C. Scheidegger and P.A. Wolseley, eds.), pp. 295–299. Kluwer Academic, Dordrecht.

Barker, W.W. and Banfield, J.F. (1996). Biologically versus inorganically mediated weathering reactions: Relationships between minerals and extracellular microbial polymers in lithobiontic communities. *Chemical Geology*, **132**, 55–69.

Barker, W.W. and Banfield, J.F. (1998). Zones of chemical and physical interaction at interfaces between microbial communities and minerals: A model. *Geomicrobiology Journal*, **15**, 223–244.

Barker, W.W., Welch, S.A. and Banfield, J.F. (1997). *Biogeochemical weathering of silicate minerals.* In: *Geomicrobiology: Interactions between Microbes and Minerals.* Reviews in Mineralogy, **35**, 391–428. Mineralogical Society of America, Washington.

Barkman, J.J. (1958). *Phytosociology and Ecology of Cryptogamic Epiphytes.* Van Gorcum, Assen, Netherlands.

Bates, J.W., Bell, J.N.B. and Massara, A.C. (2001). Loss of *Lecanora conizaeoides* and other fluctuations of epiphytes on oak in SE England over 21 years with declining SO_2 concentrations. *Atmospheric Environment*, **35**, 2557–2568.

Beck, A. (1999). Photobiont inventory of a lichen community growing on heavy-metal-rich rock. *The Lichenologist*, **31**, 501–510.

Behera, B.C., Verma, N., Sanone, A. and Makhija, U. (2005). Evaluation of antioxidant potential of the cultured mycobiont of a lichen *Usnea ghattensis. Phytotherapy Research*, **19**, 58–64.

Bell, J.N.B. and Treshow, M. (2002). *Air Pollution and Plant Life*, 2nd edn. Chichester, Wiley.

Bialonska, D. and Dayan, F.E. (2005). Chemistry of the lichen *Hypogymnia physodes* transplanted to an industrial region. *Journal of Chemical Ecology*, **31**, 2979–2991.

Blake, L., Goulding, K.W.T., Mott, C.J.B. and Johnston, A.E. (1999). Changes in soil chemistry accompanying acidification over more than 100 years under woodland and grass at Rothamsted Experimental Station, UK. *European Journal of Soil Science*, **50**, 401–412.

Branquinho, C. (2001). Lichens. In: *Metals in the Environment: Analysis by Biodiversity* (M.N.V. Prasad, ed.), pp. 117–115. Marcel Dekker, New York.

Branquinho, C., Brown, D.H. and Catarino, F. (1997). The cellular location of Cu in lichens and its effects on membrane integrity and chlorophyll fluorescence. *Environmental and Experimental Botany*, **38**, 165–179.

Branquinho, C., Catarino, F., Brown, D.H., Pereira, M.J. and Soares, A. (1999). Improving the use of lichens as biomonitors of atmospheric metal pollution. *Science of the Total Environment*, **232**, 67–77.

Brooks, R.R. (1983). Biological Methods of Prospecting for Minerals. Wiley, New York.

Brown, D.H. (1976). 17. Mineral Uptake by Lichens. In: *Lichenology: Progress and Problems* (D.H. Brown, D.L. Hawksworth and R.H. Bailey, eds.), pp. 419–439. Academic Press, London.

Brown, D.H. (1991). Lichen mineral studies—Currently clarified or confused. *Symbiosis*, **11**, 207–223.
Brown, D.H., Avalos, A., Miller, J.E. and Bargagli, R. (1994). Interactions of lichens with their mineral environment. *Cryptogamic Botany*, **4**, 135–142.
Budka, D., Przybylowicz, W.J., Mesjasz-Przybylowicz, J. and Sawicka-Kapusta, K. (2002). Elemental distribution in lichens transplanted to polluted forest sites near Krakow (Poland). *Nuclear Instruments & Methods in Physics Research Section B- Beam Interactions with Materials and Atoms*, **189**, 499–505.
Callahan, D.L., Baker, A.J.M., Kolev, S.D. and Wedd, G. (2006). Metal ion ligands in hyperaccumulating plants. *Journal of Biological Inorganic Chemistry*, **11**, 2–12.
Caviglia, A.M. and Modenesi, P. (1999). Oxidative stress and ascorbic acid contents in *Parmotrema reticulatum* and *Parmelia sulcata* thalli. *The Lichenologist*, **31**, 105–110.
Ceburnis, D. and Steinnes, E. (2000). Conifer needles as biomonitors of atmospheric heavy metal deposition: Comparison with mosses and precipitation, role of the canopy. *Atmospheric Environment*, **34**, 4265–4271.
Chisholm, J.E., Jones, G.C. and Purvis, O.W. (1987). Hydrated copper oxalate, moolooite, in lichens. *Mineralogical Magazine*, **51**, 715–718.
Church, J.M., Coppins, B.J., Gilbert, O.L., James, P.W. and Stewart, N.F. (1996). *Red Data Books of Britain and Ireland: Lichens. Volume 1.* Joint Nature Conservation Committee, Peterborough, Britain.
Collins, C.R. and Farrar, J.F. (1978). Structural resistance to mass transfer in the lichen *Xanthoria parietina. New Phytologist*, **81**, 71–83.
Coppins, B.J. and Purvis, O.W. (1987). A review of *Psilolechia. The Lichenologist*, **19**, 29–42.
Cuny, D., Denayer, F.O., de Foucault, B., Schumacker, R., Colein, P. and Van Haluwyn, C. (2004a). Patterns of metal soil contamination and changes in terrestrial cryptogamic communities. *Environmental Pollution*, **129**, 289–297.
Cuny, D., Pignata, M.L., Kranner, I. and Beckett, R. (2002). Biomarkers of pollution-induced oxidative stress and membrane damage in lichens. In: *Monitoring with Lichens—Monitoring Lichens* (P.L Nimis, C. Scheidegger and P.A. Wolseley, eds.), pp. 11–14. Kluwer, Academic, Dordrecht.
Cuny, D., Van Haluwyn, C., Shirali, P., Zerimech, F., Jerome, L. and Haguenoer, J.M. (2004b). Cellular impact of metal trace elements in terricolous lichen *Diploschistes muscorum* (Scop.) R. Sant.—Identification of oxidative stress biomarkers. *Water Air and Soil Pollution*, **152**, 55–69.
Czehura, S.J. (1977). A lichen indicator of copper mineralization, Lights Creek District, Plumas County, California. *Economic Geology*, **72**, 796–803.
Davies, B.E. (1994). Soil chemistry and bioavailability with special reference to trace elements. In: *Plants and the Chemical Elements. Biochemistry, Uptake, Tolerance and Toxicity*, (M.E. Farago, ed.), pp. 1–3. VCH, Weinheim.
Davies, L., Bates, J.W., Bell, J.N.B., James, P.W. and Purvis, O.W. (2007). Diversity and sensitivity of epiphytes to oxides of nitrogen in London. *Environmental Pollution*. **146**, 299–310.
Davison, A.W. and Barnes, J.D. (2002). Air pollutant-abiotic stress interactions. In: *Air Pollution and Plant Life* (J.N.B. Bell and M. Treshow, eds.), pp. 359–377. John Wiley, Chichester.
De Bruin, M. and Hackenitz, E. (1986). Trace element concentrations in epiphytic lichens and bark substrate. *Environmental Pollution*, **11**, 153–160.
Dolgopolova, A., Weiss, D.J., Seltmann, R. and Dulski, P. (2006). Dust dispersal and Pb enrichment at the rare-metal Orlovka-Spokoinoe mining and ore processing site: Insights from REE patterns and elemental ratios. *Journal of Hazardous Materials*, **132**, 90–97.
Easton, E.M. (1994). Lichens and rocks—A review. *Geoscience Canada*, **21**, 59–76.
Elix, J.A. (1996). Biochemistry and secondary metabolites. In: *Lichen Biology* (T.H. Nash, ed.), pp. 154–180. Cambridge University Press, Cambridge.
Ernst, G. (1995). *Vezdaea leprosa*—Spezialist am Strassenrand. *Herzogia*, **11**, 175–188.
Ernst, W.H.O. (1989). Mine vegetation in Europe. In: *Heavy Metal Tolerance in Plants: Evolutionary Aspects* (A.J. Shaw, ed.), pp. 21–37. CRC Press, Boca Raton.
Farago, M.E. (1994). Plants and the Chemical Elements. Biochemistry, Uptake, Tolerance and Toxicity. VCH, Weinheim.
Farmer, A.M. (1993). The effects of dust on vegetation—a review. *Environmental Pollution*, **79**, 63–75.
Farmer, A. (2002). Effects of particulates. In *Air Pollution and Plant Life*, (J.N.B. Bell and M. Treshow, eds.), pp. 186–199. John Wiley, Chichester.

Farmer, A.M., Bates, J.W. and Bell, J.N.B. (1991). Comparisons of three woodland sites in NW Britain differing in richness of the epiphytic *Lobarion pulmonariae* community and levels of wet acidic deposition. *Holarctic Ecology*, **14**, 85–91.

Folkeson, L. and Anderssonbringmark, E. (1988). Impoverishment of vegetation in a coniferous forest polluted by copper and zinc. *Canadian Journal of Botany*, **66**, 417–428.

Fomina, M.A., Burford, E.P. and Gadd, G.M. (2006). Fungal dissolution and transformation of minerals: Significance for nutrient and metal mobility. In: *Fungi in Biogeochemical Cycles*, (G.M. Gadd, ed.), pp. 236–266. Cambridge University Press, Cambridge.

Gadd, G.M. (1984). Effect of copper on *Aureobasidium pullulans* in solid medium: Adaptation not necessary for tolerant behaviour. *Transactions of the British Mycological Society*, **82**, 546–549.

Gadd, G.M. (1993). Tansley Review No. 47. Interactions of fungi with toxic metals. *New Phytologist*, **124**, 25–60.

Gadd, G.M., (ed.) (2006). *Fungi in Biogeochemical Cycles*. Cambridge University Press, Cambridge.

Garty, J. (1993). Lichens as biomonitors for heavy metal pollution. In: *Plants as Biomonitors. Indicators for Heavy Metals in the Terrestrial Environment* (B. Markert, ed), pp. 193–263. VCH, Weinheim.

Garty, J. (2001). Biomonitoring atmospheric heavy metals with lichens: Theory and application. *Critical Reviews in Plant Sciences*, **20**, 309–371.

Garty, J., Galun, M. and Kessel, M. (1979). Localization of heavy metals and other elements accumulated in the lichen thalli. *New Phytologist*, **82**, 159–168.

Garty, J., Tamir, O., Levin, T. and Lehr, H. (2007). The impact of UV-B and sulphur- or copper-containing solutions in acidic conditions on chlorophyll fluorescence in selected *Ramalina* species. *Environmental Pollution*, **145**, 266–273.

Gauslaa, Y. (1985). The ecology of *Lobarion pulmonariae* and *Parmelion caperatae* in *Quercus* dominated forests in south-west Norway. *The Lichenologist*, **17**, 117–140.

Gauslaa, Y. and Holien, H. (1998). Acidity of boreal *Picea abies*-canopy lichens and their substratum, modified by local soils and airborne acidic depositions. *Flora*, **193**, 249–257.

Gilbert, O.L. (1969). The effect of SO_2 on lichens and bryophytes around Newcastle upon Tyne. In: *Air Pollution, Proceedings of the First European Congress on the Influence of Air Pollution on Plants and Animals, April 22–27, 1968*, pp. 223–233. Centre for Agricultural Publishings and Documentation, Wageningen.

Gilbert, O.L. (1971). The effect of airborne fluorides on lichens. *The Lichenologist*, **5**, 26–32.

Gilbert, O.L. (1973). The effect of airborne fluorides. In: *Air Pollution and Lichens* (B.W. Ferry, M.S. Baddeley and D.L. Hawksworth, eds.), pp. 176–191. Athlone Press of the University of London, London.

Gilbert, O.L. (1976). An alkaline dust effect on epiphytic lichens. *The Lichenologist*, **8**, 173–178.

Gilbert, O.L. (1980). Effect of land-use on terricolous lichens. *The Lichenologist*, **12**, 117–124.

Gilbert, O.L. (1990). The lichen flora of urban wasteland. *The Lichenologist*, **22**, 87–101.

Gilbert, O.L. (1992). Lichen reinvasion with declining air pollution. In: *Bryophytes and Lichens in a Changing Environment* (J.W. Bates and A.M. Farmer, eds.), pp. 159–177. Clarendon Press, Oxford.

Gilbert, O.L. (2000). *Lichens*. Harper Collins, London.

Giordani, P., Modenesi, P. and Tretiach, M. (2003). Determinant factors for the formation of the calcium oxalate minerals, weddellite and whewellite, on the surface of foliose lichens. *The Lichenologist*, **35**, 255–270.

Goldberg, E.D. (1976). Rock volatility and aerosol composition. *Nature*, **260**, 128–129.

Gorbushina, A.A. (2006). Chapter 11. Fungal activities in subaerial rock-inhabiting microbial communities. In: *Fungi in Biogeochemical Cycles* (G.M. Gadd, ed.), pp. 267–288. Cambridge University, Cambridge.

Goyal, R. and Seaward, M.R.D. (1982). Metal uptake in terricolous lichens III. Translocation in the thallus of *Peltigera canina*. *New Phytologist*, **90**, 85–90.

Grime, J.P. (1979). *Plant Strategies and Vegetation Processes*. John Wiley, Chichester.

Haas, J.R., Bailey, E.H. and Purvis, O.W. (1998). Bioaccumulation of metals by lichens: Uptake of aqueous uranium by *Peltigera membranacea* as a function of time and pH. *American Mineralogist*, **83**, 1494–1502.

Haas, J.R. and Purvis, O.W. (2006). Chapter 15. Lichen Biogeochemistry. In: *Fungi in Biogeochemical Cycles* (G.M. Gadd, ed.), pp. 344–376. Cambridge University, Cambridge.

Hauck, M. (2003). Epiphytic lichen diversity and forest dieback: The role of chemical site factors. *Bryologist*, **106**, 257–269.
Hauck, M. (2005). Epiphytic lichen diversity on dead and dying conifers under different levels of atmospheric pollution. *Environmental Pollution*, **135**, 111–119.
Hauck, M., Hesse, V. and Runge, M. (2002a). Correlations between the Mn/Ca ratio in stemflow and epiphytic lichen abundance in a dieback-affected spruce forest of the Harz Mountains, Germany. *Flora*, **197**, 361–369.
Hauck, M., Hesse, V. and Runge, M. (2002b). The significance of stemflow chemistry for epiphytic lichen diversity in a dieback-affected spruce forest on Mt Brocken, northern Germany. *The Lichenologist*, **34**, 415–427.
Hauck, M. and Paul, A. (2005). Manganese as a site factor for epiphytic lichens. *The Lichenologist*, **37**, 409–423.
Hauck, M. and Runge, M. (2002). Stemflow chemistry and epiphytic lichen diversity in dieback-affected spruce forest of the Harz Mountains, Germany. *Flora*, **197**, 250–261.
Hawksworth, D.L. (1988). The variety of fungal-algal symbioses, their evolutionary significance, and the nature of lichens. *Botanical Journal of the Linnean Society*, **96**, 3–20.
Hawksworth, D.L. (1991). The fungal dimension of biodiversity: Magnitude, significance, and conservation. *Mycological Research*, **95**, 64–655.
Hawksworth, D.L. and McManus, P.M. (1989). Lichen recolonisation in London under conditions of rapidly falling sulphur dioxide levels, and the concept of zone skipping. *Botanical Journal of the Linnean Society*, **100**, 99–109.
Hay, R.W. (1994). Plant metalloenzymes. In: *Metal-Tolerant Plants: An Evolutionary Perspective* (M.E. Farago, ed.), pp. 107–148. VCH, Weinheim.
Honegger, R. (1997). Metabolic interactions at the mycobiont-photobiont interface in lichens. In: *The Mycota, Vol 5A, Plant Relationships, Part A.* (G.C. Carroll and P. Tudzynski, eds.), pp. 209–221. Springer-Verlag, New York.
Huneck, S. (2004). The significance of lichens and their metabolites. *Naturwissenschaften*, **86**, 559–570.
Huneck, S. and Yoshimura, I. (1996). *Identification of Lichen Substances.* Springer, Berlin.
Isocrono, D., Matteucci, E., Ferrarese, A., Pensi, E. and Piervittori, R. (2007). Lichen colonization in the city of Turin (N Italy) based on current and historical data. *Environmental Pollution*, **145**, 258–265.
Jahns, H. (1988). The establishment, individuality and growth of lichen thalli. *Botanical Journal of the Linnean Society*, **96**, 21–29.
James, P.W., Hawksworth, D.L. and Rose, F. (1977). Lichen communities in the British Isles: A preliminary conspectus. In: *Lichen Ecology* (M.R.D Seaward, ed.), pp. 295–413. Academic Press, London.
Jenkins, D.A., Johnson, A.H. and Freeman, C. (2000). Mynydd Parys Cu-Pb-Zn Mines: Mineralogy, microbiology and acid mine drainage. In: *Environmental Mineralogy: Microbial Interactions, Anthropogenic Influences, Contaminated Land and Waste Management.* (J.D. Cotter-Howells, L.S. Campbell, E. Valsami-Jones and M. Batchelder, eds.), pp. 161–179. Mineralogical Society of Great Britain and Ireland, London.
Jeran, Z., Byrne, A.R. and Batic, F. (1995). Transplanted epiphytic lichens as biomonitors of air—Contamination by natural radionuclides around the Zirovski Vrh Uranium-Mine, Slovenia. *The Lichenologist*, **27**, 375–385.
Jeran, Z., Mrak, T., Jaćimovića, R., Batič, F., Kastelec, D., Mavsar, R. and Simončič, P. (2007). Epiphytic lichens as biomonitors of atmospheric pollution in Slovenian forests. *Environmental Pollution*, **146**, 324–331.
Jones, D. (1988). Lichens and pedogenesis. In: *Handbook of Lichenology.* (M. Galun, ed.), Vol. 3, pp. 109–124. CRC, Boca Raton.
Jones, D., Wilson, M.J. and McHardy, W.J. (1981). Lichen weathering of rock-forming minerals—Application of scanning electron-microscopy and micro-probe analysis. *Journal of Microscopy*, **124**, 95–104.
Kameo, S., Iwahashi, H., Kojima, Y. and Satoh, H. (2000). Induction of metallothioneins in the heavy metal resistant fungus *Beauveria bassiana* exposed to copper or cadmium. *Analysis*, **28**, 382–385.

Kasama, T., Murakami, T., Ohnuki, T. and Purvis, O.W., (2001). Effects of lichens on uranium migration. In: *Scientific Basis for Nuclear Waste Management XXIV* (K.P. Hart and G.R. Lumpkin, eds.), Vol. 663, pp. 683–690. Materials Research Society, Pittsburgh, PA.

Knops, J.M.H., Nash, T.H., Boucher, V.L. and Schlesinger, W.H. (1991). Mineral cycling and epiphytic lichens — Implications at the ecosystem level. *The Lichenologist*, **23**, 309–321.

Kosman, D.J. (2004). Transition metal ion uptake in yeasts and filamentous fungi. In: *Metal Ions in Fungi* (G. Winkelmann and D.R. Winge, eds.), pp. 1–38. Marcel Dekker Inc., New York.

Kranner, I., Cram, W.J., Zorn, M., Wornik, S., Yoshimura, I., Stabentheiner, E. and Pfeifhofer, H.W. (2005). Antioxidant and photoprotection in a lichen as compared with its isolated symbiont partners. *Proceedings of the National Academy of Sciences of the United States of America*, **102**, 3141–3146.

Laaksovirta, K. and Olkkonen, H. (1977). Epiphytic lichen vegetation and element contents of *Hypogymnia physodes* and pine needles examined as indicators of air pollution at Kokkola, W Finland. *Annales Botanici Fennici*, **14**, 112–130.

Lange, O.L. (1992). Pflanzenleben unter Streß. Flechten als Pioniere der Vegetation am Extremstandorten der Erde. University of Würzburg, Würzburg.

Lange, O.L. and Ziegler, H. (1963). Der schwermetallgehalt von Flechten aus dem Acarosporetum sinopicae auf erzschlackenhalden des Harzes. I. Eisen und Kupfer. *Mitteilungen der Floristisch-soziologischen ArbeitsGemeinschaft*, **10**, 156–183.

Larsen, R.S., Bell, J.N.B., James, P.W., Chimonides, P.J., Rumsey, F.J., Tremper, A. and Purvis, O.W. (2007). Lichen and bryophyte distribution on oaks in London in relation to air pollution and bark acidity. *Environmental Pollution*, **146**, 332–340.

LeBlanc, F., Rao, D.N. and Commeau, G. (1972). The epiphytic vegetation of *Populus balsamifera* and its significance as an air pollution indicator in Sudbury, Ontario. *Canadian Journal of Botany*, **50**, 519.

Lee, M. (2000). Weathering of rocks by lichens In: *Environmental Mineralogy: Anthropogenic Influences, Contaminated Land and Waste Management*, (J.D. Cotter-Howells, L.S. Campbell, E. Valsami-Jones and M. Batchelder, eds.), pp. 77–115. Mineralogical Society of Great Britain and Ireland, London.

Lee, M. R. (1999). Organic-mineral interactions studied by controlled pressure SEM. *Microscopy and Analysis*, **70**, 9–11.

Le Roy, L.W. and Koksoy, M. (1962). The lichen—A possible plant medium for mineral exploration. *Economic Geology*, **57**, 107–111.

Loppi, S. and DeDominicis, V. (1996). Effects of agriculture on epiphytic lichen vegetation in Central Italy. *Israel Journal of Plant Sciences*, **44**, 297–307.

Loppi, S. and Pirintsos, S.A. (2000). Effect of dust on epiphytic lichen vegetation in the Mediterranean area (Italy and Greece). *Israel Journal of Plant Sciences*, **48**, 91–95.

Loppi, S. and Pirintsos, S.A. (2003). Epiphytic lichens as sentinels for heavy metal pollution at forest ecosystems (central Italy). *Environmental Pollution*, **121**, 327–332.

Loppi, S., Putorti, E., Signorini, C., Fommei, S., Pirintsos, S.A. and De Dominicis, V. (1998). A retrospective study using epiphytic lichens as biomonitors of air quality: 1980 and 1996 (Tuscany, central Italy). *Acta Oecologica. International Journal of Ecology*, **19**, 405–408.

Lounamaa, J. (1956). Trace elements in plants growing wild on different rocks in Finland. A semi-quantitative spectrographic study. *Annales Botanici Societatis Zoologicae Botanicae Fennicae 'Vanamo'*, **29**, 1–196.

Macnair, M.R. and Baker, A.J.M. (1994). Metal-tolerant plants: An evolutionary perspective, in plants and the chemical elements. In: *Biochemistry, Uptake, Tolerance and Toxicity* (M.E. Farago, ed.), pp. 31–66. VCH, Weinheim.

Markert, B. (1993). *Plants as Biomonitors. Indicators for Heavy Metals in the Terrestrial Environment*. VCH, Weinheim.

McLean, J., Purvis, O.W., Williamson, B.J. and Bailey, E.H. (1998). Role for lichen melanins in uranium remediation. *Nature*, **391**, 649–650.

Mehra, A. and Farago, M.E. (1994). Metal ions and plant nutrition. In: *Plants and the Chemical Elements. Biochemistry, Uptake, Tolerance and Toxicity*, (M.E. Farago, ed.), pp. 31–66, VCH, Weinheim.

Mikhailova, I. (2002). Transplanted lichens for bioaccumulation studies. In: *Monitoring with Lichens—Monitoring Lichens* (P.L. Nimis, C. Scheidegger and P.A. Wolseley, eds.), pp. 301–309. Kluwer Academic, Dordrecht.

Modenesi, P., Piana, M. and Pinna, D. (1998). Surface features in *Parmelia sulcata* (Lichenes) thalli growing in shaded or exposed habitats. *Nova Hedwigia*, **66**, 535–547.
Nash, T.H. III. (1972). Simplification of the Blue Mountain lichen communities near a zinc factory. *Bryologist*, **75**, 315–324.
Nash, T.H. III (1989). Metal tolerance in lichens. In: *Heavy Metal Tolerance in Plants: Evolutionary Aspects* (A.J. Shaw, ed.), pp. 119–131. CRC Press, Boca Raton.
Nash, T.H. III (1996a). *Lichen Biology.* Cambridge University Press, Cambridge.
Nash, T.H. III (1996b). Nutrients and mineral cycling. In: *Lichen Biology* (T.H. Nash, ed.), pp. 136–153. Cambridge University Press, Cambridge.
Nash, T.H., III, Gries, C., Zschau, T., Getty, S., Ameron, Y. and Zambrano, A. (2003). Historical patterns of metal atmospheric deposition to the epilithic lichen *Xanthoparmelia* in Maricopa County, Arizona, USA. *Journal De Physique IV*, **107**, 921–924.
Nieboer, E. and Richardson, D.H.S., (1980). The replacement of the nondescript term 'heavy metals' by a biologically and chemically significant classification of metal ions. *Environmental Pollution*, Ser B, **1**, 3–26.
Nieboer, E., Richardson, D.H.S. and Tomassini, F.D. (1978). Mineral uptake and release by lichens: An overview. *Bryologist*, **81**, 226–246.
Nimis, P.L., Lazzarin, G., Lazzarin, A. and Skert, N. (2000). Biomonitoring of trace elements with lichens in Veneto (NE Italy). *Science of the Total Environment*, **255**, 97–111.
Nimis, P.L. and Purvis, O.W. (2002). Monitoring lichens as indicators of pollution. An introduction. In: *Monitoring with Lichens—Monitoring Lichens*, (P.L. Nimis, C. Scheidegger and P.A. Wolseley, eds.), pp. 1–4. Kluwer Academic, Dordrecht.
Nimis, P.L., Scheidegger, C., Wolseley, P.A. (eds.) (2002). *Monitoring with Lichens—Monitoring Lichens.* Kluwer Academic, Dordrecht.
Nriagu, J.O. (1989). Global inventory of natural sources of atmospheric trace metals. *Nature*, **338**, 47–49.
Olafsdottir, E.S. and Ingolfsdottir, K. (2001). Polysaccharides from lichens: Structural characteristics and biological activity. *Planta Medica*, **67**, 199–208.
Otnyukova, T. (2007). Epiphytic lichen growth abnormalities and element concentrations as early indicators of forest decline. *Environmental Pollution*, **146**, 359–365.
Paul, A., Hauck, M. and Fritz, E. (2003). Effects of manganese on element distribution and structure in thalli of the epiphytic lichens *Hypogymnia physodes* and *Lecanora conizaeoides*. *Environmental and Experimental Botany*, **50**, 113–124.
Paul, A., Hauck, M. and Langenfeld-Heyser, M. (2004). Ultrastructural changes in soredia of the epiphytic lichen *Hypogymnia physodes* cultivated with manganese. *Environmental and Experimental Botany*, **52**, 139–147.
Pawlik-Skowrońska, B. (2000). Relationship between acid-soluble thiol peptides and accumulated Pb in the green alga *Stichococcus bacillaris*. *Aquatic Toxicology*, **50**, 221–230.
Pawlik-Skowrońska, B., di Toppi, L.S., Favali, M.A., Fossati, F., Pirszel, J. and Skowronski, T. (2002). Lichens respond to heavy metals by phytochelatin synthesis. *New Phytologist*, **156**, 95–102.
Pawlik-Skowrońska, B., Purvis, O.W., Pirszel, J. and Skowroński, T. (2006). Cellular mechanisms of Cu-tolerance in the epilithic lichen *Lecanora polytropa* growing at a copper mine. *The Lichenologist*, 267–275.
Puckett, K.J. (1976). The effect of heavy metals on some aspects of lichen physiology. *Canadian Journal of Botany*, **54**, 2695–2703.
Purvis, O.W. (1984). The occurrence of copper oxalate in lichens growing on copper sulphide-bearing rocks in Scandinavia. *The Lichenologist*, **16**, 197–204.
Purvis, O.W. (1985). The effect of mineralization on lichen communities with special reference to cupriferous substrata. PhD Thesis. Department of Geology, Royal School of Mines, Imperial College, London.
Purvis, O.W. (1993). The botanical interest of mine spoil heaps — the lichen story. *Journal of the Russell Society*, **5**, 45–48.
Purvis, O.W. (1996). Interactions of lichens with metals. *Science Progress*, **79**, 283–309.
Purvis, O.W. (1997). The species concept in lichens. In: *Species: Units of Biodiversity*, (M.F. Claridge, H.A. Dawah, and M.R. Wilson, eds.), pp 109–134. [Systematics Association Special Volume Series 54]. Chapman and Hall, London.

Purvis, O.W. (2000). *Lichens*. Life series. Natural History Museum, London.
Purvis, O.W. (2001). Metal-rich habitats. In: *Lichens and Habitat Management* (A.J. Fletcher, P.A. Wolseley and R.G. Woods, eds.), pp. 1–10. The British Lichen Society, London.
Purvis, O.W., Bailey, E.H., McLean, J., Kasama, T. and Williamson, B.J. (2004a). Uranium biosorption by the lichen *Trapelia involuta* at a uranium mine. *Geomicrobiology Journal*, **21**, 159–167.
Purvis, O.W., Chimonides, P.J., Din, V., Erotokritou, L., Jeffries, T., Jones, G.C., Louwhoff, S., Read, H. and Spiro, B. (2003). Which factors are responsible for the changing lichen floras of London? *Science of the Total Environment*, **310**, 179–189.
Purvis, O.W., Chimonides, P.J., Jeffries, T.E., Jones, G.C., Read, H. and Spiro, B. (2005). Investigating biogeochemical signatures in the lichen *Parmelia sulcata* at Burnham Beeches, Buckinghamshire, England. *The Lichenologist*, **37**, 329–344.
Purvis, O.W., Chimonides, P.D.J., Jeffries, T.E., Jones, G.C., Rusu, A.M. and Read, H. (2007). Multi-element composition of historical lichen collections and bark samples, indicators of changing atmospheric conditions. *Atmospheric Environment*, **41**, 72–80.
Purvis, O.W., Chimonides, P.J., Jones, G.C., Mikhailova, I.N., Spiro, B., Weiss, D.J. and Williamson, B.J. (2004b). Lichen biomonitoring near Karabash Smelter Town, Ural Mountains, Russia, one of the most polluted areas in the world. *Proceedings of the Royal Society of London Series B-Biological Sciences*, **271**, 221–226.
Purvis, O.W., Elix, J.A., Broomhead, J.A. and Jones, G.C. (1987). The occurrence of copper norstictic acid in lichens from cupriferous substrata. *The Lichenologist*, **19**, 193–203.
Purvis, O.W., Elix, J.A. and Gaul, K.L. (1990). The occurrence of copper-psoromic acid in lichens from cupriferous substrata. *The Lichenologist*, **22**, 345–354.
Purvis, O.W., Gilbert, O.L. and James, P.W. (1985). The influence of copper mineralization on *Acarospora smaragdula*. *The Lichenologist*, **17**, 111–114.
Purvis, O.W. and Halls, C. (1996). A review of lichens in metal-enriched environments. *The Lichenologist*, **28**, 571–601.
Purvis, O.W. and James, P.W. (1985). Lichens of the Coniston Copper Mines. *The Lichenologist*, **17**, 221–237.
Purvis, O.W., Longden, J., Shaw, G., Chimonides, P.D.J., Jeffries, T.E., Jones, G.C., Mikhailova, I.N. and Williamson, B.J. (2006). Biogeochemical signatures in the lichen *Hypogymnia physodes* in the Mid Urals. *Journal of Environmental Radioactivity*, **90**, 151–162.
Purvis, O.W., Williamson, B.J., Bartok, K. and Zoltani, N. (2000). Bioaccumulation of lead by the lichen *Acarospora smaragdula* from smelter emissions. *New Phytologist*, **147**, 591–599.
Richardson, D.H.S. (1988) Understanding the pollution sensitivity of lichens. *Botanical Journal of the Linnean Society*, **96**, 31–43.
Richardson, D.H.S. (1995). Metal uptake in Lichens. *Symbiosis*, **18**, 119–127.
Rose, C.I. and Hawksworth, D.L. (1981). Lichen recolonization in London's cleaner air. *Nature*, **289**, 289–292.
Rusu, A.M., Chimonides, P.D.J., Garcia-Sanchez, R. and Purvis, O.W. (2006a). Multi-element including rare earth content of lichens, bark, soils and waste following industrial closure. *Environmental Science & Technology*, **40**, 4599–4604.
Rusu, A.M., Jones, G.C., Chimonides, P.D.J. and Purvis, O.W. (2006b). Biomonitoring using the lichen *Hypogymnia physodes* and bark samples near Zlatna, Romania immediately following closure of a copper ore processing plant. *Environmental Pollution*, **143**, 81–88.
Saipunkaew, W., Wolseley, P.A., Chimonides, P.J. and Boonpragob, K. (2007). Epiphytic macrolichens as indicators of environmental alteration in northern Thailand. *Environmental Pollution*, **146**, 366–374.
Sanita di Toppi, L., Marabottini, R., Vattuone, Z., Musetti, R., Favali, M.A , Sorgona, A. and Badiani, M. (2005b). Cell wall immobilisation and antioxidant status of *Xanthoria parietina* thalli exposed to cadmium. *Functional Plant Biology*, **32**, 611–618.
Sanita di Toppi, L., Musetti, R., Marabottini, R., Corradi, M.G., Vattuone, Z., Favali, M.A. and Badiani, M. (2004). Response of *Xanthoria parietina* thalli to environmentally relevant concentrations of hexavalent chromium. *Functional Plant Biology*, **31**, 329–338.
Sanita di Toppi, L., Mussetti, R., Vattuone, Z., Pawlik-Skowrońska, B., Fossati, F., Bertoli, L., Badiani, M. and Favali, M.A. (2005a). Cadmium distribution and effects on ultrastructure and chlorophyll

status in photobionts and mycobionts of *Xanthoria parietina*. *Microscopy Research and Technique*, **66**, 229–238.

Sanita di Toppi, L. and Pawlik-Skowrońska, B. (eds.) (2003). *Abiotic Stresses in Plants*. Kluwer Academic, Dordrecht.

Sarrett, G., Manceau, A., Cuny, D., Van Haluwyn, C., Deruelle, S., Hazemann, J.L., Soldo, Y., Eybert-Berard, L. and Menthonnex, J.J. (1998). Mechanisms of lichen resistance to metallic pollution. *Environmental Science and Technology*, **32**, 3325–3330.

Schwartzman, D. (1993). Comment on 'weathering, plants, and the long-term carbon cycle' by Robert A. Berner. *Geochemica et Cosmochimica Acta*, **57**, 2145–2146.

Schwartzman, D. (2002). *Life, Temperature and the Earth. The Self-Organizing Biosphere*. Columbia University Press, New York.

Seaward, M.R.D. (1974). Some observations on heavy metal toxicity and tolerance in lichens. *The Lichenologist*, **6**, 158–164.

Seaward, M.R.D. and Letrouit-Galinou, M.A. (1991). Lichen recolonization of trees in the Jardin-du-Luxembourg, Paris. *The Lichenologist*, **23**, 181–186.

Shaw, A.J. (1989). Heavy Metal Tolerance in Plants: Evolutionary Aspects, CRC Press, Boca Raton.

Shotbolt, L., Ashmore, M.R. and Tipping, E. (2004). Annexe 7: Heavy metal deposition trends in five UK catchments: A reconstruction based on monitoring, historical and environmental records. Annexes to the final report to Defra EPG 1/3/188, 78–124.

Sloof, J.E. (1995). Lichens as quantitative biomonitors for atmospheric trace—Element deposition, using transplants. *Atmospheric Environment*, **29**, 11–20.

Spiro, B., Weiss, D.J., Purvis, O.W., Mikhailova, I., Williamson, B.J., Coles, B.J. and Udachin, V. (2004). Lead isotopes in lichen transplants around a Cu smelter in Russia determined by MC-ICP-MS reveal transient records of multiple sources. *Environmental Science & Technology*, **38**, 6522–6528.

Stanley, C.J. (1979). Mineralogical studies of copper, lead, zinc and cobalt mineralization in the English Lake District, Ph.D. Thesis. University of Aston, Birmingham.

Stepanenko, L.S., Krovoshchekova, O.E. and Skirina, I.F. (2002). Functions of phenolic secondary metabolites in lichens from Far East Russia. *Symbiosis*, **32**, 119–131.

Suzuki, Y. and Banfield, J.F. (1999). The geomicrobiology of Uranium. In: *Uranium*: Mineralogy, *Geochemistry and the Environment - Reviews in Mineralogy* (P.C. Burns and R. Finch, eds.), **38**, pp. 393–342. Mineralogical Society of America, Chantilly, VA.

Takala, K., Salminen, R. and Olkkonen, H. (1998). Geogenic and anthropogenic zinc in epiphytic and terricolous lichens in Finland. *Journal of Geochemical Exploration*, **63**, 57–66.

Takani M., Yajima, T., Masuda, H. and Yamauchi, O. (2002). Spectroscopic and structural characterization of copper(II) and palladium(II) complexes of a lichen substance usnic acid and its derivatives. Possible forms of environmental metals retained in lichens. *Journal of Inorganic Biochemistry*, **91**, 139–150.

Tarhanen, S. (1998). Ultrastructural responses of the lichen *Bryoria fuscescens* to simulated acid rain and heavy metal deposition. *Annals of Botany*, **82**, 735–746.

Taylor, S.R. and McLennan, S.M. (1985). *The Continental Crust: Its Composition and Evolution*. Blackwell, Oxford.

Tebo, B.M., Bargar, J.R., Clement, B.G., Dick, G.J., Murray, K.J., Parker, D., Verity, R. and Webb, S.M. (2004). Biogenic manganese oxides: Properties and mechanisms of formation. *Annual Review of Earth and Planetary Science*, **32**, 287–328.

Tomassini, F.D., Puckett, K.J., Nieboer, E., Richardson, D.H.S. and Grace, D. (1976). Determination of copper, iron, nickel and sulphur by X-ray fluorescence in lichens from the Mackenzie Valley, Northwest Territories and the Sudbury District, Ontario. *Canadian Journal of Botany*, **54**, 1591–1603.

Tømmervik, H., Johansen, B.E. and Pedersen, J.P. (1995). Monitoring the effects of air-pollution on terrestrial ecosystems in Varanger (Norway) and Nikel-Pechenga (Russia) using remote-sensing. *Science of the Total Environment*, **161**, 753–767.

Tømmervik, H., Johansen, M.E., Pedersen, J.P. and Guneriussen, T. (1998). Integration of remote sensed and in-situ data in an analysis of the air pollution effects on terrestrial ecosystems in the border areas between Norway and Russia. *Environmental Monitoring and Assessment*, **49**, 51–85.

Udachin, V., Williamson, B.J., Purvis, O.W., Spiro, B., Dubbin, W., Brooks, S., Coste, B., Herrington, R.J. and Mikhailova, I. (2003). Assessment of environmental impacts of active smelter operations and abandoned mines in Karabash, Ural mountains of Russia. *Sustainable Development*, **11**, 133–142.

USDA (2006). Air Quality Biomonitoring Program on National Forests of Northwest Oregon and Southwest Washington. http://www.fs.fed.us/r6/aq/lichen/

van Herk C.M., Mathijssen-Spiekman, E.A.M. and de Zwart, D. (2003). Long distance nitrogen air pollution effects on lichens in Europe. *The Lichenologist*, **35**, 347–359.

Wessels, J.G.H. (2000). Hydrophobins, unique fungal proteins. *Mycologist*, **14**, 153–159.

Whiting, S.N., Reeves, R.D., Richards, D., Johnson, M.S., Cooke, J A., Malaisse, F., Paton, A., Smith, J.A.C., Angle, J.S., Chaney, R.L., Ginocchio, R., Jaffre, T., Johns, R., McIntyre, T., Purvis, O.W., Salt, D.E., Schat, H., Zhao, F.J. and Baker, A.J.M. (2004). Research priorities for conservation of metallophyte biodiversity and their potential for restoration and site remediation. *Restoration Ecology*, **12**, 106–116.

Whiting, S.N., Reeves, R.D., Richards, D., Johnson, M.S., Cooke, J.A., Malaisse, F., Paton, A., Smith, J.A.C., Angle, J.S., Chaney, R.L., Ginocchio, R., Jaffre, T., Johns, R., McIntyre, T., Purvis, O.W., Salt, D.E., Schat, H., Zhao, F.J. and Baker, A.J.M. (2005). Use of plants to manage sites contaminated with metals. In: *Plant Nutritional Genomics* (M.R. Broadley and P.J. White, eds.), pp. 106–116. Oxford: Blackwell Publishing and Boca Raton (Florida): CRC Press.

Wierzchos, T. and Ascaso, C. (1994). Application of back-scattered electron imaging to the study of the lichen-rock interface. *Journal of Microscopy*, **175**, 54–59.

Williamson, B.J., McLean, J. and Purvis, O.W. (1998). Application of X-ray element mapping across the lichen-rock interface. *Journal of Microscopy*, **191**, 91–96.

Williamson, B.J., Mikhailova, I., Purvis, O.W. and Udachin, V. (2004) SEM-EDX analysis in the source apportionment of particulate matter on *Hypogymnia physodes* lichen transplants around the Cu smelter and former mining town of Karabash, South Urals, Russia. *Science of the Total Environment*, **322**, 139–154.

Wilson, M.J. (1995). Interactions between lichens and rocks: A review. *Cryptogamic Botany*, **5**, 299–305.

Wilson, M.J. and Jones, D. (1984). The occurrence and significance of manganese oxalate in *Pertusaria-corallina* (Lichenes). *Pedobiologia*, **26**, 373–379.

Wilson, M.J., Jones, D. and Russell, J.D. (1980). Glushinskite, a naturally occurring magnesium oxalate. *Mineralogical Magazine*, **43**, 837–840

Winkelmann, G., Winge, D.R. (eds.) (1994). *Metal Ions in Fungi*. Marcel Dekker Inc., New York.

Wirth, V. (1972). Die Silikatflechten. Gemeinshaften im ausseralpinen Zentraleuropa. *Dissertationes Botanicae Lehre*, **17**, 1–306, 109.

Wolseley, P.A., James, P.W., Theobald, M.R. and Sutton, M.A. (2006). Detecting changes in epiphytic lichen communities at sites affected by atmospheric ammonia from agricultural sources. *The Lichenologist*, **38**, 161–176.

Zenk, M.H. (1996). Heavy metal detoxification in higher plants: A review. *Gene*, **179**, 21–30.

Zhdanova, N.N., Zakharchenko, V.A., Vember, V.V. and Nakonechnaya, L.T. (2000). Fungi from Chernobyl: Myobiota of the inner regions of the containment structures of the damaged nuclear reactor. *Mycological Research*, **104**, 1421–1426.

CHAPTER 13

Responses of Mycorrhizal Fungi to Stress

Roger D. Finlay, Björn D. Lindahl and **Andy F. S. Taylor**

Contents

1. Introduction 202
2. Problems with Filamentous Fungi 203
3. Nutritional Stress 204
4. Fertilization and N-Deposition 204
5. Effects of Liming and Wood-Ash Additions 206
6. Toxic Metals and Oxidative Stress 209
7. Water Stress 210
8. Fragmentation and Disruption of Carbon Supply 211
9. Grazing 211
10. Microbial Interactions 212
11. Competition For Host Roots—Evolution And Specificity 214
12. Concluding Remarks 215

References 215

Abstract

Mycorrhizal fungi are exposed to all or many of the environmental stresses that other fungi may experience. These include extremes of temperature and pH, anoxia, water stress, physical fragmentation, toxic metals and other pollutants, as well as anthropogenic stresses arising from applications of fertilisers, lime and wood ash. Fungi can respond to these stresses by altering their morphology, modifying their external environment or adapting their internal metabolism, although the degree of phenotypic plasticity may vary. Since mycorrhizal fungi obtain their carbon from autotrophic host plants, the fungi may also be exposed to stress through changes in carbon allocation from hosts. The refinement and application of molecular identification methods has led to the realisation that the degree of host specificity shown by some mycorrhizal fungi may be higher than assumed in the past. This implies that the availability of compatible roots will influence fungal survival and changes in the species composition of plant communities. Restricting the supply of assimilates from compatible host roots may thus limit the growth of certain fungi. Other types of stress may arise from interactions with competing mycorrhizal symbionts, other fungi, soil animals or bacteria. These different types of stress are reviewed from physiological,

British Mycological Society Symposia Series
Published by Elsevier Ltd.

ecological and evolutionary perspectives in an attempt to identify potentially fruitful lines of future research.

1. INTRODUCTION

Stress is often defined in terms of growing conditions that are harmful to an organism. Jennings (1993) reviewed different aspects of stress tolerance in fungi. One definition of stress mentioned by Jennings, includes "situations that restrict or prevent the growth and reproduction of the vast majority of fungi" suggesting that extreme environments and extremophiles are involved. An alternative way of considering stress is that stressful environments are those that differ significantly from normal culture conditions. This highlights the problematic, anthropocentric nature of the word "stress". Many natural environments are considered "nutrient-poor" usually because crop plants or trees do not grow at rates considered "optimal" by humans. Competition for nutrients or other resources is a type of natural stress that has been discussed below. Another possible definition is that stress occurs when conditions change, however, since most natural environments are subject to high spatial and temporal variability, the latter two definitions imply that stress is a normal phenomenon for most organisms—the rule rather than the exception.

Mycorrhizal fungi are exposed to all or many of the environmental stresses that other fungi may experience. These include a range of physical, chemical and biological effects involving extremes of temperature, pH, oxygen, water or nutrient availability, physical fragmentation, toxic metals and other pollutants, as well as anthropogenic stresses arising from applications of fertilisers, lime and wood ash, or CO_2 enrichment. Possible outcomes of stress interactions involving mycorrhizal fungi are illustrated schematically in Figure 1. Stress factors may act directly on the mycorrhizal fungi themselves (a), but since mycorrhizal fungi form symbiotic associations with autotrophic host plants they may also be subject to indirect forms of stress through reduced allocation of host assimilates (b) caused by plant responses to environmental changes (c). Conversely, mycorrhizal symbionts may confer a greater degree of stress tolerance on their host plants (d) and a stress stimulus usually elicits a response that may or may not confer a greater degree of tolerance on the fungus itself (e). Stress may be induced by natural environmental extremes or competition with other organisms, or anthropogenically induced through human modification of an environment. Stress may be acute or chronic, and both the effects of stress and responses of mycorrhizal fungi to it can operate on different time scales. The responses may involve homeostatic regulation of their internal or external environment (f), which may take place on an instantaneous or phenological time scale, short-term changes in fungal populations and community structures (g) or long-term changes over evolutionary time that involve adaptations to different suites of edaphic parameters and have led to characteristic vegetation types (h). Whether or not this last level of interaction, which concerns the role of stress in speciation,

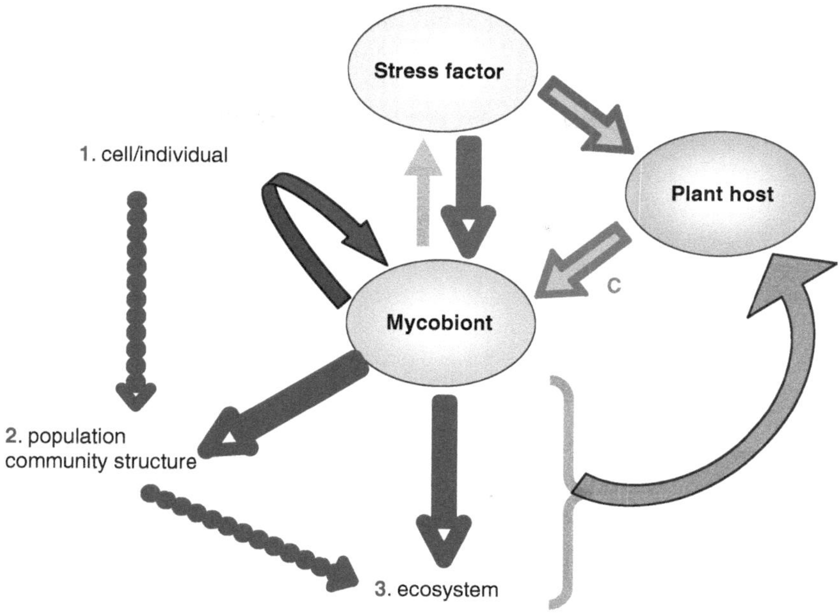

Figure 1 Stress interactions involving mycorrhizal fungi. Stress factors may act directly on mycorrhizal fungi (a) but may also arise indirectly through reduced allocation of host assimilates (b), caused by plant responses to environmental changes (c). Conversely mycorrhizal fungi may confer a greater degree of stress tolerance on their host plants (d) and stress stimuli may eleicit responses which may or may not confer a greater degree of tolerance on the fungus itself (e). Stress may be acute or chronic and both the effects of stress and the responses of mycorrhizal fungi to it may operate on different time scales. The responses may involve homeostatic regulation of their internal or external environment (f) which may take place on an instantaneous or phenological time scale, short-term changes in fungal population and community structure (g), or long-term changes over evolutionary time that involve adaptations to different suites of edaphic parameters and have led to characteristic vegetation types (h). **(See Colour Section)**

can be classified as involving "stress", as discussed within the context of this book, can be debated; but the different types of mycorrhizal symbiosis currently in existence undoubtedly play important roles in the different types of ecosystem they dominate, influencing a range of ecosystem processes (Read and Perez-Moreno, 2003). Ultimately responses of mycorrhizal fungi to environmental stress at levels f, g & h will all influence the plant host and, in natural systems, multiple mycobionts and plant hosts will co-exist.

2. PROBLEMS WITH FILAMENTOUS FUNGI

Simple definitions of stress that can be applied to larger organisms occupying a well-defined habitat are less easily applicable to filamentous fungi for a number of reasons. The environment to which filamentous fungi are exposed is less easily defined since they can simultaneously occupy discrete, nutritionally diverse substrates, re-distributing nutrients and carbon over large distances. Fungi can also modify their environment by excreting protons, antibiotics, enzymes, siderophores and water, and by sequestration of toxic compounds. This local modification of the environment in the immediate vicinity of the mycelium will alter the stresses that the fungus is actually exposed to (Lindahl and Olsson, 2004). Although fungal hyphae may extend long distances through soil, they are microscopic in diameter and can occupy microsites that may be environmentally distinct from the bulk substrate. As Jennings (1993) has argued, the changed niche of a fungus exposed to stress may be masked by features of the bulk substrate, making the definition of a particular stress factor difficult.

3. NUTRITIONAL STRESS

Nutritional stress is a normal feature of most environments in which organisms compete for growth limiting resources. The modern concept of a growth limiting nutrient, familiar in agronomy and plant biology, was first put forward by Justus von Liebig in 1840. "Liebig's Law of the Minimum" states that the growth of a plant is limited by the one essential mineral that is in the (relative to need) shortest supply. This is a qualitative version of principles used in fertiliser application in modern agriculture and Liebig is credited with the discovery of nitrogen as an essential plant nutrient. This concept was subsequently extended by Shelford (1913) to include not only the lower limit of an essential nutrient required, but also the maximum limit of tolerance, and the notion that organisms have an ecological maximum and minimum with a range in between which represents the limits of tolerance is a basic feature of modern niche theory today. These theories are not only relevant to the role of ectomycorrhizal (ECM) fungi in accessing organic nitrogen in boreal forest ecosystems that are poor in mineral nitrogen (Read, 1989) but also to mycorrhizal responses to excess nitrogen in the form of anthropgenic nitrogen deposition (Wallenda and Kottke, 1998).

4. FERTILIZATION AND N-DEPOSITION

Most ECM fungi appear to be remarkably sensitive to changes in the abiotic environment in which they grow (Erland and Taylor, 2002). In particular, chronic and/or drastic increases in mineral N availability, as a result of either N deposition or from fertilizer additions, can cause dramatic losses in species richness and substantially alter ECM community structures (reviewed by Wallenda and Kottke, 1998). Given the N-limited environment under which the ECM associations are most extensively developed, and important (Read, 1991), responses to N availability are to be expected. However, not all ECM fungi react negatively to N additions (Peter *et al.*, 2001). While members of the genus *Cortinarius*, the most species rich genus with approximately 2000 spp., are usually very sensitive to

N-additions and are often absent from fertilised forest plots, species of the genus *Lactarius* often increase in abundance following fertilisation.

It is still unclear whether the commonly observed loss of diversity in ECM fungi following fertilisation with N is due to a direct stress effect on the fungi or an indirect effect *via* the host plant, reducing the supply of C to the fungi. The development of mycelium by some ECM is known to be negatively affected by elevated soil N levels (Wallander and Nylund, 1992; Arnebrant, 1994). Wallander (1995) put forward the theory that ECM fungi could not restrict N uptake and mycelial growth was reduced because C that would normally be used for hyphal extension was diverted into the assimilation of the N taken up. This work has focussed on the cost of metabolising ammonium and little account has been taken of direct mineral N toxicity on the fungi.

Nitrogen fertilisation of forests usually involves the application of approximately 100–200 kg N ha^{-1} as ammonium nitrate. It has been noted on several occasions that nitrate can be toxic to some ECM fungi (Smith and Read, 1997). In recent experiments (Taylor, unpublished), examining the use of different N sources by ECM fungi, mycorrhizal formation and development by *Suillus variegatus*, a common ECM associate of *Pinus sylvestris*, was strongly negatively influenced by the presence of nitrate (Figure 2). This was the case even when ammonium or an organic sources, in this case bovine serum albumin (BSA), was present as an alternative N source.

The chronic addition of N from aerial N-deposition has also been linked to the reduction in sporocarp production in European forests (Arnolds, 1991; Baar and

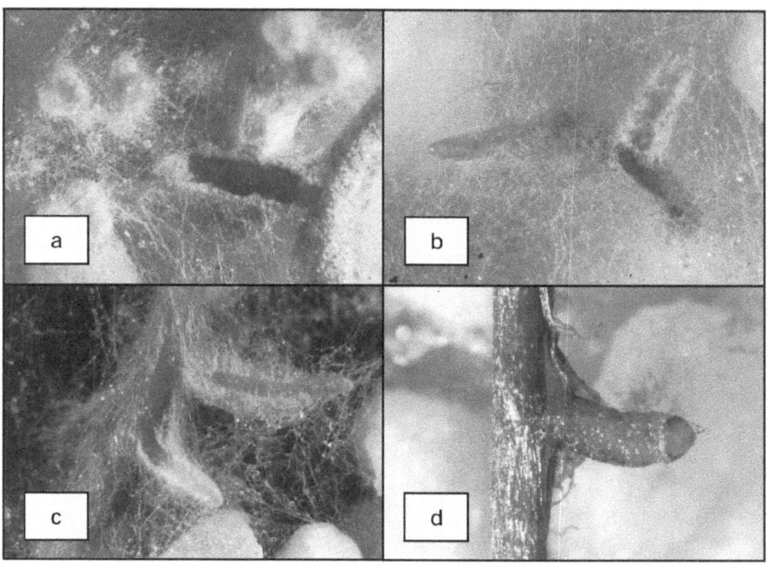

Figure 2 Normal development of tuberculate mycorrhizas by *Suillus variegatus* on scots pine when grown with an organic nitrogen source (a). Disruption of development when grown with (b) ammonium/nitrate; (c) organic N/nitrate, and (d) nitrate alone. **(See Colour Section)**

ter Braak, 1996). Wallenda and Kottke (1998) summarised the available data and concluded that there seemed little doubt that ECM fungal species richness, as measured by sporocarp production, was negatively affected by increasing N-deposition. They also concluded that "specialist" species (especially the symbionts of conifers) were more adversely affected than "generalist" species, which are able to form mycorrhizas with a wide range of host plants. Taylor *et al.* (2000) examined the hypothesis that pollutant N-deposition contributes to a reduction in diversity of ECM communities by investigating below-ground (ECM fungi on root tips) and above-ground communities (sporocarp production) in beech and spruce forests along a N–S transect in Europe, ranging from polluted sites in central Europe to relatively unpolluted sites in Scandinavia. A clear negative relationship was found between the diversity of ECM fungi in spruce forests and N-deposition. However, no negative effects on ECM fungal diversity were found in the beech forests, where there was, in fact, a positive correlation between extractable mineral N in the soil and species richness on root tips. These results highlight the context dependent response of mycorrhizal fungi to different types of stress.

In arbuscular mycorrhizal (AM) fungi, the effects of fertilisation have been shown to depend upon the initial nutrient status of the ecosystems, as well as the particular mycorrhizal species involved (Treseder and Allen, 2002). Johnson *et al.* (2003) demonstrated that the composition of the plant community may influence the structure of the AM fungal community, but it is also known that the diversity of AM fungal communities can influence the diversity and productivity of plant communities (van der Heijden *et al.*, 1998). Studies by Jumpponen *et al.* (2005) showed minimal changes in AM colonisation in response to nitrogen enrichment but did reveal changed patterns of community composition. Santos *et al.* (2006) demonstrated a negative correlation between numbers of AM sequence groups and soil mineral nitrogen along a gradient of fertilisation in a grassland.

5. EFFECTS OF LIMING AND WOOD-ASH ADDITIONS

The slow acidification of soils under many vegetation types, in particular under some coniferous species, is a natural process and plants have developed a variety of mechanisms to tolerate or avoid the resulting stress (Marschner, 1991). However, the deposition of many anthropogenic pollutants has greatly increased the rate of soil acidification (Falkengren-Grerup *et al.*, 1987). In areas of northern Europe, where the parent material has little buffering capacity, lowering of soil pH has been recognised as a potentially important threat to the long-term survival of forest ecosystems (Falkengren-Grerup and Eriksson, 1990; Hahn and Marschner, 1998). Increasing soil acidity increases the rates of mineral dissolution and mobility of many potentially toxic metals, in particular Al, as well as causing leaching of base cations (Finlay, 1995). However, leaching of these base cations may be reduced by the presence of an extensive extraradical ECM mycelium (Ahonen-Jonnarth *et al.*, 2003). While increasing acidity and toxic metal availability are likely to affect the mycorrhizal fungi in forests, proposed counter measures such as liming, which are employed to de-acidify the soil (Nowotny *et al.*, 1998), may be expected to induce the greatest stress response in the fungi and other soil biota. In most boreal and temperate forests, where the mycorrhizal fungi are adapted to the prevailing acidic soil conditions (Read, 1991), it seems inevitable that liming will constitute a stress that significantly alters community composition and structure. In contrast, the effects of moderate reductions in soil pH are likely to have less drastic effects.

A number of negative "side-effects" have been noted after liming (Kreutzer, 1995), of these most apparently related to the response of the soil biota to the dramatic increases in soil pH that are a concomitant consequence of liming. Increased rates of decomposition (Persson *et al.*, 1989; Priha and Smolander, 1994) lead to increased loss of C, higher levels of ammonium availability can also lead to increased nitrification (Andersson and Valeur, 1994) and leaching of nitrate to groundwater (Kreutzer, 1995). While these responses may partly reflect physiological plasticity within existing organisms, it is also well documented that liming can dramatically alter fungal-community structure and composition.

A number of early studies on ECM fungi found shifts in community structure after liming (Erland and Söderström, 1991; Lehto, 1994; Andersson and Söderström, 1995). The small number of more recent studies, using morphological (Taylor and Finlay, 2003) and/or molecular identification (Jonsson *et al.*, 1999; Bakker *et al.*, 2000) of ECM fungi on root tips, have confirmed that liming can radically alter species composition. However, there are still too few studies to predict how an ECM community will respond to the addition of lime, but it is clear that the reaction is dependent upon the species present and that the effects of the lime are long lasting.

Piloderma fallax is a common ECM fungus in acidic-boreal forest soils and shows a particularly strong negative reaction to increasing soil pH. When the mycelium of this fungus was challenged with organic matter collected from either control or plots that had been limed 15 years previously (Figure 3a), only very limited growth occurred in the limed material (Figure 3b). Mycelial allocation of carbon was measured following labelling of the host seedlings with ^{14}C-labelled CO_2 (Figure 3c) and was significantly higher to the control substrate than the limed substrate. The mycelia of ECM fungi are important in the colonisation of new root tips and hence a major factor in the competitive capability of ECM fungi. Any perturbation, such as liming, that affects the growth of ECM fungi in soil must, therefore, alter the competitive interactions between species. Those taxa favouring conditions with a higher soil pH will increase at the expense of those dependent upon low soil pH. Liming of a spruce forest in southern Sweden with 8.75 ton ha^{-1} of dolomite resulted in the almost total replacement of the typical ECM community by taxa not normally found under the prevailing conditions. A similar situation was reported by Taylor and Brand (1992) from a limed spruce forest in southern Germany. It is unclear whether the taxa favoured by liming occur naturally in these forests, surviving in the forests in soil microsites with high pH, but below our detection levels, or whether they migrate to the site

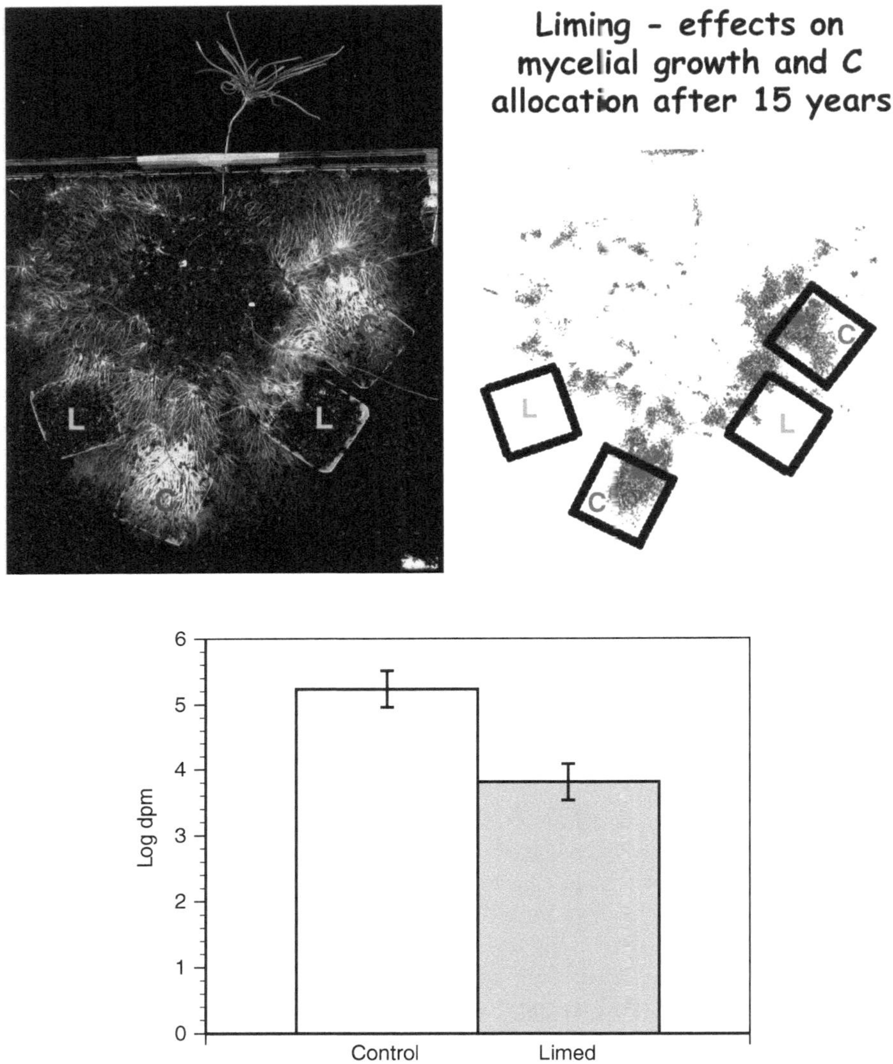

Figure 3 Effects of liming on carbon allocation by the extraradical mycelium of the ectomycorrhizal fungus *Piloderma fallax*. (a) Microcosm containing a *Pinus sylvestris* seedling with intact extraradical mycelium colonising patches of control (C) and limed (L) substrate taken from the fermentation horizon of a forest 15 years after the addition of the lime. (b) Electronic autoradiograph showing the distribution of label following exposure of the seedling to $^{14}CO_2$. (c) Histogram showing the relative distribution of mean levels of radioactivity (Log_{10} cpm) in control (white) and limed (grey) substrate following sample oxidation and quantification using liquid scintillation spectrometry. Error bars indicate $\pm$ 1.96 SE. **(See Colour Section)**

as spores from surrounding sites with high pH. It is possible that they survive in the forests in soil microsites with high pH.

The application of wood ash to forest ecosystems has been proposed as a means whereby nutrients removed during whole-tree harvesting can be, at least in part, replaced. Wood ash is produced from burning biofuels and it is costly to dispose it off in landfill sites. Returning the ash to forests therefore represents both a means by which essential nutrients may be replaced and a convenient solution for the disposal of this by-product. However, wood ash contains the macronutrients Ca, P, K and Mg and the micronutrients B, Cu and Zn, but it can be highly alkaline (pH 10–13). Wood ash may also contain high levels of heavy metals (Steenari and Lindkvist, 1999).

There have been remarkably few studies examining the effect of wood-ash addition on mycorrhizal fungi. Mahmood *et al.* (2000) used molecular typing to study the effects of granulated wood ash (3 and 6 ton ha^{-1}) on the below-ground ECM community in a Swedish spruce forest 7 years after the treatment. A small shift in ECM community was detecting with some species apparently increasing in the presence of the ash. One unidentified *Piloderma* species rapidly colonised the ash granules, whereas another *Piloderma* species, *P. fallax*, was shown to strongly avoid ash patches. Erland and Söderstöm (1991) examined short-term effects of non-granulated wood ash on ECM-colonisation in *Pinus sylvestris* plots treated with 7.5 tons ha^{-1} wood ash. Unfortunately, the level of resolution did not allow a detailed examination of species responses but they did record a significant rise in the humus pH of the ashed plots (6.4) compared to the control plots (pH 3.8). Taylor and Finlay (2003) recorded little effect of the addition of 4.25 ton ha^{-1} of wood ash on the species richness of the ECM community in a spruce forest in southern Sweden. However, some species either increased or decreased in response to ash additions. In summary, the levels of nutrients in wood ash are unlikely to have major impacts upon the ECM community, but the potential shifts in soil pH as a result of wood-ash additions could be expected to have measurable effects.

6. TOXIC METALS AND OXIDATIVE STRESS

Responses of mycorrhizal fungi to heavy metals are dealt with by Colpaert in Chapter 11 in this volume and will therefore not be discussed at length here. There is an extensive literature concerning effects of metals on mycorrhizal fungi, as well as the responses of the fungi, that include modification of the plant-host responses. Fomina *et al.* (2005) investigated solubilisation of toxic-metal minerals and metal tolerance by ericoid and ECM fungi. In general, metal-tolerant fungi grew and solubilised toxic-metal minerals better than non-tolerant isolates. Metal dissolution by fungi may take place through proton-promoted or ligand promoted mechanisms and organic acids provide both a source of protons for solubilisation and metal-chelating anions to complex the metal cations. Increased production of organic acids, particularly oxalic acid, by ECM fungi has been demonstrated in response to elevated concentrations of Al and Cu and may play a role in protecting plants from the toxic effects of these metals (Ahonen-Jonnarth *et al.*, 2000). Toxic metals may also cause oxidative stress and several studies of mycorrhizal fungal responses suggested that the fungi may be able to regulate genes providing protection against reactive oxygen species (ROS). Lanfranco *et al.* (2005) present evidence of a functional AM CuZn superoxide dismutase, which may provide protection against localised host defence responses involving ROS. Other studies (Ott *et al.*, 2002; Schützendübel and Polle, 2002) suggest that ECM fungi improve protection against toxic metal induced oxidative stress through strongly induced glutathione synthesis.

7. WATER STRESS

Mycorrhizal fungi are known to confer a greater degree of tolerance to water stress in plants, but the exact mechanisms are not fully understood and there appear to be varying effects of water stress on different mycorrhizal associations (Augé, 2001). Some of the effects are nutritionally mediated since supply of nutrients is limited in dry soil by the increasing tortuosity of the diffusion path and the mycorrhizal hyphal contribution to nutrient uptake will become increasingly important as soil dries. Increased levels of root colonisation by AM fungi in response to soil drying are more common than decreased levels and this effect may be related to reductions in plant P levels due to decreased diffusion of soil P or lowered P uptake. The increased size and improved P nutrition of mycorrhizal plants often introduce confounding factors in studies of the effect of mycorrhizal symbiosis on water balance, however several studies suggest that there may be effects which do not involve these factors and it is likely that a range of physical, nutritional and cellular effects is involved (Ruiz-Lozano, 2003). The enhanced tolerance of AM plants to water deficit may involve modulation of drought-induced plant genes (Ruiz-Lozano *et al.*, 2006). Although there is no evidence of any effects on accumulation of late embyogenesis-abundant proteins such as dehydrins or on the induction of *p5cs* genes involved in the biosynthesis of proline, several experiments suggest that expression of aquaporin genes may be altered by AM symbiosis. Porcel *et al.* (2006) showed that genes encoding plasma membrane aquaporins (water channel proteins that facilitate and regulate the passive movement of water down a water potential gradient) were down regulated in roots of soybean and lettuce under drought stress, and that the down regulation was more severe in AM plants than in non-mycorrhizal plants. This is suggested to be a mechanism to decrease membrane-water permeability and to allow cellular-water conservation. It is also known that AM fungi may increase the tolerance of plants to salt stress (Al-Karaki *et al.*, 2001). Aquaporins also appear to be important in tolerance of salt stress and Ouiziad *et al.* (2006) found reductions in transcript levels of a plasmalemma aquaporin gene in tomato plants subjected to salt stress. Interestingly, down regulation of the aquaporin genes in the study by Porcel *et al.* (2006) occurred at an earlier stage than in non-mycorrhizal plants.

8. FRAGMENTATION AND DISRUPTION OF CARBON SUPPLY

Mycorrhizal fungi may be subjected to stress through physical fragmentation of the mycelium, either by human activity or through grazing by animals. Comparisons between disrupted and non-disrupted treatments in both upland grassland ecosystems and boreal forest ecosystems have provided a better picture of the *in situ* functioning of mycorrhizal mycelia under natural conditions and the contribution of mycorrhizal mycelium to total soil respiration.

Johnson *et al.* (2001, 2002) used rotated and non-rotated mesh cylinders which could be penetrated by mycelia but not roots and demonstrated that the mycelia of AM fungi provide a rapid pathway for flux of carbon to the soil and atmosphere. Disruption of the mycelium-reduced colonisation of non-mycorrhizal roots, mycelial transport of ^{33}P-labelled phosphorus to plants and flow of carbon from host plants to the mycelium.

The effects of disruption of carbon flow below ground to roots and mycorrhizal fungi in a boreal forest were examined in a tree girdling experiment by Högberg *et al.* (2001). Girdling reduced soil respiration by 54% within 1–2 months and by up to 37% in five days, suggesting that flux of current assimilates to roots and mycorrhizal fungi is a key driver of soil respiration. Marked reductions in the numbers of ECM fungal fruiting bodies suggested there was a direct effect on the mycorrhizal fungi. One natural process which may restrict carbon flow to roots is herbivory causing reduced photosynthetic capacity. Kuikka *et al.* (2003) demonstrated that artificial defoliation of pine trees to simulate pine sawfly attack reduced both the numbers of fruiting bodies and the number of ECM species forming fruiting bodies.

9. GRAZING

One of the principal natural causes of fragmentation is grazing by animals and Taylor and Alexander (2005) argue that this may have been a major selection pressure in the evolution of ECM fungi. There is fossil evidence of Paleozoic mites in the early Devonian period, 360–400 million years ago (Shear and Kukalová-Peck, 1990), and some of the modern descendents of these are fungivorous. Arbuscular mycorrhizal fungi existed at this time but a large proportion of their mycelial biomass is within the plant roots, protected from grazing, and in extant AM taxa hyphal healing seems to be widespread (Giovannetti *et al.*, 1999; de la Providencia *et al.*, 2005). In ECM fungi the fungal mantle is exposed to grazing on the root surface and no hyphal healing mechanisms have been reported. However, different physical and chemical defence mechanisms have been reported that may restrict grazing. Many ECM species produce thick-walled, pointed, setae-like structures that are often melanised and may form a physical barrier for grazing. Thin-walled, swollen cells called cystidia are also present on the fungal mantle of many species, especially those within the Russulaceae. The cellular contents of these cells appear to be specialised and contain chemical deterrents to grazing such as the biologically inactive precursor stearoylvelutinal that, upon injury to the cell, is rapidly converted to strongly antibiotic and pungent sesquiterpenoid dialdehydes such as isovellerral (Mier *et al.*, 1996). These compounds are active against a range of arthropod grazers (Stadler and Sterner, 1998).

Interactions between hyphae and potential grazers may also result in the death of the grazing animals, although the underlying mechanisms are not known. This may be of significance with regard to nutitional stress since it has been shown that nitrogen can be transferred from soil animals to the mycorrhizal fungus and subsequently to the host plant in both Collembola (Klironomos and Hart, 2001) and nematodes (Perez-Moreno and Read, 2001). Grazing by Collembola in an upland grassland ecosystem has been shown to disrupt the flow of carbon through mycorrhizal mycelial networks (Johnson *et al.*, 2005). The consequences of this for the fungi themselves have not yet been fully investigated but grazing animals may affect the interactions of mycorrhizal fungi with other fungi. Tiunov and Scheu (2005) recently demonstrated that interactions between AM fungi and fungal saprotrophs are influenced in turn by the presence of Collembola.

10. MICROBIAL INTERACTIONS

Mycorrhizal fungi are subject to different types of stress through interactions with other fungi and bacteria. These include indirect interactions involving competition for nutrients and other resources, as well as direct biotic interactions that may be either stimulatory or antagonistic. Bacteria and fungi involved in decomposition may compete for plant derived substrates, but mutualistic strategies have also evolved involving the utilisation of fungal exudates (see de Boer *et al.*, 2005), or even endosymbiotic relationships (Bertaux *et al.*, 2003; Jargeat *et al.*, 2004).

Interactions between microorganisms may occur on different spatial scales. On the smallest scale, bacteria living in close proximity to mycorrhizal hyphae affect the fungi by altering their common chemical environment. The mycorrhizal mycelium routes labile carbon into the soil environment, and the large surface area of the mycelium may constitute an attractive environment for bacterial colonisation (Toljander *et al.*, 2005). Some bacteria may directly exert stress on the mycelium (Riedlinger *et al.*, 2006), interacting antagonistically with the hyphae and subsequently utilising the mycelium as a nutrient source (de Boer *et al.*, 2005). Other bacteria may compete for nutrients in the hyphosphere. Such competition is, however, not likely to involve much nutritional stress to the fungus, as immobilisation of resources in bacteria is only transitory, and in most cases, the resources will eventually become available to the fungus. Bacteria associated with mycorrhizal mycelium may also act to reduce environmental stress in the hyphal vicinity (Lindahl and Olsson, 2004), increasing nutrient availability or controlling antagonistic microbial populations. There, thus, seems to be an intricate interplay between the fungal and bacterial soil communities. The identity and vigour of the fungal mycelium that is present in the soil greatly influence the composition of the bacterial community (Timonen *et al.*, 1998; Tornberg *et al.*, 2003), and the bacterial community may in turn both stimulate and inhibit mycelial growth (Riedlinger *et al.*, 2006). These positive and negative mycorrhizosphere interactions may ultimately have applications in forestry and agriculture (Johansson *et al.*, 2004) but, at present, our knowledge about them is extremely limited due to their spatial complexity and small scale.

In contrast to bacteria, fungal individuals may occupy large volumes of soil, and their capacity to store and translocate resources implies that they often compete for control of the substrate rather than directly for resources within the substrate. Wu *et al.* (1999) demonstrated that an ECM individual may displace the mycelium of another species from the soil in competition for root tips and also replace the inferior competitor on already colonised roots. Lindahl *et al.* (1999) showed that ECM fungi that are successful in competition with saprotrophic fungi may gain nutrients from the saprotrophic mycelium. If the balance in inoculum potential (= energy availability) is shifted in favour of the saprotroph, however, the mycorrhizal mycelium may be replaced by the saprotroph and outcompeted from the soil (Lindahl *et al.*, 2001). Similarily, Leake *et al.* (2001) showed that a saprotrophic fungus may exert high levels of stress on an ECM mycelium, and that a seemingly localised antagonistic interaction may reduce the vigour of the entire mycelium and the amount of carbon allocated to extraradical hyphae (Figure 4). On the spatial scale of an ecosystem, Lindahl *et al.* (2006)

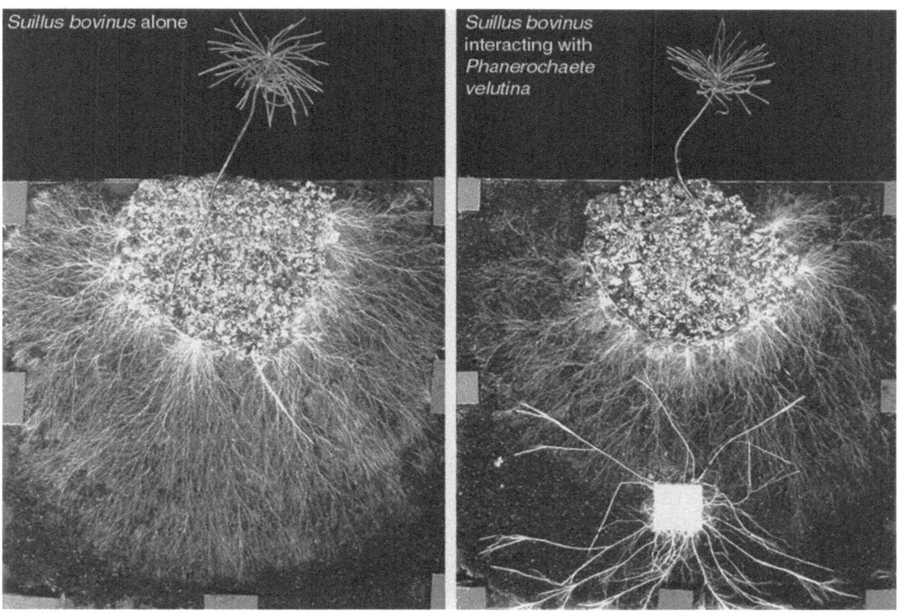

Figure 4 (a) Microcosm containing mycelium of *Suillus bovinus* mycorrhizal with *Pinus sylvestris*, forming an almost complete covering of the surface of the peat. (b) Microcosm in which the mycelium of *S. bovinus* has interacted with mycelial cords of the wood-decomposer fungus *Phanerochaete velutina* growing from a wood block. Note the difference in the density of cover and extent of mycorrhizal mycelium compared with Figure 2a. (From Leake *et al.*, 2001, with permission). **(See Colour Section)**

recently showed that the mycorrhizal and saprotrophic communities in a boreal forest were spatially separated with saprotrophs occupying relatively fresh litter on the surface of the forest floor and mycorrhizal fungi occupying the underlying more degraded litter, humus and mineral soil. This observation supports the hypothesis that these two functional groups of fungi may compete with each other for space, and that their different energy acquisition strategies determine their realised niches in the soil.

Mycorrhizal fungi may thus live under constant stress in the form of antagonistic interference from other fungal individuals. Toljander *et al.* (2006) studied how species diversity of wood-rotting fungi affected the decomposing efficiency of the community. Under stable environmental conditions, decomposition rates and metabolic efficiency decreased with the number of species added to the system, indicating a high-metabolic cost associated with interactions. Regardless of initial diversity, the fungal community was subsequently reduced to one or two species, showing that positive feedbacks during competitive colonisation of a substrate eventually lead to dominance by the most competitive individual (Lindahl and Olsson, 2004). The addition of an external environmental stress, constituted by fluctuating temperature, permitted a higher degree of co-existence of different species, presumably due to interspecific differences in temperature optima. Furthermore, intermediate levels of species diversity increased the decomposing capacity of the community under fluctuating temperatures, indicating a higher degree of stress tolerance of a mixed community than of single species (Toljander *et al.*, 2006).

11. COMPETITION FOR HOST ROOTS—EVOLUTION AND SPECIFICITY

Successful growth of mycorrhizal fungi depends upon having access to roots of compatible host plants that are able to provide them with photosynthetically derived carbon compounds. Recent phylogenetic analyses of free-living, saprotrophic and mycorrhizal homobasidiomycetes (Hibbett *et al.*, 2000) suggest that mycorrhizal symbionts have evolved repeatedly from saprotrophic precursors and there may also have been multiple reversals to a free-living condition, supporting the view that mycorrhiza are unstable, evolutionarily dynamic associations.

Over 400 achlorophyllous plant species in 87 genera are known to be parasitic upon fungi, exploiting them as a principal source of carbon (Leake, 2005). In the Orchidaceae, over 100 species are fully myco-heterotrophic but a further 30,000 species have dust-like seeds with no carbon reserves and are initially myco-heterotrophic, depending upon fungi until they develop green leaves. The identities of the fungi involved with mycoheterotrophs are being unravelled using DNA based identification and many have now been shown to be AM (Bidartondo *et al.*, 2002), or ECM (Leake *et al.*, 2004), deriving their own carbon from fully photosynthetic plant hosts. The myco-heterotrophic plants are therefore effectively epiparasitic on photosynthetic plant hosts *via* a shared mycelial network. Epiparasitic plants have consistently been revealed to have extremely high levels of fungal specificity, irrespective of which of the major groups of mycorrhizal fungi they associate with. In each case only a very narrow range of fungi is involved and the fungal distribution appears to control the plant distribution, providing a clear example of the way in which mycorrhizal-community structure may control the success of plant species. Different explanations have been advanced to account for the extreme fungal specificity. Specialisation might allow plants to adapt to particular fungi to enable efficient capture of carbon; alternatively most fungi may develop resistance to "cheating" plants from which they derive no carbon, forcing the plants to specialise on fungi without this resistance. From the fungal perspective, there is clearly a selective advantage to be obtained from having a relatively broad host range to minimise disruptions to carbon supply caused by the lack of available hosts.

12. CONCLUDING REMARKS

Mycorrhizal fungi are similar to other filamentous fungi in the range of responses they exhibit to different stress factors. Different types of mycorrhizal fungi occupy different habitats with a wide range of environmental extremes but they share the common feature that they derive their carbon directly from plant hosts and are therefore subject to both direct and plant-mediated stresses. Responses to both abiotic and biotic stresses include alteration of morphology, modification of the external environment and altered internal metabolism. The concept of stress is problematical in the case of filamentous fungi since mycelia may have large dimensions, extending over many metres, and are able to re-allocate their biomass and transport nutrients and carbon between patchily distributed resources. At the same time, the individual hyphae are microscopic in diameter and may experience a local environment that is dissimilar from the bulk environment, which can be easily measured. Responses to different types of stress can occur on different spatial scales and at different time scales. Many responses at the hyphal or mycelial level probably involve autoregulatory responses of the type discussed by Ugalde (2006) and ultimately some type of programmed cell death (Lu, 2006) to allow re-allocation of resources and mycelial differentiation. These processes have so far only been studied in non-mycorrhizal, model fungi, however mycorrhizal fungi play such a central role in plant soil interactions influencing plant health and yield that the ultimate rewards of applying available knowledge to mycorrhizal fungi should be very high indeed.

REFERENCES

Ahonen-Jonnarth, U., Göransson, A. and Finlay, R.D. (2003). Growth and nutrient uptake of ectomycorrhizal *Pinus sylvestris* seedlings treated with elevated Al concentrations. *Tree Physiology*, **23**, 157–167.

Ahonen-Jonnarth, U., van Hees, P.A.W., Lundström, U. and Finlay, R.D. (2000). Production of organic acids by mycorrhizal and non-mycorrhizal *Pinus sylvestris* seedlings exposed to elevated concentrations of aluminium and heavy metals. *New Phytologist*, **146**, 557–567.

Al-Karaki, G.N., Hammad, R. and Rusan, M. (2001). Response of two tomato cultivars differing in salt tolerance to inoculum with mycorrhizal fungi under salt stress. *Mycorrhiza*, **11**, 43–47.
Andersson, S. and Söderström, B. (1995). Effects of lime ($CaCO_3$) on ectomycorrhizal colonisation of *Picea abies* (L.) Karst. seedlings planted in a spruce forest. *Scandinavian Journal of Forest Research*, **10**, 149–154.
Andersson, S. and Valeur, I. (1994). Influence of lime on soil respiration, leaching of DOC, and C/S relationships in the mor humus of a haplic podsol. *Environment International*, **20**, 81–88.
Arnebrant, K. (1994). Nitrogen amendments reduce the growth of extramatrical mycelium. *Mycorrhiza*, **5**, 7–15.
Arnolds, E. (1991). Decline of ectomycorrhizal fungi in Europe. *Agriculture Ecosystems & Environment*, **35**, 209–244.
Augé, R.M. (2001). Water relations, drought and vesicular-arbuscular mycorrhizal symbiosis. *Mycorrhiza*, **11**, 3–42.
Baar, J. and ter Braak, C.J.F. (1996). Ectomycorrhizal sporocarp occurrence as affected by manipulation of litter and humus layers in Scots pine stands of different age. *Applied Soil Ecology*, **4**, 61–73.
Bakker, M.R., Garbaye, J. and Nys, C. (2000). Effect of liming on the ectomycorrhizal status of oak. *Forest Ecology and Management*, **126**, 121–131.
Bertaux, J., Schmid, M., Prevost-Boure, N.C., Churin, J.L., Hartmann, A., Garbaye, J. and Frey-Klett, P. (2003). *In situ* identification of intracellular bacteria related to *Paenibacillus* spp. in the mycelium of the ectomycorrhizal fungus *Laccaria bicolor* S238N. *Applied and Environmental Microbiology*, **69**, 4243–4248.
Bidartondo, M.I., Redecker, D., Hijri, I., Wiemken, A., Bruns, T.D., Dominguez, L., Sersic, A., Leake, J.R. and Read, D.J. (2002). Epiparasitic plants specialized on arbuscular mycorrhizal fungi. *Nature*, **419**, 389–392.
de Boer, W., Folman, L.B., Summerbell, R.C. and Boddy, L. (2005). Living in a fungal world: Impact of fungi on soil bacterial niche development. *FEMS Microbiology Reviews*, **29**, 795–811.
de la Providencia, I.E., de Souza, F.A., Fernández, F., Delmas, N.S. and Declerck, S. (2005). Arbuscular mycorrhizal fungi reveal distinct patterns of anastomosis formation and hyphal healing mechanisms between different phylogenic groups. *New Phytologist*, **165**, 261–271.
Erland, S. and Söderström, B. (1991). Effects of lime and ash treatments on ectomycorrhizal infection of *Pinus sylvestris* L. seedlings planted in a pine forest. *Scandinavian Journal of Forest Research*, **6**, 519–526.
Erland, S. and Taylor, A.F.S. (2002). Diversity of ectomycorrhizal fungal communities in relation to the abiotic environment. In: *Mycorrhizal Ecology* (M.G.A. van der Heijden and I.R. Sanders, eds.), *Ecological Studies*, Vol. 157, pp. 263–200. Springer-Verlag, Heidelberg.
Falkengren-Grerup, U. and Eriksson, H. (1990). Changes in soil, vegetation and forest yield between 1947 and 1988 in beech and oak sites of southern Sweden. *Forest Ecology & Management*, **38**, 37–53.
Falkengren-Grerup, U., Linnermark, N. and Tyler, G. (1987). Changes in acidity and cation pools of south Sweden soils between 1949 and 1985. *Chemosphere*, **16**, 2239–2248.
Finlay, R.D. (1995). Interactions between soil acidification, plant growth and nutrient uptake in ectomycorrhizal associations of forest trees: Problems and progress. In: *Effects of Acid Deposition and Ozone on the Terrestrial Environment in Sweden*, (H. Staaf and G. Tyler, eds.), *Ecological Bulletin*, Vol. 44, pp. 197–214. Munksgaard, Copenhagen.
Fomina, M.A., Alexander, I.J., Colpaert, J.V. and Gadd, G.M. (2005). Solubilization of toxic metal minerals and metal tolerance of mycorrhizal fungi. *Soil Biology & Biochemistry*, **37**, 851–866.
Giovannetti, M., Azzolini, D. and Citernesi, A.S. (1999). Anastomosis formation and nuclear and protoplasmic exchange in arbuscular mycorrhizal fungi. *Applied and Environmental Microbiology*, **65**, 5571–5575.
Hahn, G. and Marschner, H. (1998). Effect of acid irrigation and liming on root growth of Norway spruce. *Plant Soil*, **199**, 11–22.
van der Heijden, M.G.A., Klironomos, J.N., Ursic, M., Moutoglis, P., Streitwolf-Engel, R., Boller, T., Wiemken, A., Sanders, I.R. (1998). Mycorrhizal fungal diversity determines plant biodiversity, ecosystem variability and productivity. *Nature*, **396**, 69–72.
Hibbett, D.S., Gilbert, L.-B. and Donaghue, M.J. (2000). Evolutionary instability of ectomycorrhizal symbioses in basidiomycetes. *Nature*, **407**, 506–508.

Högberg, P., Nordgren, A., Buchmann, N., Taylor, A.F.S., Ekblad, A., Högberg, M.N., Nyberg, G., Ottosson-Löfvenius, M. and Read, D.J. (2001). Large-scale forest girdling shows that current photosynthesis drives soil respiration. *Nature*, **411**, 789–792.

Jargeat, P., Cosseau, C., Ola'h, B., Jauneau, A., Bonfante, P., Batut, J. and Bécard, G. (2004). Isolation, free-living capacities, and genome structure of "Candidatus Glomeribacter gigasporarum," the endocellular bacterium of the mycorrhizal fungus *Gigaspora margarita*. *Journal of Bacteriology*, **186**, 6876–6884.

Jennings, D.H. (ed.) (1993). *Stress Tolerance of Fungi*. Marcel Dekker, New York.

Johansson, J., Paul, L.R. and Finlay, R.D. (2004). Microbial interactions in the mycorrhizosphere and their significance for sustainable agriculture. *FEMS Microbiology Ecology*, **48**, 1–12.

Johnson, D., Krsek, M., Wellington, E.M.H., Stott, A.W., Cole, L., Bardgett, R.D., Read, D.J. and Leake, J.R. (2005). Soil invertebrates disrupt carbon flow through fungal networks. *Science*, **309**, 1047.

Johnson, D., Leake, J.R., Ostle, N., Ineson, P. and Read, D.J. (2002). *In situ* (CO_2)-^{13}C pulse-labelling of upland grassland demonstrates a rapid pathway of carbon flux from arbuscular mycorrhizal mycelia to the soil. *New Phytologist*, **153**, 327–334.

Johnson, D., Leake, J.R. and Read, D.J. (2001). Novel in-growth core system enables functional studies of grassland mycorrhizal mycelial networks. *New Phytologist*, **152**, 555–562.

Johnson, D., Vandenkoornhuyse, P., Leake, J.R., Gilbert, L., Booth, R., Grime, J., Young, J.P.W. and Read, D.J. (2003). Plant communities affect arbuscular mycorrhizal fungal diversity and community composition in grassland microcosms. *New Phytologist*, **161**, 503–515.

Jonsson, T., Finlay, R.D., Kokalj, S. and Erland, S. (1999). Effects of liming on ectomycorrhizal community structure in a spruce forest. *Mycological Research*, **103**, 501–508.

Jumpponen, A., Trowbridge, J., Mandyam, K. and Johnson, L. (2005). Nitrogen enrichment causes minimal changes in arbuscular mycorrhizal colonization but shifts community composition—Evidence from rDNA data. *Biology and Fertility of Soils*, **41**, 217–224.

Klironomos, J.N. and Hart, M.M. (2001). Food-web dynamics—Animal nitrogen swap for plant carbon. *Nature*, **410**, 651–652.

Kreutzer, K. (1995). Effects of forest liming on soil processes. *Plant and Soil*, **168–169**, 447–470.

Kuikka, K., Härmä, E., Markkola, A.M., Rautio, P., Roitto, M., Saikkonen, K., Ahonen-Jonnarth, U., Finlay, R. and Tuomi, J. (2003). Severe defoliation of scots pine reduces reproductive investment by ectomycorrhizal symbionts. *Ecology*, **84**, 2051–2061.

Lanfranco, L., Novero, M. and Bonfante, P. (2005). The mycorrhizal fungus *Gigaspora margarita* possesses a CuZn superoxide dismutase that is up-regulated during symbiosis with legume hosts. *Plant Physiology*, **137**, 1319–1330.

Leake, J.R. (2005). Plants parasitic on fungi: Unearthing the fungi in myco-heterotrophs and debunking the 'saprophytic' plant myth. *Mycologist*, **19**, 113–122.

Leake, J.R., Donnelly, D.P., Saunders, E.M., Boddy, L. and Read, D.J. (2001). Rates and quantities of carbon flux to ectomycorrhizal mycelium following ^{14}C pulse labeling of *Pinus sylvestris* seedlings: Effects of litter patches and interaction with a wood-decomposer fungus. *Tree Physiology*, **21**, 71–82.

Leake, J.R., McKendrick, S.L., Bidartondo, M. and Read, D.J. (2004). Symbiotic germination and development of the myco-heterotroph *Monotropa hypopitys* in nature and its requirement for locally distributed *Tricholoma* spp. *New Phytologist*, **163**, 405–423.

Lehto, T. (1994). Effects of liming and boron fertilization on mycorrhizas of *Picea abies*. *Plant and Soil*, **163**, 65–68.

Lindahl, B.D., Ihrmark, K., Trumbore, S., Högberg, P., Stenlid, J. and Finlay, R.D. (2007). Spatial separation of litter decomposition and mycorrhizal nitrogen uptake in boreal forests. *New Phytologist*, **173**, 611–620.

Lindahl, B. and Olsson, S. (2004). Fungal translocation—Creating and responding to environmental heterogeneity. *Mycologist*, **18**, 79–88.

Lindahl, B., Stenlid, J. and Finlay, R.D. (2001). Effects of resource availability on mycelial interactions and ^{32}P transfer between a saprotrophic and an ectomycorrhizal fungus in soil microcosms. *FEMS Microbiology Ecology*, **38**, 43–52.

Lindahl, B., Stenlid, J., Olsson, S. and Finlay, R.D. (1999). Translocation of ^{32}P between interacting mycelia of a wood decomposing fungus and ectomycorrhizal fungi in microcosm systems. *New Phytologist*, **144**, 183–193.

Lu, B.C.K. (2006). Programmed cell death in fungi. In: *The Mycota, Growth, Differentiation and Sexuality* (U. Kües and R. Fischer, eds.), Vol. I. Springer, Berlin.

Mahmood, S., Finlay, R.D., Erland, S. and Wallander, H. (2000). Solubilisation and colonisation of wood ash by ectomycorrhizal fungi isolated from a wood ash fertilised spruce forest. *FEMS Microbiology Ecology*, **35**, 151–161.

Marschner, H. (1991). Mechanisms of adaptation of plants to acid soils. *Plant and Soil*, **134**, 1–20.

Mier, N., Canete, S., Klaebe, A., Chavant, L. and Fournier, D. (1995). Insecticidal properties of mushroom and toadstool carpophores. *Phytochemistry*, **41**, 1293–1299.

Nowotny, I., Dähne, J., Klingelhöfer, D. and Rothe, G.M. (1998). Effect of artificial soil acidification and liming on growth and nutrient status of mycorrhizal roots of Norway spruce (*Picea abies* [L.] Karst.), *Plant and Soil*, **199**, 29–40.

Ott, T., Fritz, E., Polle, A. and Schützendübel, A. (2002). Characterisation of antioxidative systems in the ectomycorrhiza-building basidiomycete *Paxillus involutus* (Bartsch) Fr. and its reaction to cadmium. *FEMS Microbiology Ecology*, **42**, 359–366.

Ouziad, F., Wilde, P., Schmelzer, E., Hildebrandt, U. and Bothe, H. (2006). Analysis of expression of aquaporins and Na+/H+ transporters in tomato colonized by arbuscular mycorrhizal fungi and affected by salt stress. *Environmental and Experimental Botany*, **57**, 177–186.

Perez-Moreno, J. and Read, D.J. (2001). Nutrient transfer from soil nematodes to plants: A direct pathway provided by the mycorrhizal mycelial network. *Plant Cell and Environment*, **24**, 1219–1226.

Persson, T., Lundkvist, H., Wiren, A., Hyvönen, R. and Wessen, B. (1989). Effects of acidification and liming on carbon and nitrogen mineralization and soil organisms in mor humus. *Water Air & Soil Pollution*, **54**, 351–364.

Peter, M., Ayer, F. and Egli, S. (2001). Nitrogen addition in a Norway spruce stand altered macromycete sporocarp production and below-ground ectomycorrhizal species composition. *New Phytologist*, **49**, 311–325.

Porcel, R., Aroca, R., Azcón, R. and Ruiz-Lozano, J.M. (2006). PIP aquaporin gene expression in arbuscular mycorrhizal *Glycine max* and *Lactuca sativa* plants in relation to drought stress tolerance. *Plant Molecular Biology*, **60**, 389–404.

Priha, Q. and Smolander, A. (1994). Fumigation-extraction and substrate-induced respiration derived microbial biomass C, and respiration rate in limed soil of Scots pine sapling stand. *Biology & Fertility of Soils*, **17**, 301–308.

Read, D.J. (1987). In support of Frank's organic nitrogen theory. *Angewandte Botanik*, **61**, 25–37.

Read, D.J. (1991). Mycorrhizas in ecosystems–Nature's response to the Law of the minimum. In: *Frontiers in Mycology*. (D.L. Hawksworth, ed.), pp. 101–130. CAB International, Wallingford, UK.

Read, D.J. and Perez-Moreno, J. (2003). Mycorrhizas and nutrient cycling in ecosystems — A journey towards relevance? *New Phytologist*, **157**, 475–492.

Riedlinger, J., Schrey, S.D., Tarkka, M.T., Hampp, R., Kapur, M. and Fiedler, H-P. (2006). Auxofuran, a novel metabolite that stimulates the growth of fly agaric, is produced by the mycorrhiza helper bacterium *Streptomyces* strain ACH 505. *Applied and Environmental Microbiology*, **72**, 3550–3557.

Ruiz-Lozano, J.M. (2003). Arbuscular mycorrhizal symbiosis and alleviation of osmotic stress. New perspectives for molecular studies. *Mycorrhiza*, **13**, 309–317.

Ruiz-Lozano, J.M., Porcel, R. and Aroca, R. (2006). Does the enhanced tolerance of arbuscular mycorrhizal plants to water deficit involve modulation of drought.induced plant genes? *New Phytologist*, **171**, 693–698.

Santos, J.C., Tehler, A. and Finlay, R.D. (2006). Molecular analysis of arbuscular mycorrhizal fungi colonising a semi-natural grassland along a fertilisation gradient. *New Phytologist*, **172**, 159–168.

Schützendübel, A. and Polle, A. (2002). Plant responses to abiotic stresses. Heavy metal-induced oxidative stress and protection by mycorrhization. *Journal of Experimental Botany*, **53**, 1351–1365.

Shear, W.A. and Kukalová-Peck, J. (1990). The ecology of Palaeozoic terrestrial arthropods: The fossil evidence. *Canadian Journal of Zoology*, **68**, 1807–1834.

Shelford, V.E. (1913: 1912). Ecological Succession. *Biological Bulletin*, **22**, 1 and **21**, 127.

Smith, S.E. and Read, D.J. (1997). *Mycorrhizal Symbiosis*. Academic Press, London.
Stadler, M. and Sterner, O. (1998). Production of bioactive secondary metabolites in the fruit bodies of macrofungi as a response to injury. *Phytochemistry*, **49**, 1013–1019.
Steenari, B.-M. and Lindqvist, O. (1999). Fly ash characteristics in co-combustion of wood with coal, oil or peat. *Fuel*, **78**, 479–488.
Taylor, A.F.S. and Alexander, I. (2005). The ectomycorrhizal symbiosis: Life in the real world. *Mycologist*, **19**, 102–112.
Taylor, A.F.S. and Brand, F. (1992). Reaction of the natural Norway spruce mycorrhizal flora to liming and acid irrigation. In: *Mycorrhizas in Ecosystems*. (D.J. Read, D.H. Lewis, A.H. Fitter and I.J. Alexander, eds.), CAB International, Wallingford, UK.
Taylor, A.F.S. and Finlay, R.D. (2003). Effects of liming and ash application on below ground ectomycorrhizal community structure in two Norway spruce forests. *Water Soil and Air Pollution: Focus*, **3**, 63–76.
Taylor, A.F.S., Martin, F. and Read, D.J. (2000). Fungal diversity in ectomycorrhizal communities of Norway spruce (*Picea abies* (L.) Karst.) and Beech (*Fagus sylvatica* L.) in forests along north-south transects in Europe. In: *Carbon and Nitrogen Cycling in European Forest Ecosystems* (E.-D. Schulze, ed.), *Ecological Studies*, Vol. 142, pp 343–365. Springer-Verlag, Heidelberg.
Timonen, S., Jorgensen, K.S., Haahtela, K. and Sen, R. (1998). Bacterial community structure at defined locations of *Pinus sylvestris Suillus bovinus* and *Pinus sylvestris Paxillus involutus* mycorrhizospheres in dry pine forest humus and nursery peat. *Canadian Journal of Microbiology*, **44**, 499–513.
Tiunov, A.V. and Sheu, S. (2005). Arbuscular mycorrhiza and Collembola interact in affecting community composition of saprotrophic microfungi. *Oecologia*, **142**, 636–642.
Toljander, J.F., Artursson, V., Paul, L.R., Jansson, J.K. and Finlay, R.D. (2005). Attachment of different soil bacteria to arbuscular mycorrhizal fungi is determined by hyphal vitality and fungal species. *FEMS Microbiology Letters*, **254**, 34–40.
Toljander, Y.K., Lindahl, B.D., Holmer, L. and Högberg, N. (2006). Environmental fluctuations facilitate species co-existence and increase decomposition in communities of wood decay fungi. *Oecologia*, **148**, 625–631.
Tornberg, K., Bååth, E. and Olsson, S. (2003). Fungal growth and effects of different wood decomposing fungi. *Biology and Fertility of Soils*, **37**, 190–197.
Treseder, K.K. and Allen, M.F. (2002). Direct nitrogen and phosphorus limitation of arbuscular mycorrhizal fungi: A model and field test. *New Phytologist*, **155**, 507–515.
Ugalde, U. (2006). Autoregulatory signals in mycelial fungi. In: *The Mycota, Growth, Differentiation and Sexuality* (U. Kües and R. Fischer, eds.), Vol. I, pp. 203–213. Springer, Berlin.
Wallander, H. (1995). A new hypothesis to explain allocation of dry-matter between mycorrhizal fungi and pine-seedlings in relation to nutrient supply. *Plant and Soil*, **169**, 243–248.
Wallander, H. and Nylund, J.-E. (1992). Effects of excess nitrogen and phosphorus starvation on the extramatrical mycelium of ectomycorrhizas on *Pinus sylvestris* L. *New Phytologist*, **119**, 405–411.
Wallenda, T. and Kottke, I. (1998). Nitrogen deposition and ectomycorrhizas. *New Phytologist*, **139**, 169–187.
Wu, B., Nara, K. and Hogetsu, T. (1999). Competition between ectomycorrhizal fungi colonizing *Pinus densiflora*. *Mycorrhiza*, **9**, 151–159.

CHAPTER 14

Regulation of Protein Synthesis in Yeast by Oxidative Stress

Daniel Shenton, Claire Mascarenhas and **Chris M. Grant**

Contents

1.	Introduction	222
2.	Translation Initiation is a Key Regulated Step in Gene Expression	222
3.	Regulation of Translation Initiation in Yeast	223
4.	Regulation of Translation Elongation	226
5.	mRNA-Specific Translational Control	227
6.	Adaptation to Oxidant Stress	229
7.	Future Perspectives	231
	Acknowledgement	231
	References	231

Abstract

All organisms must be able to respond to changes in their external environment. With the availability of genome sequences, much attention has focused on analysing the changes in gene expression/transcription profiles during these adaptive responses. The translation of mRNA into protein is a fundamental component of the gene expression pathway. However, relatively little is known regarding the role of translational control mechanisms in response to stress conditions. Cells typically respond to stress conditions by invoking complex regulatory mechanisms including global inhibition of translation. This reduction in protein synthesis may prevent continued gene expression during potentially error-prone conditions as well as allow for the turnover of existing mRNAs and proteins whilst gene expression is reprogrammed to deal with the stress. The initiation phase of translation has long been thought to be the main target of regulation and represents a key control point for eukaryotic gene expression. Recent data have also implicated control at the post-initiation phase of translation in response to oxidants. This chapter reviews recent studies on yeast and how these findings are leading to a better understanding of the role of translational control in eukaryotic gene expression.

British Mycological Society Symposia Series
Published by Elsevier Ltd.

1. INTRODUCTION

All aerobic organisms are exposed to reactive oxygen species (ROS), such as H_2O_2, the superoxide anion and the hydroxyl radical, during the course of normal aerobic metabolism or following exposure to radical-generating compounds. These ROS cause wide-ranging damage to macromolecules eventually leading to cell death (Halliwell and Gutteridge, 1989; Gutteridge, 1993). The yeast *Saccharomyces cerevisiae*, like all organisms, contains effective antioxidant defence mechanisms, which detoxify ROS as they are generated and maintain the intracellular redox environment in a reduced state (Temple *et al.*, 2005). Protective enzymes include catalases, superoxide dismutases and glutathione peroxidases, which are present in different subcellular compartments such as the mitochondrion and vacuole. Non-enzymic defences typically consist of small molecules that can act as free radical scavengers and include glutathione (GSH), uric acid and vitamins C and E (Jamieson, 1998). Yeast cells can respond to oxidative stress by altering global transcription patterns including genes coding for antioxidants and other metabolic enzymes (Gasch *et al.*, 2000; Causton *et al.*, 2001). However, recent data have shown that exposure to oxidative stress conditions causes a rapid and reversible inhibition of protein synthesis (Shenton and Grant, 2003; Dunand-Sauthier *et al.*, 2005; Swaminathan *et al.*, 2006). Thus, it is unclear how changes in the transcriptome are translated into the cellular proteome.

2. TRANSLATION INITIATION IS A KEY REGULATED STEP IN GENE EXPRESSION

The initiation phase of protein synthesis is rate limiting and is a target of extensive regulation. It is a complex process involving more than 30 polypeptide factors interacting with ribosomal subunits, methionyl-tRNA (Met-$tRNA_i^{met}$) and mRNAs (Kapp and Lorsch, 2004). Most studied forms of regulation involve protein phosphorylation and the interaction of accessory proteins with translation factors (Hinnebusch, 2000). Eukaryotic initiation factor 2 (eIF2) is one such regulated factor and plays a key role in the response to oxidative stress. eIF2 is a guanine nucleotide-binding factor, which in its GTP-bound form interacts with the initiator Met-$tRNA_i^{met}$ to form a ternary complex that is competent for translation initiation (Fig. 1). Following each round of initiation, eIF2 is released from the ribosome as a binary complex with GDP. GDP is removed and replaced by GTP in a guanine nucleotide exchange reaction catalysed by eIF2B. Met-$tRNA_i^{met}$ can only bind the eIF2/GTP complex, so translational control can be regulated by the activity of eIF2B (Fig. 1). In both yeast and mammals this is achieved by phosphorylation of the alpha subunit of eIF2 at a conserved serine (Ser51) residue (Pavitt *et al.*, 1998; Harding *et al.*, 2000). Phosphorylation converts eIF2 from a substrate to a competitive inhibitor of the guanine nucleotide exchange factor eIF2B and the resulting decrease in eIF2B activity leads to reduced ternary complex levels, which inhibits translation initiation (Pavitt *et al.*, 1998).

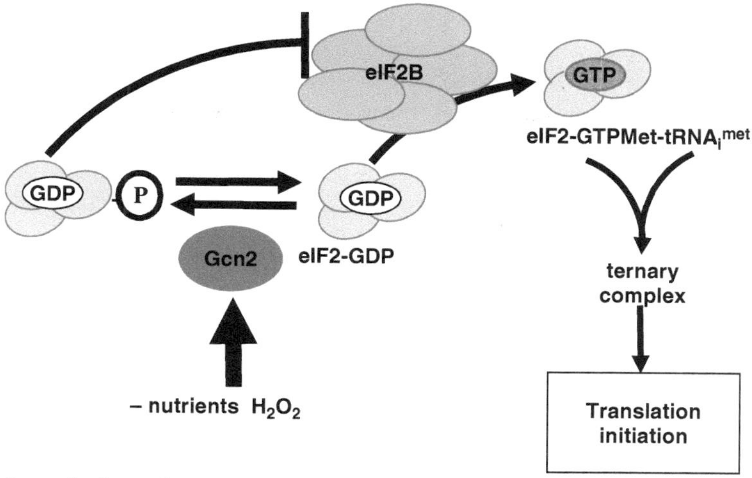

Figure 1 *Control of translation initiation by eIF2.* Gcn2 phosphorylates eIF2 in response to nutrient starvation and oxidative stress. Phosphorylation of eIF2 converts it into a competitive inhibitor of the eIF2B guanine nucleotide exchange factor. Decreased eIF2B activity generates less eIF2 in the GTP bound form, resulting in decreased ternary complex levels and inhibition of translation initiation.

Four mammalian kinases have been identified which inhibit translation initiation by phosphorylating eIF2α. GCN2 (the amino acid control kinase), PKR (the double-stranded RNA-activated protein kinase), HRI (the haem-regulated inhibitor) and PERK/PEK (the PKR-like endoplasmic reticulum (ER) eIF2α kinase) are regulated independently in response to various different cellular stresses (Dever, 2002; Proud, 2005). For example, PERK has been found in all multicellular eukaryotes and is a component of the unfolded protein response (UPR). Consistent with its central role in the ER stress response, cells lacking PERK fail to phosphorylate eIF2α and do not down-regulate protein synthesis during ER stress conditions (Bertolotti *et al.*, 2000). Attenuating protein synthesis may act to reduce the burden of newly synthesised ER client proteins on the ER folding machinery. Additionally, eIF2α phosphorylation induces translation of specific mRNAs, such as that encoding the metazoan activating transcription factor 4 (*ATF4*) (Lu *et al.*, 2004; Vattem and Wek, 2004). *ATF4* mediates the "integrated stress response" whose targets include genes encoding proteins involved in amino acid metabolism and resistance to oxidative stress, ultimately protecting against the deleterious consequences of ER oxidation (Harding *et al.*, 2003).

3. REGULATION OF TRANSLATION INITIATION IN YEAST

In the yeast *S. cerevisiae*, Gcn2 is the sole eIF2 kinase and phosphorylates eIF2α in response to diverse conditions including nutrient starvation, and sodium or rapamycin exposure (reviewed in Hinnebusch (2005)). As in mammalian cells, the resulting decrease in ternary complex levels reduces the bulk rate of protein synthesis. Paradoxically, translation of the *GCN4* mRNA is activated in response to low ternary complex in a mechanism involving four short upstream open reading frames (Hinnebusch, 2005). Gcn4 is a transcription factor that activates gene expression of many targets including amino acid biosynthetic genes (Natarajan *et al.*, 2001). Thus, analogous to the mammalian integrated stress response, activation of Gcn4 serves to overcome the imposed starvation, which initially led to the translational control. More recently, we have used microarray analysis combined with polysome analysis to demonstrate that lowering ternary complex levels results in widespread translational reprogramming, identifying a fundamental role for translational control in the adaptation to nutrient limitation (Smirnova *et al.*, 2005).

In mammalian cells, eIF2α is phosphorylated by PERK in response to oxidative stress as part of the integrated stress response (Harding *et al.*, 2003). Yeast does not contain PERK and hence oxidative stress had not been thought to inhibit protein synthesis. However, recent data indicate that moderating protein synthesis in response to ROS through eIF2 α phosphorylation is a highly conserved mechanism in yeasts. The fission yeast *Schizosaccharomyces pombe* has fewer eIF2 kinases than mammalian cells, containing only Gcn2 and two HRI-related protein kinases called Hri1 and Hri2 (Zhan *et al.*, 2004). Hri2 is the primary kinase that is activated in response to heat shock, arsenite or cadmium stress, whereas, Gcn2 responds to nutrient downshift, amino acid starvation and osmotic stress (Zhan *et al.*, 2004). Both Hri2 and Gcn2 are activated and phosphorylate eIF2α in response to oxidative stress induced by exposure to H_2O_2. The role of these eIF2 kinases during oxidative stress conditions was further confirmed by the sensitivity of a mutant lacking *GCN2* and *HRI2* to H_2O_2. More recently, phosphorylation of eIF2α and inhibition of protein synthesis in fission yeast has been shown to be under the control of the stress-activated protein kinase (SAPK) pathway (Dunand-Sauthier *et al.*, 2005).

We have analysed the regulation of protein synthesis in *S. cerevisiae* in response to oxidative stress induced by exposure to H_2O_2. Our data show that H_2O_2 causes a dose-dependent inhibition of protein synthesis (Shenton *et al.*, 2006). Maximum inhibition is observed at concentrations >1 mM H_2O_2 which results in approximately 90% inhibition of protein synthesis within 15 min. Translational activity has been further investigated by examining the distribution of ribosomes in response to H_2O_2. Polysomes are ribosomes that are actively translating mRNAs. They can be separated on sucrose density gradients and quantified by measuring absorbance at 254 nm. There is a dramatic shift of ribosomes from the polysomal region into the monosome or 80S peak following treatment with H_2O_2, which is indicative of decreased translation initiation (Fig. 2). Western blot analysis has confirmed that H_2O_2 induces phosphorylation of eIF2α through activation of Gcn2. Furthermore, the inhibition of translation initiation is entirely dependent on the presence of Gcn2 since it is absent in a *gcn2* mutant (Fig. 2). Regulating translation initiation via phosphorylation of eIF2α appears therefore to be a common response mechanism to ROS in eukaryotic cells.

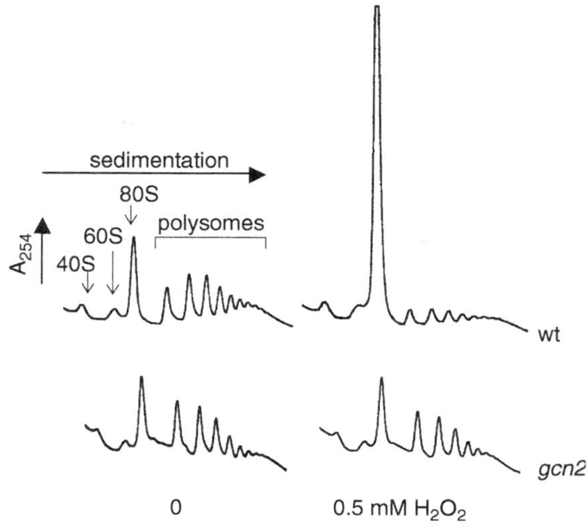

Figure 2 *Regulation of translation initiation by hydrogen peroxide.* H_2O_2-treatment specifically inhibits translation initiation. Polyribosome traces are shown for the wild-type strain treated with 0.5 mM H_2O_2 for 15 min. The peaks that contain the small ribosomal subunit (40S), the large ribosomal subunit (60S) and both subunits (80S) are indicated by arrows. The polysome peaks generated by 2, 3, 4, 5, etc. 80S ribosomes on a single mRNA are bracketed. An increase in monosomes is characteristic of a block in translation initiation. Loss of *GCN2* restores the polysome profile indicating that translation initiation is inhibited by a Gcn2-dependent mechanism.

The exact mechanism by which oxidative stress activates eIF2 kinases in eukaryotic cells is not known. Yeast Gcn2 phosphorylates eIF2α in response to various conditions including nutrient starvation and exposure to sodium or rapamycin. Activation in response to amino acid starvation is the best characterised regulatory mechanism (Hinnebusch, 2005). Depletion of amino acids leads to an accumulation of uncharged tRNA, which activates the Gcn2 protein kinase through its histidyl-tRNA synthetase (HisRS)-related domain. It is likely that other stress conditions ultimately affect the levels of uncharged tRNA in the cell. For example, volatile anaesthetics inhibit amino acid uptake (Palmer *et al.*, 2005) and Gcn2 is activated by glucose starvation partly through an effect on vacuolar amino acid pools (Yang *et al.*, 2000). The signals activating Gcn2 in response to rapamycin or NaCl are not well understood. Rapamycin appears to work by blocking Tor-mediated phosphorylation of Gcn2 at Ser577 (Cherkasova and Hinnebusch, 2003). However, activation of Gcn2 by rapamycin and NaCl still requires the HisRS-related domain of Gcn2, as well as Gcn1 and Gcn20, which are thought to mediate the activation of Gcn2 by uncharged tRNA (Narasimhan *et al.*, 2004). Similarly, the inhibition of translation initiation in response to H_2O_2 requires Gcn1 and Gcn20 (Shenton *et al.*, 2006). Oxidative stress may conceivably cause an accumulation of uncharged tRNA through a variety of mechanisms. Free amino acids and amino acids in proteins are highly susceptible to oxidation by ROS (Stadtman and Levine, 2003). Oxidised amino acids can be detected in yeast cells, and for example, oxidised phenylalanine (*m*- and *o*-Tyr) is elevated following exposure to similar concentrations of H_2O_2 (0.2–2 mM) as those that activate Gcn2 (Poljak *et al.*, 2003). Although the levels of oxidised amino acids detected in yeast cells exposed to H_2O_2 are relatively low, representing about one modification per 10^3 phenylalanine residues, the total oxidative load on the amino acid pool may trigger an amino acid starvation response. Alternatively, the proteins and nucleic acids that are required for tRNA-aminoacylation may be susceptible to oxidation resulting in an accumulation of uncharged tRNA and activation of Gcn2. For example, oxidative damage to RNAs, including tRNAs, has been implicated in the pathogenesis of Alzheimer's disease (Ding *et al.*, 2005; Honda *et al.*, 2005) and ROS can affect the stability and activity of aminoacyl-tRNA synthetases (Takahashi and Goto, 1990).

4. REGULATION OF TRANSLATION ELONGATION

Deletion of *GCN2* abrogates phosphorylation of eIF2α and results in translational resistance to amino acid starvation (Hinnebusch, 2005). Similarly, no inhibition of translation initiation is observed in a *gcn2* mutant in response to oxidative stress since loss of *GCN2* prevents the accumulation of monosomes in response to H_2O_2 (Fig. 2). However, in contrast to amino acid starvation, protein synthesis is still inhibited in response to H_2O_2 in the absence of Gcn2 (Shenton *et al.*, 2006). Loss of *GCN2* increases protein synthesis during oxidative stress conditions but does not restore it to wild-type levels. These data indicate that protein synthesis is regulated at the initiation phase via Gcn2 and is also inhibited at the post-initiation phase in response to ROS. We have confirmed directly that H_2O_2 inhibits the rate of ribosomal transit on mRNAs by measuring transit times and by analysing the rate of ribosomal run-off from mRNAs. The average mRNA transit time is increased by exposure to H_2O_2 confirming that oxidative stress can influence the rate of elongating ribosomes. Attenuation of elongating ribosomes should promote increased association of mRNAs with polysomes resulting in a decreased rate of ribosomal run-off. Ribosomal run-off occurs in untreated cells resulting in a loss of approximately 60% of polysomes within 2 min. In contrast, polysomes are maintained following treatment with H_2O_2 in agreement with the idea that oxidative stress promotes the association of ribosomes with mRNA (Shenton *et al.*, 2006). These data indicate that oxidative stress causes an inhibition of protein synthesis at the post-initiation phase, but as yet, it is unclear whether this affects translation elongation and/or termination.

Regulating mRNA expression levels by modulating translation elongation or termination is relatively poorly understood. It is known that cells can alter the bulk rate of protein synthesis in response to different growth conditions or hormones by changing the overall rates of elongation and/or termination (reviewed in Proud (2000)). There are very few characterised examples where the expression of individual mRNAs is regulated via alterations in ribosomal transit times (reviewed in Mathews *et al.* (2000)). One good example is provided by the tyrosine aminotransferase mRNA where the rate of ribosomal transit is increased fivefold in response to dibutyryl cyclic AMP (Roper and Wicks, 1978). Attenuating elongating ribosomes in response to stress conditions, as opposed to ribosomal initiation, offers the advantage that ribosomes remain bound to mRNAs and can rapidly resume proteins synthesis once the stress is removed or detoxified. For an oxidative stress condition, it would also prevent continued protein synthesis during potentially error-prone conditions.

Oxidative stress in mammalian cells promotes phosphorylation and oxidative modification of translation elongation factor 2 (eEF2) (Ayala *et al.*, 1996; Patel *et al.*, 2002). eEF2 is a crucial translation factor which catalyses the translocation of the deacylated tRNA in the ribosomal P-site and the peptidyl-tRNA in the ribosomal A-site into the E- and P- sites, respectively. Phosphorylation of eEF2 is thought to inhibit translation by reducing the affinity of eEF2 for GTP and by decreasing ribosome binding (Nairn and Palfrey, 1987; Dumont-Miscopein *et al.*, 1994). Interestingly, this regulatory mechanism appears to be conserved in yeast. eEF2 is phosphorylated by the mitogen-activated kinase, Rck2, in response to osmotic stress conditions, resulting in translation inhibition (Teige *et al.*, 2001). Rck2 is itself a substrate for phosphorylation by the Hog1 MAP kinase indicating a link between MAP kinase signalling and the control of protein synthesis (Teige *et al.*, 2001). Both Hog1 and Rck2 are phosphorylated in response to oxidative stress indicating that this signalling pathway can respond to diverse stress conditions (Bilsland *et al.*, 2004). Most recently, Rck2 has been implicated in the attenuation of elongating ribosomes during oxidative stress conditions (Swaminathan *et al.*, 2006). Exposure to oxidative (*tert*-butyl hydroperoxide) or osmotic (NaCl) stress was found to cause a pronounced dissociation of polysomes in an *rck2* mutant. Microarray analysis indicated that a number of weakly transcribed mRNAs associate more avidly with polysomes during stress conditions, consistent with a role for Rck2 in mRNA–polysome association. The authors speculated that Rck2 targets the interchange of mRNAs between polysomes and stress granules (Swaminathan *et al.*, 2006). Stress granules are thought to act as storage factors which sort mRNAs for degradation or later translation following the relief of stress conditions.

5. mRNA-SPECIFIC TRANSLATIONAL CONTROL

The use of expression-profiling techniques, such as microarray analysis, has enabled detailed quantitative comparisons of the levels of all cellular mRNAs in many organisms (Gasch *et al.*, 2000; Causton *et al.*, 2001). More recently, this technology has been extended to analyse protein synthesis by comparing polysome and monosome mRNA pools (Beilharz and Preiss, 2004). We have used microarray technology to examine the nature and extent of mRNAs, which are translationally regulated in response to two stresses that lower eIF2B activity (Smirnova *et al.*, 2005). Amino acid starvation leads to an accumulation of non-aminoacylated-tRNAs and subsequent activation of Gcn2, whereas, exposure to the fusel alcohol butanol, inhibits eIF2B activity. Surprisingly, even though the stresses impact upon the same translation initiation factor, eIF2B, they have quite different outcomes in terms of the specific mRNAs that are translationally controlled. This creates a highly specific stress response that facilitates adaptation to the particular stress condition (Smirnova *et al.*, 2005). We have suggested that although these stresses both act via eIF2B, there must be other stress-specific modulatory inputs on the translational pathway that ultimately change which mRNAs are translationally selected following stress. The impact of oxidative stress on ribosomal initiation and transit is therefore, particularly interesting, since it indicates that control of the translational machinery at a stage subsequent to initiation may be used to modulate the proteomic output.

The environmental stress response (ESR) cluster encompasses approximately 900 genes that are transcriptionally activated or repressed by a large number of stress conditions including heat shock, osmotic stress, oxidative stress and starvation (Gasch *et al.*, 2000; Causton *et al.*, 2001). The genes, which are transcriptionally induced as part of the ESR, encode products that are thought to protect against and/or detoxify the stress agent as well as repair the resulting cellular damage. This includes genes encoding products that are involved in stress defences and detoxification, protein modification and degradation, energy generation and metabolism. A large number of genes involved in growth-related processes are repressed including genes encoding products involved in protein synthesis, transcription, metabolism, and cellular transport. This presumably indicates a requirement to down-regulate energy-consuming processes during stress conditions. In addition to these general responses to stress conditions, each particular stress can invoke a unique pattern of transcriptional regulation, which appears to be specialised for the specific conditions (Gasch *et al.*, 2000). We have used microarray analysis of polysome- and monosome-associated mRNA pools to examine how these changes in transcript levels are regulated at the translational level. In particular, we identified genes which are translationally up- or down-regulated following exposure to a low (0.2 mM) or a high (2.0 mM) concentration of H_2O_2 (Shenton *et al.*, 2006). Interestingly, these two stress conditions induce distinct patterns of regulation.

The low concentration induced several antioxidants and stress-protective molecules, as might be expected. These include a cytosolic catalase (Ctt1) and an atypical 2-Cys peroxiredoxin (Gpx2) which can reduce H_2O_2 directly, thioredoxin reductase (*TRR1*), which provides the reducing power for the thioredoxin system, sulphiredoxin (*SRX1*), which can reduce cysteine-sulphinic acid residues that are formed in peroxiredoxins following oxidative stress, and a glutathione transferase (*GTT2*) and two GS-X pumps (*YCF1, YBT1*) which form part of the glutathione conjugation/removal system of cells that is active against a broad range of toxic substrates. The most prominent class of genes up-regulated by 0.2 mM H_2O_2 includes genes encoding various cellular transporters. Strikingly, nine ABC transporters were identified which are known to confer considerable multiple drug resistance. A number of metabolic genes are up- or down-regulated in response to low concentrations of H_2O_2 consistent with significant metabolic reconfiguration occurring during oxidative stress conditions. The high concentration of H_2O_2 did not significantly affect the translation of antioxidants or other stress protective molecules. It resulted in prominent up-regulation of genes involved in ribosome biogenesis and rRNA processing. This is in contrast to the large number of similar genes that are transcriptionally repressed as part of the ESR (Gasch and Werner-Washburne, 2002). It is unclear why genes which are transcriptionally repressed by oxidants may be translationally activated but it may represent a means of "fine-tuning" expression levels. These data indicate that there may be a requirement to replace ribosomal proteins and rRNA, which become damaged by oxidative stress. In contrast to the low peroxide treatment, there was no significant up-regulation of metabolic genes in response to high concentrations of H_2O_2. However, many genes affecting several aspects of metabolism are down-regulated in response to 2.0 mM H_2O_2 in agreement with the idea that metabolic reconfiguration is required in response to oxidative stress. Surprisingly, a significant number of the genes that are translationally down-regulated in response 2.0 mM H_2O_2 are increased at the transcript level. These data indicate that certain genes are increased at the transcriptional level in response to H_2O_2, but remain poorly translated. Increasing transcript levels in the absence of active translation may provide a source of mRNAs, which can become rapidly translated once the stress is removed (Shenton *et al.*, 2006).

6. ADAPTATION TO OXIDANT STRESS

In *S. cerevisiae* there are a number of cellular responses that ensure the survival of the cell following exposure to oxidants. Cells can adapt to oxidative stress, becoming more resistant to a subsequent high dose following exposure to a low, non-lethal dose of a particular ROS (including H_2O_2, superoxide anion) or product of oxidation (Collinson and Dawes, 1992; Jamieson, 1992; Turton *et al.*, 1997). Similar adaptive responses have been shown in higher eukaryotes (Wiese *et al.*, 1995). Whilst there has been considerable research on cellular antioxidant defence systems, less is known about how this adaptation occurs. Its nature depends on the treatment. Heat shock induces cells to become resistant to most oxidants, but oxidative stress does not have the reverse effect. Cells adapted to H_2O_2 treatment become resistant to menadione (a superoxide generator), but not vice versa (Temple *et al.*, 2005). This hierarchical response to stress may indicate the existence of a number of different adaptation systems, which have overlapping components. In most cases, the adaptive response depends on gene activation and *de novo* protein synthesis.

In yeast cells, low concentrations of H_2O_2 (0.2 mM) induce resistance to a subsequent challenge with a high lethal (2 mM) dose of H_2O_2 (Grant *et al.*, 1997; and Fig. 3A). The high concentration of H_2O_2 inhibits yeast cell growth for several hours, but cells can recover following relief of the stress and extended incubation periods. Both concentrations of peroxide inhibit protein synthesis, but the high concentration inhibits $>96\%$ of total protein synthesis (Fig. 3B). Similarly, both concentrations reduce cell viability, with the greatest effect seen with the 2 mM concentration of H_2O_2 (Fig. 3B). Adaptation may be explained by the pattern of proteins that are produced in response to different concentrations of oxidants. Metabolic labelling and two-dimensional gel electrophoresis have

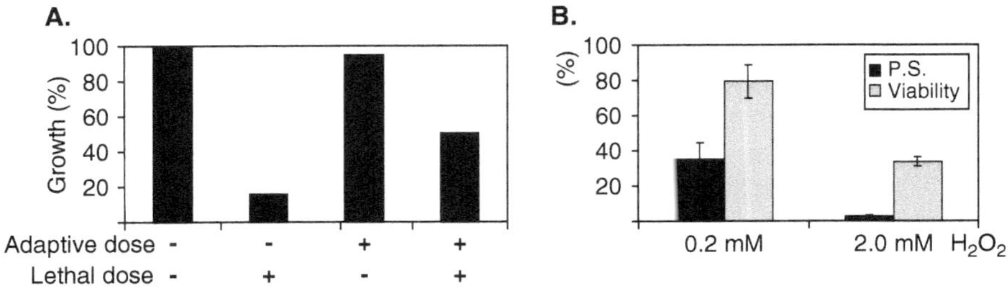

Figure 3 *Adaptive response to hydrogen peroxide.* (A) A wild-type strain of *S. cerevisiae* was either pretreated with 0.2 mM H_2O_2 for 1 h (adaptive dose) then challenged with 2 mM H_2O_2 (lethal dose) to test for lethality, or exposed to 2 mM H_2O_2 without pretreatment. Growth was monitored by absorbance at 600 nm. (B) The adaptive and lethal doses of H_2O_2 inhibit protein synthesis (P.S.) and reduce viability to different extents. Protein synthesis was measured by pulse labelling with [35S]-cysteine/methionine for 5 min. Data are shown following treatment with H_2O_2 for 15 min relative to untreated control cultures (100%). Cells were diluted and plated on Yeast extract, Peptone, dextrose (YEPD) to monitor cell viability, which is expressed as percent survival relative to untreated control cultures (100%).

been used to examine the changes in gene expression that occur in response to low adaptive concentrations of H_2O_2 (Godon *et al.*, 1998). This study reported that a substantial number of proteins are induced or repressed as a result of H_2O_2 treatment. As expected, the H_2O_2-responsive targets include a number of stress proteins and antioxidants, which can act to protect against the oxidant. Similarly our data have shown that a range of stress-protective molecules is translationally induced at similar concentrations of H_2O_2 (Shenton *et al.*, 2006). Low concentrations of oxidants appear to promote changes in gene expression that provide the crucial functions which are required to detoxify subsequent, higher doses of oxidants.

Relatively few studies have characterised the changes in protein levels that occur in response to high lethal doses of oxidants. Extensive transcriptional reprogramming is evident from transcriptome studies (Gasch and Werner-Washburne, 2002; Temple *et al.*, 2005), but it is as yet unclear how these changes are translated into the cellular proteome. Polysome/microarray analysis indicates that numerous mRNAs increase or decrease their association with polysomes in response to a lethal concentration of H_2O_2. For at least three of these mRNAs, increased ribosome association does not correlate with increased protein production (Shenton *et al.*, 2006). One possible explanation is that these mRNAs are inhibited at the elongation phase of protein synthesis. This would increase their association with ribosomes without increasing protein production. These mRNAs may become rapidly activated following detoxification of the oxidant and relief of the translational block. Perhaps not surprisingly therefore, the difference between low adaptive and high lethal concentrations of oxidants, appears to be that higher concentrations of oxidants promote a more pronounced inhibition of protein production. Inhibiting protein synthesis at multiple steps may leave it primed such that mRNA specific translation can be rapidly resumed following relief of the oxidative stress.

7. FUTURE PERSPECTIVES

In summary, oxidative stress elicits complex translational reprogramming that is fundamental for adaptation to the stress. It is important to understand how gene expression is moderated in response to oxidants since oxidative stress is a major biological problem that underlies numerous disease processes. Whilst the yeast system has enabled a clearer understanding of the translational responses to oxidative stress, important questions remain to be addressed. It is clear that different ROS invoke different protective responses. Establishing how the translational regulatory systems overlap and interact to achieve mRNA-specific translational regulation will be a major challenge.

ACKNOWLEDGEMENT

Work in the authors' lab is supported by the Biotechnology and Biological Sciences Research Council (BBSRC).

REFERENCES

Ayala, A., Parrado, J., Bougria, M., and Machado, A. (1996). Effect of oxidative stress, produced by cumene hydroperoxide, on the various steps of protein synthesis. Modifications of elongation factor-2. *The Journal of Biological Chemistry*, **271**, 23105–23110.

Beilharz, T.H. and Preiss, T. (2004). Translational profiling: The genome-wide measure of the nascent proteome. *Briefings in Functional Genomics and Proteomics*, **3**, 103–111.

Bertolotti, A., Zhang, Y., Hendershot, L.M., Harding, H.P. and Ron, D. (2000). Dynamic interaction of BiP and ER stress transducers in the unfolded-protein response. *Nature Cell Biology*, **2**, 326–332.

Bilsland, E., Molin, C., Swaminathan, S., Ramne, A. and Sunnerhagen, P. (2004). Rck1 and Rck2 MAPKAP kinases and the HOG pathway are required for oxidative stress resistance. *Molecular Microbiology*, **53**, 1743–1756.

Causton, H.C., Ren, B., Koh, S.S., Harbison, C.T., Kanin, E., Jennings, E.G., Lee, T.I., True, H.L., Lander, E.S. and Young, R.A. (2001). Remodeling of yeast genome expression in response to environmental changes. *Molecular Biology of the Cell*, **12**, 323–337.

Cherkasova, V.A. and Hinnebusch, A.G. (2003). Translational control by TOR and TAP42 through dephosphorylation of eIF2alpha kinase GCN2. *Genes & Development*, **17**, 859–872.

Collinson, L.P. and Dawes, I.W. (1992). Inducibilty of the response of yeast cells to peroxide stress. *Journal of General Microbiology*, **138**, 329–335.

Dever, T.E. (2002). Gene-specific regulation by general translation factors. *Cell*, **108**, 545–556.

Ding, Q., Markesbery, W.R., Chen, Q., Li, F. and Keller, J.N. (2005). Ribosome dysfunction is an early event in Alzheimer's disease. *Journal of Neuroscience*, **25**, 9171–9175.

Dumont-Miscopein, A., Lavergne, J.P., Guillot, D., Sontag, B. and Reboud, J.P. (1994). Interaction of phosphorylated elongation factor EF-2 with nucleotides and ribosomes. FEBS Letters, **356**, 283–286.

Dunand-Sauthier, I., Walker, C.A., Narasimhan, J., Pearce, A.K., Wek, R.C. and Humphrey, T.C. (2005). Stress-activated protein kinase pathway functions to support protein synthesis and translational adaptation in response to environmental stress in fission yeast. *Eukaryotic Cell*, **4**, 1785–1793.

Gasch, A.P., Spellman, P.T., Kao, C.M., Carmel-Harel, O., Eisen, M.B., Storz, G., Botstein, D. and Brown, P.O. (2000). Genomic expression programs in the response of yeast cells to environmental changes. *Molecular Biology of the Cell*, **11**, 4241–4257.

Gasch, A.P. and Werner-Washburne, M. (2002). The genomics of yeast responses to environmental stress and starvation. *Functional & Integrative Genomics*, **2**, 181–192.

Godon, C., Lagniel, G., Lee, J., Buhler, J.M., Kieffer, S., Perrot, M., Boucherie, H., Toledano, M.B. and Labarre, J. (1998). The H_2O_2 stimulon in *Saccharomyces cerevisiae*. *Journal of Biological Chemistry*, **273**, 22480–22489.
Grant, C.M., Maciver, F.H. and Dawes, I.W. (1997). Mitochondrial function is required for resistance to oxidative stress in the yeast *Saccharomyces cerevisiae*. FEBS Letters, **410**, 219–222.
Gutteridge, J.M.C. (1993). Free radicals in disease processes: A compilation of cause and consequence. *Free Radical Research Communications*, **19**, 141–158.
Halliwell, B. and Gutteridge, J.M.C. (1989). *Free Radicals in Biology and Medicine*, 2nd edn. Oxford University Press, Oxford.
Harding, H.P., Novoa, I., Zhang, Y., Zeng, H., Wek, R., Schapira, M. and Ron, D. (2000). Regulated translation initiation controls stress-induced gene expression in mammalian cells. *Molecular Cell*, **6**, 1099–1108.
Harding, H.P., Zhang, Y., Zeng, H., Novoa, I., Lu, P.D., Calfon, M., Sadri, N., Yun, C., Popko, B., Paules, R., *et al.* (2003). An integrated stress response regulates amino acid metabolism and resistance to oxidative stress. *Molecular Cell*, **11**, 619–633.
Hinnebusch, A.G. (2000). Mechanism and regulation of initiator methionyl-tRNA binding to ribosomes. In: *Translational Control of Gene Expression* (N. Sonenberg, J.W.B. Hershey and M.B. Mathews, eds.). Cold Spring Harbor Laboratory Press, New York, pp. 185–243.
Hinnebusch, A.G. (2005). Translational regulation of gcn4 and the general amino acid control of yeast. *Annual Review of Microbiology*, **59**, 407–450.
Honda, K., Smith, M.A., Zhu, X., Baus, D., Merrick, W.C., Tartakoff, A.M., Hattier, T., Harris, P.L., Siedlak, S.L., Fujioka, H., *et al.* (2005). Ribosomal RNA in Alzheimer disease is oxidized by bound redox-active iron. *Journal of Biological Chemistry*, **280**, 20978–20986.
Jamieson, D.J. (1992). *Saccharomyces cerevisiae* has distinct adaptive responses to both hydrogen peroxide and menadione. *Journal of Bacteriology*, **174**, 6678–6681.
Jamieson, D.J. (1998). Oxidative stress responses of the yeast *Saccharomyces cerevisiae. Yeast*, **14**, 1511–1527.
Kapp, L.D. and Lorsch, J.R. (2004). The molecular mechanics of eukaryotic translation. *Annual Review of Biochemistry*, **73**, 657–704.
Lu, P.D., Harding, H.P. and Ron, D. (2004). Translation reinitiation at alternative open reading frames regulates gene expression in an integrated stress response. *Journal of Cell Biology*, **11**, 27–33.
Mathews, M.B., Sonenberg, N. and Hershey, J.W.B. (2000). Origins and principles of translational control. In: *Translational Control of Gene Expression* (M.B. Mathews, N. Sonenberg, and J.W.B. Hershey, eds.). Cold Spring Harbor, Cold Spring Harbor laboratory Press, New York, pp. 1–31.
Nairn, A.C. and Palfrey, H.C. (1987). Identification of the major M^r 100,000 substrate for calmodulin-dependent protein kinase III in mammalian cells as elongation factor-2. *The Journal of Biological Chemisty*, **262**, 17299–17303.
Narasimhan, J., Staschke, K.A. and Wek, R.C. (2004). Dimerization is required for activation of eIF2 kinase Gcn2 in response to diverse environmental stress conditions. *The Journal of Biological Chemisty*, **279**, 22820–22832.
Natarajan, K., Meyer, M.R., Jackson, B.M., Slade, D., Roberts, C., Hinnebusch, A.G. and Marton, M.J. (2001). Transcriptional profiling shows that Gcn4p is a master regulator of gene expression during amino acid starvation in yeast. *Molecular and Cellular Biology*, **21**, 4347–4368.
Palmer, L.K., Shoemaker, J.L., Baptiste, B.A., Wolfe, D. and Keil, R.L. (2005). Inhibition of translation initiation by volatile anesthetics involves nutrient-sensitive GCN-independent and -dependent processes in yeast. *Molecular Biology of the Cell*, **16**, 3727–3739.
Patel, J., McLeod, L.E., Vries, R.G., Flynn, A., Wang, X. and Proud, C.G. (2002). Cellular stresses profoundly inhibit protein synthesis and modulate the states of phosphorylation of multiple translation factors. *European Journal of Biochemistry*, **269**, 3076–3085.
Pavitt, G.D., Ramaiah, K.V., Kimball, S.R. and Hinnebusch, A.G. (1998). eIF2 independently binds two distinct eIF2B subcomplexes that catalyze and regulate guanine-nucleotide exchange. *Genes & Development*, **12**, 514–526.
Poljak, A., Dawes, I.W., Inglese, B.A., Duncan, M.W., Smythe, G.A. and Grant, C.M. (2003). Oxidative damage to proteins in yeast cells exposed to adaptive levels of H_2O_2. *Redox Report*, **8**, 371–377.

Proud, C.G. (2000). Control of the elongation phase of protein synthesis. In: *Translational Control of Gene Expression* (N. Sonenberg, J.W.B. Hershey and M.B. Mathews, eds.). Cold Spring Harbor, Cold Spring Harbor Laboratory Press, New York, pp. 719–739.

Proud, C.G. (2005). eIF2 and the control of cell physiology. *Seminars in Cell & Development Biology,* **16,** 3–12.

Roper, M.D. and Wicks, W.D. (1978). Evidence for acceleration of the rate of elongation of tyrosine aminotransferase nascent chains by dibutyryl cyclic AMP. *Proceedings of the National Academy of Sciences of the United States of America,* **75**, 140–144.

Shenton, D. and Grant, C.M. (2003). Protein S-thiolation targets glycolysis and protein synthesis in response to oxidative stress in the yeast *Saccharomyces cerevisiae. The Biochemical Journal,* **374,** 513–519.

Shenton, D., Smirnova, J.B., Selley, J.N., Carroll, K., Hubbard, S.J., Pavitt, G.D., Ashe, M.P. and Grant, C.M. (2006). Global translational responses to oxidative stress impact upon multiple levels of protein synthesis. *The Journal of Biological Chemistry,* **281**, 29011–29021.

Smirnova, J.B., Selley, J.N., Sanchez-Cabo, F., Carroll, K., Eddy, A.A., McCarthy, J.E., Hubbard, S.J., Pavitt, G.D., Grant, C.M. and Ashe, M.P. (2005). Global gene expression profiling reveals widespread yet distinctive translational responses to different eukaryotic translation initiation factor 2B-targeting stress pathways. *Molecular and Cellular Biology,* **25**, 9340–9349.

Stadtman, E.R. and Levine, R.L. (2003). Free radical-mediated oxidation of free amino acids and amino acid residues in proteins. *Amino Acids,* **25**, 207–218.

Swaminathan, S., Masek, T., Molin, C., Pospisek, M. and Sunnerhagen, P. (2006). Rck2 is required for reprogramming of ribosomes during oxidative stress. *Molecular Biology of the Cell,* **17**, 1472–1482.

Takahashi, R., and Goto, S. (1990). Alteration of aminoacyl-tRNA synthetase with age: Heat-labilization of the enzyme by oxidative damage. *Archives of Biochemistry and Biophysics,* **277**, 228–233.

Teige, M., Scheikl, E., Reiser, V., Ruis, H. and Ammerer, G. (2001). Rck2, a member of the calmodulin-protein kinase family, links protein synthesis to high osmolarity MAP kinase signaling in budding yeast. *Proceedings of the National Academy of Sciences of the United States of America,* **98**, 5625–5630.

Temple, M.D., Perrone, G.G. and Dawes, I.W. (2005). Complex cellular responses to reactive oxygen species. *Trends in Cell Biology,* **15**, 319–326.

Turton, H.E., Dawes, I.W. and Grant, C.M. (1997). *Saccharomyces cerevisiae* exhibits a yAP-1-mediated adaptive response to malondialdehyde. *Journal of Bacteriology,* **179**, 1096–1101.

Vattem, K.M. and Wek, R.C. (2004). Reinitiation involving upstream ORFs regulates ATF4 mRNA translation in mammalian cells. *Proceedings of the National Academy Sciences of the United States of America,* **101**, 11269–11274.

Wiese, A.G., Pacifici, R.E. and Davies, K.J. (1995). Transient adaptation of oxidative stress in mammalian cells. *Archives of Biochemistry and Biophysics,* **318**, 231–240.

Yang, R., Wek, S.A. and Wek, R.C. (2000). Glucose limitation induces GCN4 translation by activation of Gcn2 protein kinase. *Molecular and Cellular Biology,* **20**, 2706–2717.

Zhan, K., Narasimhan, J. and Wek, R.C. (2004). Differential activation of eIF2 kinases in response to cellular stresses in *Schizosaccharomyces pombe. Genetics,* **168**, 1867–1875.

CHAPTER **15**

Cell Differentiation as a Response to Oxidative Stress

Wilhelm Hansberg, Jesús Aguirre, Mauricio Ríos-Momberg, Pablo Rangel, Leonardo Peraza, Yésika Montes de Oca and **Nallely Cano-Domínguez**

Contents

1. Dioxygen has Shaped Life on Earth 236
2. ROS and Cell Differentiation in Eukaryotic Microorganisms 237
   2.1 Increased ROS Levels and Antioxidant Enzymes During Cell Differentiation 238
   2.2 NADPH Oxidases in Cell Differentiation 238
   2.3 Ras, ROS and Cell Differentiation 240
3. ROS and Cell Differentiation in *Neurospora* 240
   3.1 Exposure to a High Oxygen Tension Induces the Conidiation Process; Hyphal Adhesion is the First Differentiated State 240
   3.2 A Hyperoxidant State at the Start of Hyphal Adhesion 241
   3.3 Adhesion of Hyphae in Liquid Cultures 244
   3.4 The Adhered Hyphae Produce the Aerial Hyphae 247
   3.5 A Hyperoxidant State at the Start of the Formation of Aerial Hyphae 249
   3.6 A Hyperoxidant State at the Start of Formation of Conidia 250
4. The Development of *N. crassa* in Null Mutant Strains Affected in Genes Encoding Antioxidant and Pro-Oxidant Enzymes 251
   4.1 CAT-3 Null Mutants 251
   4.2 CAT-2 Null Mutant Strains 252
   4.3 The Development of Mutant Strains Affected in *sod-1* and *ras-1* Genes 254
   4.4 Development in NADPH-Oxidase Null Mutant Strains 255
References 255

Abstract

Photosynthetic dioxygen (O_2) discharge and accumulation in the atmosphere compelled cells to adapt or to die in the presence of a reactive compound and to protect themselves from it and from the inevitable formation of more reactive oxygen species (ROS). At the same time, by using O_2 as final electron acceptor, early heterotrophs could obtain more energy from reduced carbon

British Mycological Society Symposia Series
Published by Elsevier Ltd.

and attain faster growth. However, fast growing aerobes were made more dependent on the availability of reduced carbon. In the fight for food, microorganisms colonized different niches including other organisms, which they parasitized or ate. With time, life-damaging ROS became signals used by cells to regulate growth and proliferation, cell differentiation and cell death.

We have proposed that cell differentiation is a response to oxidative stress. In different eukaryotic microorganisms, the presence of ROS and the induction of antioxidant mechanisms are associated with development. Moreover, the superoxide-producing NADPH-oxidases proved essential for some developmental processes. The small GTPase RAS-1 also regulates the production of ROS and different mutations in the corresponding gene affect cell differentiation. The process of asexual reproduction (conidiation) in *Neurospora crassa* involves three morphogenetic transitions, each preceded by a hyperoxidant state. An accentuated redox imbalance, the oxidation and degradation of oxidized protein, oxidative inactivation of different enzyme activities, and increased production of ROS characterize the hyperoxidant state. Null mutants in catalase genes show increased levels of development. Strains without superoxide dismutase express a conidiation rhythm, similar to the "band" strain, which contains a semi-dominant *ras-1* mutation. Mutation of NADPH-oxidase genes reduced or completely abolished different developmental processes. Together, these results support our view on cell differentiation as a response to oxidative stress.

1. DIOXYGEN HAS SHAPED LIFE ON EARTH

Photosynthetic O_2 discharged into the atmosphere and its subsequent accumulation, which started approximately 2.2 billion years ago, was probable the most significant event in biology that shaped life and its evolution and forever changed the surface chemistry of Earth (Canfield, 2005). The electron affinity of O_2 makes it reactive and capable of generating ROS by its sequential univalent reduction or by the capture of radiation energy. Due to their irreversible nature, most ROS reactions with cellular components disrupt the organization of the cell. Primeval cells, originated in an anoxic environment and specialized to conduct a reduced set of redox reactions (Raymond and Segre, 2006), evolved to live in the presence of a poisonous gas. Adaptation occurred by mechanisms that shielded cells from O_2, or through the efficient reduction of entering O_2 and the development of mechanisms to cope with or to avoid formation of ROS. Since the generation of an oxidant atmosphere, the threat of damaging effects caused by ROS has been prevalent for all organisms.

Atmospheric O_2 was poisonous to cells but at the same time it was highly beneficial; it led to the formation of an ozone layer in the stratosphere that shielded cells from the solar, high-energy ultraviolet (UV) radiation that is damaging to nucleic acids. Absorption of UV light in the stratosphere was essential for life outside bodies of water, the colonization of land and its global dispersal. Furthermore, by using O_2 as final electron acceptor, early heterotrophs could obtain four times more energy from reduced carbon sources. A large supply of energy allowed fast growth and the acquisition of new abilities, with the consequence of an unprecedented and dramatic increase in biomass production on Earth.

However, fast growing aerobes were made more dependent on the availability of reduced carbon, indispensable for O_2 reduction and for avoidance of ROS generation and ROS neutralization. The increased need for reduced carbon sources resulted in the colonization of all kinds of niches including other organisms, which were parasitized or eaten. The fight for food produced the complex food chains that characterize current modes of life. With time, some organisms evolved strategies to use the life-damaging capacity of ROS. Devices were developed for the controlled production of ROS avoiding, at the same time, their deleterious effects. Controlled ROS production could then be used to kill or control the growth of other organisms, as can be observed in many host–parasite interactions (Daub and Ehrenshaft, 2000; Fang, 2004; Chauhan *et al.*, 2006). Because O_2 and its unavoidable ROS have been of such importance for survival, mechanisms were selected to exquisitely detect the presence of O_2 and ROS and to cope with, or to avoid ROS through different antioxidant responses (Blokhina *et al.*, 2003; Giaccia *et al.*, 2004). Most interestingly, ROS became signals used by cells to regulate growth and proliferation (Burdon, 1995; Esposito *et al.*, 2004), cell differentiation (Aguirre *et al.*, 2005; Gapper and Dolan, 2006; Li *et al.*, 2006), cell ageing (Droge, 2002; Esposito *et al.*, 2004) and cell death (Simon *et al.*, 2000; Martindale and Holbrook, 2002).

2. ROS AND CELL DIFFERENTIATION IN EUKARYOTIC MICROORGANISMS

ROS are generated by absorption of radiation energy and by partial reduction of O_2 (Halliwell and Gutteridge, 1999). UV "A" and blue light can excite O_2 unpaired electrons to form singlet oxygen ($^1O_2$). $^1O_2$ is unstable and decays to the triplet ground state of O_2 or reacts with a great variety of cellular compounds. Partial reduction of O_2 sequentially forms superoxide ($O_2\cdot^-$), hydrogen peroxide (H_2O_2) and hydroxyl radical ($HO\cdot$). $O_2\cdot^-$ and H_2O_2 are not much more reactive than O_2, but they can generate $^1O_2$ and $HO\cdot$ through different reactions; $HO\cdot$ being one of the most reactive known species. Thus, enzymatic disposal of $O_2\cdot^-$ through the superoxide dismutases and of H_2O_2 through the catalases, peroxidases, and peroxiredoxins is important to avoid $^1O_2$ and $HO\cdot$ formation. On the other hand, ROS and other reactive species of oxygen combined with nitrogen, such as nitric oxide and peroxynitite, have been implicated in different physiological processes.

We have proposed that cell differentiation is a response to oxidative stress (Hansberg and Aguirre, 1990; Hansberg, 1996; Aguirre *et al.*, 2005). According to this idea, growth and differentiated states are stable conditions in which low levels of ROS are maintained by a balance between ROS generation and elimination. A switch between these states occurs when a transient increase in ROS levels, beyond the cellular capability to neutralize them, is produced (Fig. 1).

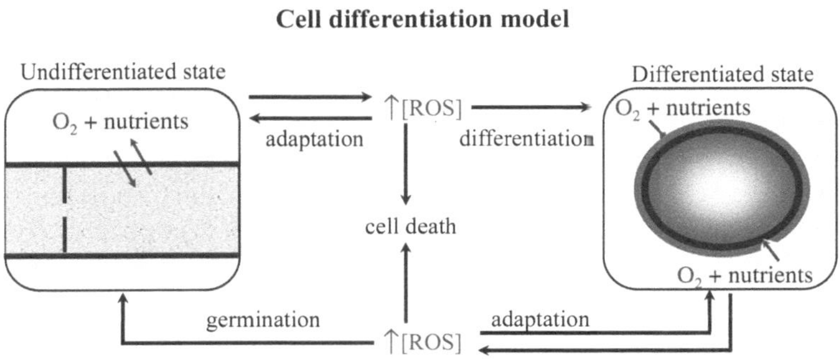

Figure 1 Cell differentiation model. The transition between undifferentiated and differentiated states, which are both stable states, occurs through a transient and unstable state in which generation of ROS surpasses the antioxidant capacity of the cell. This hyperoxidant state has three possible outcomes: (i) The cell compensates with a source of reducing power and returns to the previous stable state, thereby adapting to a more oxidizing condition; (ii) The cell differentiates by insulating itself from environmental oxygen; and (iii) When adaptation or cell differentiation cannot take place, the reduced internal medium equilibrates with the oxidizing external medium and the cell dies. Cell death enables other cells to either adapt or differentiate. The transition from a differentiated state to the undifferentiated state occurs through a germination process, which is characterized by breakage of insulation, development of a hyperoxidant state, and compensation of this state with the reducing power derived from nutrient utilization. ROS levels are maintained by a balance between the rate of ROS generation by mitochondria, NADPH oxidases and other enzymes, and the rate of ROS decomposition by antioxidant enzymes and reaction of ROS with other antioxidant mechanisms. This model predicts increased levels of ROS at the start of transitions between stable states, activation during cell differentiation of antioxidant responses and of mechanisms for insulation from environmental dioxygen. It further predicts the inhibition of cell differentiation by antioxidant activities, an increased cell differentiation by pro-oxidant conditions, such as the lack of antioxidant enzymes, and the requirement of ROS generating processes for cell differentiation. The data reviewed in this Chapter support this model. **(See Colour Section)**

2.1 Increased ROS Levels and Antioxidant Enzymes During Cell Differentiation

In different eukaryotic microorganisms, the presence of increased ROS levels is associated with cell differentiation. As a consequence, in several of these systems there is an up-regulation of antioxidant enzymes, such as superoxide dismutases, catalases, catalase-peroxidase, peroxidases and peroxiredoxins. Also, in several of these cases, antioxidants inhibit development, supporting the idea that ROS are needed for the developmental process to occur. This subject has been reviewed recently (Aguirre *et al.*, 2005) and will not be discussed in detail here.

2.2 NADPH Oxidases in Cell Differentiation

It has been recognized that animal cells use different NADPH-oxidases (NOX) to produce superoxide in a regulated manner and control fundamental processes such as cell defence, cell proliferation and apoptosis (Lambeth, 2004). The phagocytic NOX consists of the membrane-associated catalytic core gp91^{phox} and p22^{phox} subunits (cytochrome *b558*). The assembly of the cytosolic regulatory proteins p47^{phox}, p67^{phox} and Rac1/Rac2 with the cytochrome *b558* results in NOX activation. The presence of NOX enzymes in microbial eukaryotes and their roles in regulation of development were not recognized until recently. NoxA from *Emericella nidulans* (anamorph: *Aspergillus nidulans*), the first gp91^{phox} microbial orthologue characterized was found to produce ROS and was essential for differentiation of the sexual fruiting bodies (cleistothecia). In addition, the derepression of *noxA* was shown to correlate with a premature and abundant production of cleistothecia (Lara-Ortiz *et al.*, 2003). The phylogenetic analysis of *noxA* homologues shows that the number of *nox* genes in fungi ranges from none in unicellular and some dimorphic fungi, to one, two and three present in different fungal species. An interesting correlation between the presence of *nox* genes and the ability to develop multicellular fruiting bodies was observed (Lara-Ortiz *et al.*, 2003; Aguirre *et al.*, 2005). The significance of this correlation is strengthened by the fact that in *Podospora anserina* (Malagnac *et al.*, 2004) and *N. crassa* (Cano-Domínguez, Álvarez-Delfín, Hansberg and Aguirre, unpublished), the disruption of *nox1* genes prevents the development of sexual fruiting bodies. In both fungi, the inactivation of a second *nox-2* gene results in the production of sexual spores that are non-viable or unable to germinate (Malagnac *et al.*, 2004) (Cano-Domínguez *et al.*, unpublished). In the endophytic fungus *Epichloë festucae*, mutants with the *noxA* gene deleted increase their pattern of hyphal branching and dramatically change the interaction with the plant host from mutualistic to antagonistic, whereas the deletion of a second *noxB* gene does not produce any visible phenotype (Tanaka *et al.*, 2006). The role of these *nox* genes in *E. festucae* sexual development has yet to be analyzed.

Dictyostelium discoideum contains three *nox* (*noxA*, *noxB* and *noxC*) and single p22^{phox} and p67^{phox} homologues. Notably, inactivation of *noxA*, *noxB*, *noxC* or *p22^{phox}* results in arrested development and lack of asexual spores, without affecting viability, growth or the phagocytosis of bacteria (Lardy *et al.*, 2005). These results indicate that using NOX enzymes to regulate cell-differentiation processes is a common strategy in eukaryotes.

A distinct group of putative *nox* genes has been recently identified in algae. With a gp91^{phox} gene homologue, initially isolated from the red macroalga *Chondrus crispus*, additional homologues in EST and genome databases from unicellular red and diatom algae were identified. *C. crispus* produces superoxide in response to cell-free extracts from its pathogenic green alga and the gp91^{phox} gene was induced during infection by the green alga and in response to fatty-acid derived hydroperoxides. In contrast to other gp91^{phox} homologues, putative NOX from algae contain an insertion of approximately 300 amino acids between two NADPH-binding sites, which predict four additional transmembrane-spanning domains (Herve *et al.*, 2006). The function of these NOX has not been analyzed.

2.3 Ras, ROS and Cell Differentiation

Small GTPases, mainly from the Ras and Rho families, are involved in perceiving and controlling the production of ROS. Ras has been recognized as a target of ROS (Lander *et al.*, 1995), through the promotion of GDP dissociation from Ras caused by ROS (Heo and Campbell, 2006) or by controlling the activity of Ras regulators by ROS (Accorsi *et al.*, 2001).

On the other hand, Ras and Rac can regulate intracellular ROS levels after different stimulations (Sundaresan *et al.*, 1996). Permanently active versions of these proteins are involved in malignant transformation of cells through continuous mitogenic signalling, mediated by ROS (Irani *et al.*, 1997). Both effects are suppressed by dominant negative forms of Rac, supporting a mechanism for ROS generation in which Rac lies downstream of Ras. Rac proteins can form complexes with different mammalian NOXs and control superoxide formation by these (Hordijk, 2006). ROS generated by NOX complexes are *bona fide* signalling molecules, regulating the activation of MAP kinase cascades (Torres and Forman, 2003) or inhibiting phosphatases and thus promoting tyrosine phosphorylation (Salmeen *et al.*, 2003). A NOX protein is proposed to act upstream of a MAPK pathway in *Podospora anserina* (Malagnac *et al.*, 2004). In *Colletotrichum trifolii* it was shown that Ras may directly activate Rac to generate ROS and to activate a MAPK pathway, but through parallel pathways (Chen and Dickman, 2004).

ROS are associated with developmental transitions in *N. crassa* and NOX proteins are required for proper fungal morphogenesis (Aguirre *et al.*, 2005). Further work in the coming years is expected to elucidate the interplay between ROS, NOX, Ras-like proteins and MAP kinase pathways.

3. ROS AND CELL DIFFERENTIATION IN *Neurospora*

In this section we review our results demonstrating that a hyperoxidant state is developed at the start of each morphogenetic transition of the conidiation process in *N. crassa*. We also describe unpublished results about the production of a polysaccharide in the culture medium that is related to the process of hyphal adhesion.

3.1 Exposure to a High Oxygen Tension Induces the Conidiation Process; Hyphal Adhesion is the First Differentiated State

A synchronous process of conidiation is started when an aerated liquid culture is filtered and the resulting mycelial mat is exposed to air (Toledo *et al.*, 1986; Springer and Yanofsky, 1989). Hyphae in direct contact with air adhere to each other. The adhesion process takes place in the course of 40 min, but most hyphae are already adhered after 20 min (Toledo *et al.*, 1986; Toledo and Hansberg, 1990). By contrast, hyphae in the inner part of the mycelial mat rapidly become microoxic and do not adhere to each other.

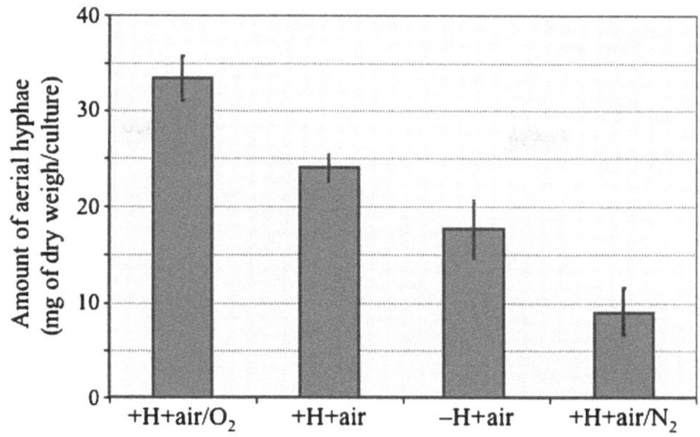

Figure 2 Amount of aerial hyphae formed under different O_2 concentrations in a humid or dry atmosphere. Values are mean values of four experiments with four cultures for each condition in every experiment. Symbols: +H+air/O_2, humidified mixture of equal amounts of air and O_2; +H+air, humidified air; -H+air, non-humidified air; +H+air/N_2, humidified mixture of equal amounts air and N_2. Humidification was done by bubbling the gas through a 5 cm high water column at a rate of 50 ml/min.

It has been wrongly stated that desiccation of mycelia induces the conidiation process. It is not the lack of water that starts conidiation but the sudden exposure to an increased oxygen tension. When exposed to atmospheres with increased or decreased O_2 concentrations, mycelial mats respectively show increased or decreased amounts of aerial hyphae after 15 h of exposure. By contrast, desiccation of a mycelial mat by exposure to a flux of dry air results in lower amounts of aerial hyphae than when exposed to humidified air (Fig. 2).

Dioxygen dissolves poorly in water and its solubility is limited by the concentration of other solutes and by temperature. The maximal O_2 concentration that is attained in a nutrient liquid medium at 30°C is not much higher than 200 μM. Because of active respiration, dissolved O_2 concentration rapidly becomes the growth-limiting factor in shaken cultures. Under such conditions, the amount of O_2 dissolved in the medium is equated with the amount of O_2 consumed by respiration and as a result, O_2 is almost undetectable. However, when the culture is filtered, the small amount of water surrounding hyphae on the surface of the mycelial mat becomes saturated with O_2 because of a sudden increase in the water surface that is in direct contact with air. Thus, hyphae at the mat surface are exposed to an increased O_2 tension (Toledo *et al.*, 1991) and rapidly adhere to each other. What follows is an account of the rapid and critical changes that occur between the instant that hyphae are exposed to the air and 1 h later.

3.2 A Hyperoxidant State at the Start of Hyphal Adhesion

Low-level chemiluminescence (LLCL) was used to directly and continuously follow the generation of excited species during the process of *N. crassa* conidiation (Hansberg *et al.*, 1993). Singlet oxygen and excited carbonyls emit light when their electrons return to their ground state, and this light can be detected with a photon counting system.

LLCL emission is detected as soon as a mycelial mat is exposed to air. LLCL starts at a high level and continuously decreases during the following 2 h. If a current of N_2 is passed over the mycelial mat, at 20 min after start, there is an initial abrupt decline in LLCL followed by a faster decrease in LLCL. When instead of N_2, O_2 (95%) is passed over the culture, an immediate increase in LLCL is observed and the decrease in LLCL is retarded. Thus, oxygen-dependent LLCL indicates ROS production as soon as hyphae are exposed to air.

This high rate of ROS production should be reflected in the redox state of cells. In fact, as soon as hyphae are exposed to air, a redox imbalance is detected: there is a dramatic fall in the reduced-to-oxidized ratio of NAD and NADP coenzymes during the first 6 min due to a rapid decrease of NADH and NADPH and a concomitant increase in NAD and NADP (Toledo *et al.*, 1991). The NAD(P)-reductive charge (the ratio of the sum of coenzymes in a reduced state to the total NAD(P)(H) content) drops from a value of 0.6 to 0.3. After this descent in NAD(P)-reductive charge, the reduced levels of these coenzymes return to their starting values after 30 min of air exposure. Such redox imbalance of the NAD(P)(H) coenzymes does not occur in a mycelial mat exposed to an N_2-enriched atmosphere. On the contrary and as expected from an anoxic or microoxic state, concentration of reduced coenzymes increases under N_2. Reductive-charge in the non-adhered hyphae, which are in a microoxic state, is maintained at a value of 0.6, even though changes are detected in the levels of NAD(P)(H) (Toledo *et al.*, 1991). Adding a carbon source to the starving mycelial mat does not abolish the observed loss of NAD(P)-reductive charge in the adhered hyphae, indicating that the lack of reduced carbon *per se* does not cause the observed redox change.

Glutathione (GSH) is one of the major safeguards of the cellular reduced state. GSH in cells is at a millimolar concentration and the ratio of GSH to glutathione disulphide (GSSG) is maintained around 100:1 by glutathione reductase, which uses NADPH to reduce GSSG. Thus, a large change in NAD(P)-reductive charge should be reflected in the amount and GSH/GSSG ratio. Indeed, with the loss of NAD(P)-reducing power there is also an increase in GSSG concentration during the first 10 min of air exposure with a concurrent 30% loss of GSH during the first 15 min, producing a GSH/GSSG ratio of 80 between 3 and 15 min of air exposure. Thereafter, during the next 15 min, the levels of GSH and GSSG and its ratio recover to initial values (Toledo *et al.*, 1995). Moreover, there is a rapid increase in GSSG in the medium surrounding the hyphae, which is maximal between 6 and 10 min of air exposure and thereafter declines to original values at 30 min (Toledo *et al.*, 1991). Under an N_2-enriched atmosphere, no extracellular GSSG is detected and adding glucose to an air exposed mycelial mat does not avoid GSSG secretion. Thus, there is an accentuated redox imbalance of both NAD(P)H and GSH during the first 20 min in the air-exposed hyphae.

Increase in ROS and loss in reductive power is in turn reflected in an increased oxidation of proteins, as measured by protein carbonyl content. Carbonyls in total protein of the air-exposed hyphae increase two fold during the first 15–20 min; thereafter values decrease during the next 15–20 min to almost reach their initial values (Toledo and Hansberg, 1990). Oxidized proteins are labile to the activity of endogenous proteases and thus are rapidly degraded, allowing the carbonyl content in total protein to be reduced 30–40 min after air exposure. In contrast to the adhered hyphae, carbonyl content in total protein from non-adhered hyphae decrease between 20 and 45 min to almost undetectable levels (Toledo *et al.*, 1994).

Does oxidation of proteins affect metabolic pathways? We have studied the nitrogen assimilation pathway, which is essential for ammonium assimilation and for the distribution of nitrogen to different compounds. Activities of the glutamine synthetase (GS) and the anabolic, NADPH dependent, glutamate dehydrogenase (GDH(NADPH)) rapidly decrease in the air-exposed hyphae (Cárdenas and Hansberg, 1984b). Maximal loss in GS activity (70%) is seen after 20 min; thereafter activity recovers up to the original values at 45 min air exposure (Toledo *et al.*, 1994). Enzyme loss correlates with increase in total protein carbonyl content and enzyme recovery with diminishing protein carbonyl content.

The GS, immuno-precipitated from a cell extract of the adhered hyphae after 30 min of air exposure, shows the characteristic oxidized α and β polypeptides that are observed when the purified GS is treated with ascorbate + Fe^{3+} or H_2O_2 + Fe^{2+} (Toledo *et al.*, 1994), indicating that the same oxidation takes place *in vivo* and *in vitro* and indicating that the hydroxyl radical is the oxidizing agent. The same oxidized polypeptides are observed in a cell extract when it is incubated with NADH or NADPH and air, or in a cell extract from a mycelium that was starved for reduced carbon in an aerated liquid culture (Aguirre and Hansberg, 1986). Under the same conditions but in the absence of air, no oxidized GS polypeptides are detected. By contrast, in the non-adhered microoxic hyphae, although starved of carbon and nitrogen, GS activity does not change and oxidized GS polypeptides are not detected in the cell extract (Toledo *et al.*, 1994).

Loss of GDH(NADPH) activity also occurs in the adhered hyphae (Cárdenas and Hansberg, 1984b). The GDH(NADPH) activity decreases as soon as cells are exposed to air, during the first 30 min. However, the activity does not recover as observed for GS, but continues to decrease for 1.5 h (Toledo *et al.*, 1994).

The amount of GDH(NADPH) polypeptide immunoprecipitated from a cell extract of the adhered hyphae decreases with time, but no oxidized polypeptides could be detected after 20 min of air exposure, as it is observed when the purified enzyme is subjected to an oxidizing reaction and then immunoprecipitated (Aguirre *et al.*, 1989). This could indicate that oxidized polypeptides are rapidly degraded.

In contrast to the anabolic enzyme, the catabolic NAD-dependent GDH activity increases during the first hour of air exposure (Toledo *et al.*, 1994). This indicates that under highly oxidative conditions, only some enzymes are oxidatively inactivated. Some enzymes may be resistant to oxidation, or may be oxidatively modified but not affected in their activity, and some may in fact be activated by an oxidative modification.

As a direct consequence of protein oxidation, which is an irreversible process (except for incomplete thiol oxidation), increased proteolysis takes place. In fact, oxidized GS in cell extracts is completely degraded, whereas the non-oxidized enzyme is preserved (Aguirre and Hansberg, 1986). Protein degradation in the air-exposed mycelial mat accounts for 37% of total protein in 10 h (Cárdenas and Hansberg, 1984b). Most of it occurs during the first 20 min (25%) and is confined to the air-exposed adhering hyphae in which increased carbonyl content in total protein is observed (Toledo *et al.*, 1994). Protein degradation is again at low values after 30 min. Thus, increased protein oxidation as well as oxidation of GS and augmented proteolysis take place during the first 20 min of air exposure and then recover during the next 20–30 min.

Since proteins are actively degraded during the first 20 min of air exposure, amino acids are probably drained to the tricarboxylic acid cycle. Thus, it makes physiological sense to rapidly inhibit the nitrogen assimilation pathway, which consumes ATP and NADPH, and to increase catabolic GDH activity, which produces NADH. In this way, the loss of reductive charge would be partially compensated.

The morphological consequence of the dramatic changes described above is the adhesion of hyphae. Hyphae adhere to each other because their cell wall is modified and an additional layer is deposited which glues neighbouring hyphae together. Under the electron microscope, this extra layer was observed as crossing filaments between the cell walls of adjacent hyphae. Adhered hyphae cannot be separated by strong agitation in water, or water with salts, and tearing hyphae apart only caused breakage but not the dissociation of hyphae. Thus, hyphal adhesion seems an irreversible process (Fig. 3).

3.3 Adhesion of Hyphae in Liquid Cultures

An increase in O_2 tension can be induced by the sudden increase in agitation of a liquid culture or when the air to liquid ratio in a culture is increased. Under these conditions, adhesion of hyphae is also observed (Toledo *et al.*, 1994). Adhesion of hyphae in liquid correlates with protein oxidation; the higher the air-to-liquid ratio, the more hyphae adhere and greater levels of carbonyl content in total protein are observed (Michán *et al.*, 2003).

The reason for hyphal adhesion is probably related to the presence in the medium of a polysaccharide (PS) that sticks hyphae together. In fact, Congo red, which is known to bind to PSs permanently, binds to the cell wall of hyphae, and more dye is fixed when hyphae adhere to each other.

When the wild-type (Wt) strain was grown with a high air-to-liquid ratio (5 or 25), the rate of glucose consumption was 1.04 mg per mg total protein per ml culture per hour during exponential growth. However, during the pre-stationary growth phase, when glucose in the medium becomes limiting, the glucose consumption rate increased considerably (50 to 100% increase), rapidly depleting glucose from the medium in about 2 h. Concomitant to depletion of the carbon source, a PS appeared in the medium and then aggregation of hyphae was observed at the onset of the stationary growth phase.

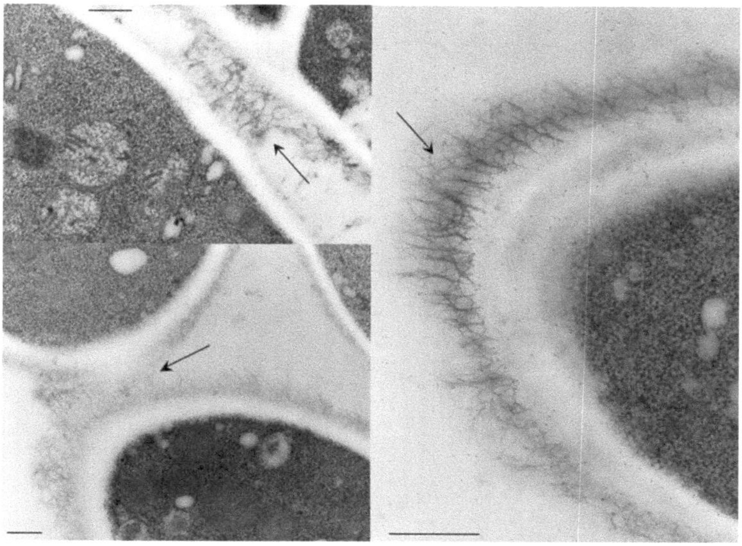

Figure 3 An additional layer is observed on the cell wall of the adhered hyphae. Photographs of the adhered hyphae taken with the electron microscopy; the bar indicates 500 nm. The samples were fixed with 0.5% glutaraldehyde in 0.1 M cacodilate buffer, pH 7.2, for 1 h at room temperature. After washing three times with cacodilate buffer, samples were dehydrated with ethanol solutions from 30 to 70% at room temperature. Thereafter samples were impregnated overnight in LR white resin in 70% ethanol. The resin was allowed to polymerize at 60°C for 48 h and then 90 nm thick cuts were made. Cuts were layered on a nickel 300 meshes covered with formvar and stained with 2% uranile acetate for 30 min and thereafter with lead citrate. Observations were done in a JEOL 1200 EX II transmission electron microscope at 60 KV.

By contrast, when Wt strain was grown in a low air-to-liquid ratio (1.08), glucose consumption rate during exponential growth was lower (0.87), exponential growth was prolonged for 5 h and there was no apparent increase in glucose consumption rate during pre-stationary growth phase (Fig. 4). In these cultures, there was no PS in the medium (Fig. 5) and no adhesion of hyphae at the entrance into the stationary growth phase took place. Hyphae finally adhered in these cultures after prolonged incubation.

The PS in the medium can be isolated by ethanol precipitation, and consists of two fractions: a fraction that became insoluble after centrifugation and had the consistency of a gel, and a fraction that was water-soluble. The insoluble fraction was collected from the medium by centrifugation at 15,300*g* for 50 min, and the soluble fraction was precipitated from the supernatant with 0.5 M KCl and collected by centrifugation at 15,300*g* for 10 min. The two fractions were quantified in cultures grown at high (5.25) or low (1.08) air-to-liquid ratio. PS concentration in the culture with low aeration was very low and only increased slightly during the stationary growth phase. In contrast, PS was abundant in the culture grown at a high air-to-liquid ratio; the soluble fraction was present in a higher amount in

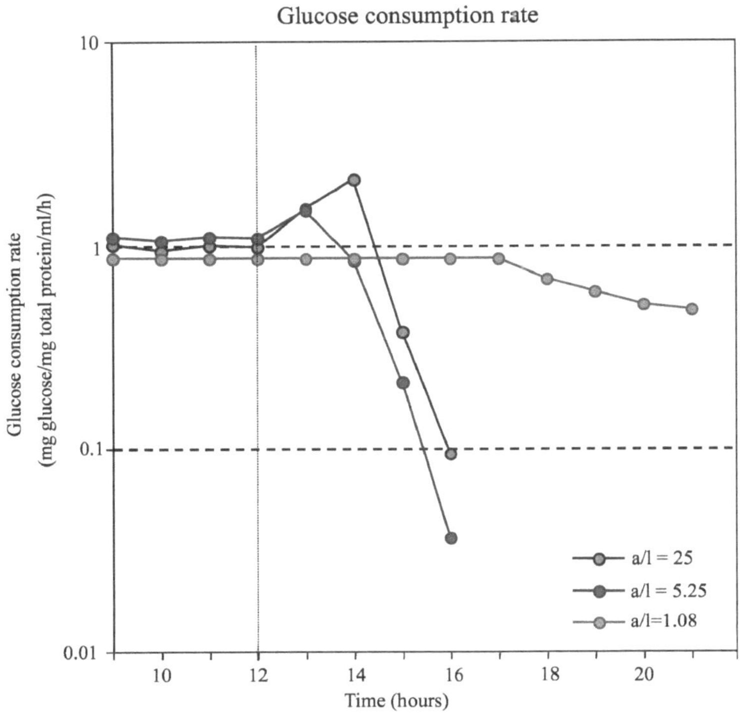

Figure 4 Glucose consumption rate in liquid cultures grown at different air to liquid ratios. Cultures in erlenmeyer flasks of 125 ml containing 5 (a/l=25), 20 (a/l=5.25) or 60 (a/l=1.08) ml of Vogel's minimal medium supplemented with 100 mM glucose and inoculated with 10^6 conidia/ml were incubated at 30°C in a shaker at 250 rpm. At times indicated, glucose in the medium was measured with the glucose oxidase/peroxidase assay (glucose oxidase 10 U/ml in 100 mM phosphate buffer, pH 7, using 2.2′-azino-di-[3 ethilbenzotiazolidine-sulphonic acid]$(NH_4)_2$ as substrate of peroxidase); protein in the acetone-dehydrated mycelium was determined by the method of bradford. **(See Colour Section)**

the pre-stationary growth phase; the insoluble fraction at the entrance into stationary growth phase, but both were low after 24 h growth (Fig. 6). These results suggest a precursor/product relationship between the soluble and the insoluble PS fractions. In fact, when the soluble PS was incubated with Fe^{2+} plus H_2O_2 or only Fe^{2+} (or Fe^{3+}), the soluble PS became insoluble (Fig. 7). An oxidation reaction probably caused the soluble PS to polymerize into an insoluble PS. Disappearance of PS during the stationary growth phase is due to PS binding to the cell wall. Moreover, adding PS to an exponentially growing culture at a high air-to-liquid ratio enhanced the aggregation of hyphae.

Our interpretation of these results is that a culture grown at a high air-to-liquid ratio develops oxidative stress when the carbon source in the medium becomes limiting. The remaining carbon source is consumed rapidly to compensate stress and to synthesize PS. By binding and polymerizing at the cell wall, PS promotes adhesion of hyphae, and by structuring water molecules, bound PS

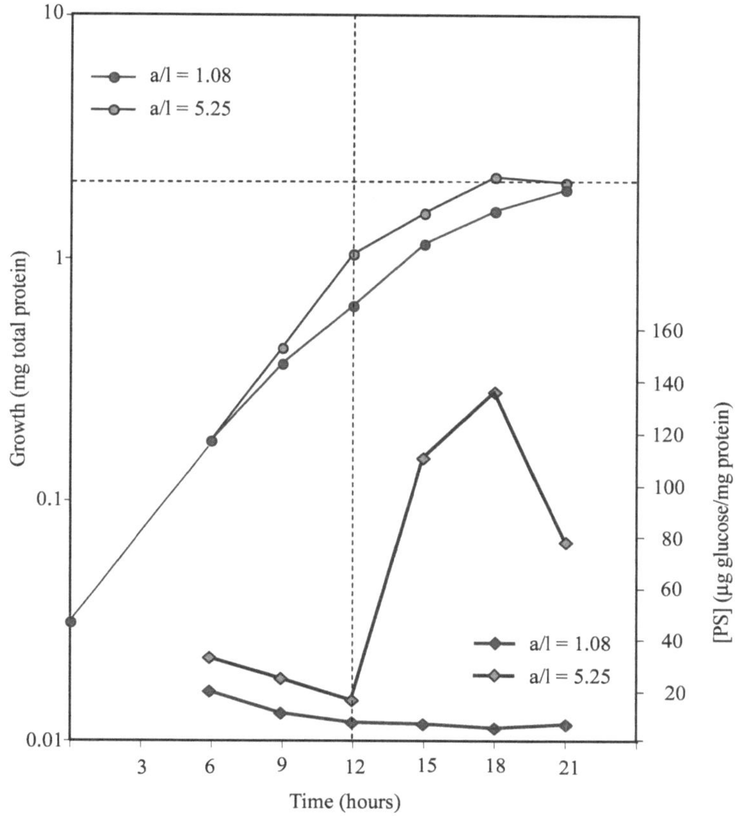

Figure 5 Polysaccharide (PS) isolated from the medium in cultures grown at low and high air to liquid ratios. PS was isolated from the medium of 20 (a/l=5.25) or 60 (a/l=1.08) ml cultures in 125 ml erlenmeyer flasks at the times Indicated; protein was determined in the acetone dehydrated mycelium by the method of bradford. PS was isolated by precipitation from the medium, adding two volumes of ethanol and storing overnight at -20°C. Precipitated PS was collected by centrifugation at 14,000g for 36 min, washed with distilled water, suspended in a third of the previous volume and precipitated again with ethanol. The collected pellet was dried at room temperature, re-suspended in 0.5 ml 0.5 N H_2SO_4 and incubated 15 h at 100°C in sealed ampoules. Thereafter H_2SO_4 was evaporated under vacuum, the dried powder was re-suspended in distilled water, pH was neutralized with 5 N KOH, and glucose was determined as indicated in Figure 3. **(See Colour Section)**

could diminish entry of O_2 to the cell. Besides, active respiration of closely attached hyphae can diminish surrounding O_2 concentration and thereby compensate oxidative stress.

3.4 The Adhered Hyphae Produce the Aerial Hyphae

Adhered hyphae constitute the active cell structure that forms the aerial hyphae. The non-adhered hyphae in the mycelial mat do not participate directly in aerial

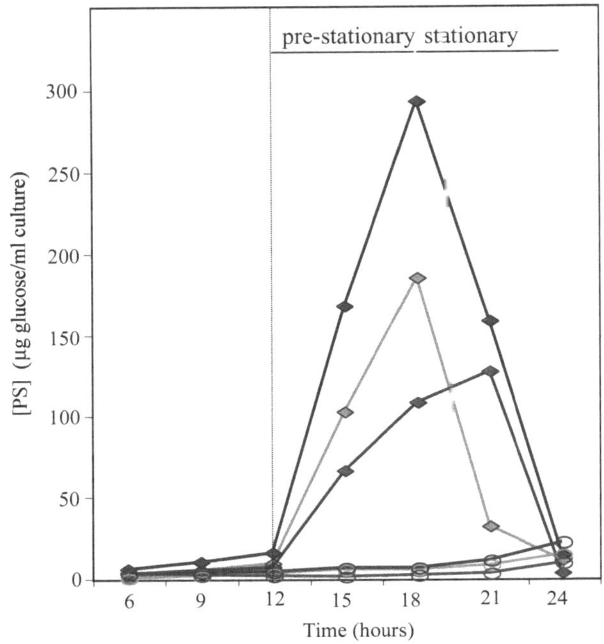

Figure 6 Soluble and insoluble fractions of polysaccharide Isolated from the medium in cultures with low (a/l=1.08) and high (a/l=5.25) air-to-liquid ratios. The insoluble fraction was collected from the medium by centrifugation at 15,300*g* for 50 min, and the water-soluble fraction was precipitated from the supernatant with 0.5 M KCl and collected by centrifugation at 15,300*g* for 10 min. Both fractions were hydrolyzed as described in (Figure 5) and glucose quantified as indicated in Figure 3. **(See Colour Section)**

hyphae formation (Toledo *et al.*, 1986). Adhered hyphae remain viable after 15 h of air exposure under starving conditions; vital dyes are excluded from these hyphae (Toledo *et al.*, 1986) and hyphae rapidly resume growth when placed in an aerated nutrient medium (Toledo *et al.*, 1994). By contrast, non-adhered hyphae, when placed in an aerated nutrient medium, first develop oxidative stress with a rapid loss of GS and GDH(NADPH) activities and cell adhesion. These enzymes have first to be synthesized again before these hyphae can start growing, which occurs after 6 h incubation (Toledo *et al.*, 1994).

Moreover, adhered hyphae in a mycelial mat incorporate radioactive labelled amino acids in higher amounts than the non-adhered hyphae; 2 times more when labelled at the start of exposure to air; 12 times more when labelled after 3.5 h. Fifty percent of the label in the adherent mycelium precipitates with acid whereas only 15% precipitates in the non-adherent mycelium (Toledo *et al.*, 1986).

A definitive demonstration that the adhered hyphae produce the aerial hyphae comes from mycelial mats where hyphae of an auxotrophic strain is

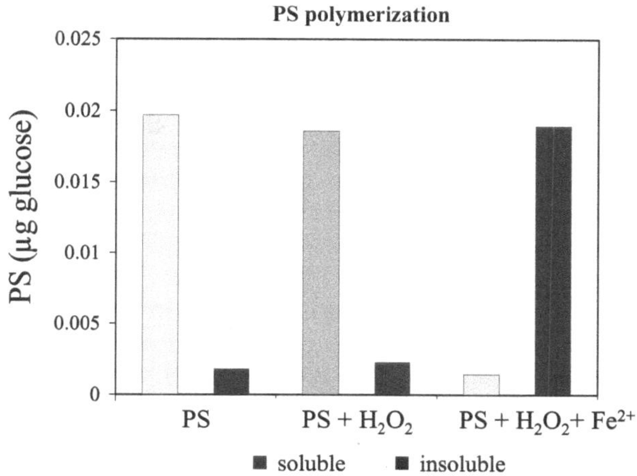

Figure 7 Polymerization of the PS soluble fraction. PS isolated from the medium by ethanol precipitation was centrifuged to eliminate the PS insoluble fraction. Aliquots of soluble PS were incubated for 30 min at room temperature in the presence of 0.7 mM $FeSO_4$ and/or H_2O_2 and thereafter centrifuged 10 min in a microfuge at full speed. Glucose was quantified in the resulting pellets and supernatants. Fe^{3+} can substitute Fe^{2+}; 1 mM EDTA or 1,10-phenantroline inhibits the formation of a precipitate. Fe ions are required for PS precipitation; H_2O_2 enhances precipitation, but is not required.

stratified on top of a different auxotrophic strain that was previously labelled with a radioactive amino acid. Conidia have mainly or exclusively the genetic background from the auxotroph layered on top, and the amount of label transported to the aerial hyphae is ten times less in a stratified mat than in a non-stratified labelled mycelial mat (Toledo *et al.*, 1986).

3.5 A Hyperoxidant State at the Start of the Formation of Aerial Hyphae

Using a photon counting system, it is observed that the initial high emission of light from the mycelial mat reaches a minimum at 2.5–3 h of air exposure. Thereafter, LLCL increases again to give a second peak with a maximum approximately 6.5 h after onset of air exposure (Hansberg *et al.*, 1993). Thus, ROS in the adhered hyphae increase in concentration again during the formation of aerial hyphae.

Even before a rise in ROS is detected, a redox imbalance is observed in the adhered hyphae (Toledo *et al.*, 1995). After having reached initial values 1 h after air exposure, the NADH/NAD and NADPH/NADP ratios fall to a one third lower reductive charge within 1 h. This imbalance is observed even earlier in the GSH redox state (Toledo *et al.*, 1995). After having recovered from the first hyperoxidant state to initial values 30 min after air exposure, the amount of GSH + GSSG is reduced 40% and the GSH/GSSG ratio falls to a third within 1 h.

As a consequence, protein carbonyls in the adhered hyphae increase again after 1.5 h of air exposure giving a new peak of protein oxidation at 3 h and decrease during the next 1.5 h (Toledo and Hansberg, 1990). Concomitant to protein oxidation, GS and GDH(NADPH) activities decline steadily between 1 and 4 h of air exposure. In contrast, the catabolic GDH(NAD) activity increases at the same time.

In summary, with the adhesion of hyphae in the mycelial mat, a stable condition is reached that is similar to the one of growing hyphae. However, half an hour later, adhered hyphae start losing their reductive power and hyphae arrive at a critical point in which its redox state is at a minimum, proteins are oxidized, GS and GDH are inactivated and catabolic GDH activity increases, and ROS production augments.

The appearance of aerial hyphae seems to be the morphological consequence of this second hyperoxidant state. Adhered hyphae serve as a substrate for the growth of aerial hyphae. Amino acids coming from proteolysis are utilized for protein synthesis but also as carbon sources and excess nitrogen is excreted as ammonium ions (Cárdenas and Hansberg, 1984b). Most of the radioactive label that is incorporated in the adhered hyphae is transported to the aerial hyphae. Because the pool of glutamine is essential for aerial growth (Cárdenas and Hansberg, 1984a), radioactive-labelled glutamine is conveyed from the adhered hyphae to the aerial hyphae (Toledo *et al.*, 1986) and is catabolized there by the glutamine transaminase-ω-amidase pathway; glutamate is formed by transamination of α-ketoglutarate (Cárdenas and Hansberg, 1984b). Both, glutamine and alanine transaminases are particularly active in the aerial hyphae (Cárdenas and Hansberg, 1984b).

3.6 A Hyperoxidant State at the Start of Formation of Conidia

A similar pattern of events to those observed as soon as hyphae are exposed to air (first hyperoxidant state) is observed in the adhered hyphae just before the start of aerial hyphae development (second hyperoxidant state) and again in aerial hyphae before conidia are formed (third hyperoxidant state).

To measure LLCL in the aerial hyphae, light coming from the adhered hyphae has to be shaded with a black filter paper. Aerial hyphae grow through the paper and give a peak of LLCL between 9.5 and 11 h of exposure to air, when aerial hyphae start forming conidia.

Aerial hyphae can be separated from mycelial mat and used for biochemical measurements. In contrast to growing hyphae and the adhered hyphae, in the aerial hyphae the NADP(H) pool is 2.8 times higher than the NAD(H) pool, reflecting an extensive use of the internal carbon sources. Both NADH/NAD and NADPH/NADP ratios decrease between 6 and 9 h to a very low value and remain at low values thereafter. The amount of GSH in aerial hyphae decreases at the same time and the GSH/GSSG ratio abates to low values (Toledo *et al.*, 1995).

This redox imbalance is reflected in the oxidation of proteins. Carbonyls in proteins of aerial hyphae increase steadily between 6 and 8.5 h of air exposure and decreases thereafter to initial values (Toledo and Hansberg, 1990). The GS activity in the aerial hyphae decreases sharply between 7 and 8.5 h of air exposure and is definitively lost (Toledo *et al.*, 1994). There is no anabolic

GDH(NADPH) activity in the aerial hyphae, but the catabolic GDH(NAD) is particularly high (Cárdenas and Hansberg, 1984b).

In summary, in the aerial hyphae after 9h of our exposure, the redox state has decreased, proteins are oxidized and GS is inactivated, and ROS continue to increase. At this critical moment conidia are formed in a burst between 9 and 15 h and continue to form thereafter at a lower rate (Toledo and Hansberg, 1990).

Thus, formation of conidia from growing hyphae involves three morphogenetic transitions: growing hyphae to adhered hyphae; adhered hyphae to aerial hyphae; and aerial hyphae to conidia. A hyperoxidant state develops at the start of each of these morphogenetic transitions (Aguirre and Hansberg, 1986; Hansberg and Aguirre, 1990; Hansberg, 1996).

During germination of conidia, a hyperoxidant state develops in which singlet oxygen is produced (Lledías *et al.*, 1999). Catalase-1, which is highly accumulated in conidia, is oxidized by singlet oxygen during germination. From the first minutes of germination to approximately 1 h, carbonyl content in total protein increases, proteolysis augments and there is a high LLCL. Moreover, oxidative stress is required for conidia germination (Lledías *et al.*, 1999).

4. THE DEVELOPMENT OF *N. crassa* IN NULL MUTANT STRAINS AFFECTED IN GENES ENCODING ANTIOXIDANT AND PRO-OXIDANT ENZYMES

The data reviewed indicate that a hyperoxidant state develops at the start of all three morphogenetic transitions of the conidiation process. A hyperoxidant state also develops during the transition from a differentiated state, the conidium, to the growth state. If cell differentiation is a response to oxidative stress, then mutant strains lacking antioxidant enzymes should undergo augmented oxidative stress and this should promote cell differentiation. In contrast, mutants affected in pro-oxidant enzymes should show a diminished oxidative stress and thus show inhibited cell differentiation. Here we discuss the roles of different antioxidant and pro-oxidant enzymes on asexual and sexual development.

4.1 CAT-3 Null Mutants

Fungi have a battery of enzymes for H_2O_2 removal, which comprises various peroxiredoxins, peroxidases and catalases, and a catalase-peroxidase. The K_M for H_2O_2 between these enzymes varies by four orders of magnitude, being very low for peroxiredoxins, low for peroxidases, high for catalase-peroxidase and very high for catalases. Thus, catalase-peroxidase and catalases are important when H_2O_2 concentration is in the millimolar range.

Most euascomycetes have 2 large-subunit catalases and 1–4 small-subunit catalases; *N. crassa* has large-subunit catalases CAT-1 and CAT-3 but only one small-subunit catalase, CAT-4, which is probable non-peroxisomal as it lacks a peroxisome targeting signal; *Botryotina fuckeliana* (anamorph: *Botrytis cinerea*) and *Hypocrea jecorina* (anamorph: *Trichoderma reesei*) have orthologues of *N. crassa*

CAT-1 and CAT-3 and four additional small-subunit catalases, one of which has a peroxisomal signal (Salas and Hansberg, unpublished).

The induction of catalases is an expected consequence of a hyperoxidant state. There is a stepwise increase in catalase activity during the conidiation process. Conidia have 60 times more catalase activity than hyphae growing in liquid medium, which corresponds mostly to CAT-1 (Michán *et al.*, 2002). Different catalase activities are observed during conidiation, corresponding to different enzymes as well as to CAT-1 modification (Michán *et al.*, 2002). Activities of CAT-1, CAT-3 and a catalase-peroxidase (CAT-2) can be identified by their molecular weight and charge. The first two are homo-tetramers; CAT-2 is a homo-dimer.

CAT-3 activity is present throughout growth and is induced during the pre-stationary growth phase, at the start of conidiation and under different stress conditions (Michán *et al.*, 2002). The enzyme has four N-glycosylation sites (Díaz, Valdés, Rudiño-Piñera, Horjales, Hansberg, unpublished), it has a signal peptide and is partially secreted (Michán *et al.*, 2002). Strains without CAT-3 were isolated by Repeat-induced point mutation (RIP) (Cambareri *et al.*, 1989) in the *cat-3* gen (Michán *et al.*, 2003). Colonies of $cat\text{-}3^{RIP}$ mutants are sensitive to H_2O_2: they do not survive a 10 min treatment with 15 mM H_2O_2, while 85% of the Wt colonies survive this treatment. The lack of CAT-3 activity in the $cat\text{-}3^{RIP}$ strains is not compensated for by other catalases. In fact, Paraquat or H_2O_2 induces CAT-3 activity in the Wt but no other catalases are induced in the mutant strains. In support of our hypothesis, $cat\text{-}3^{RIP}$ and Δ *cat-3* strains produce more aerial hyphae and conidia than the Wt, forming longer and more densely packed aerial hyphae. Precocious conidiation and increased number of conidia is observed in $cat\text{-}3^{RIP}$ colonies formed in solid medium. Compared to Wt, these $cat\text{-}3^{RIP}$ colonies are subjected to a higher oxidative stress during conidiation, as judged by a higher amount of oxidized proteins and increased production of carotenes (Michán *et al.*, 2003). Conidiating $cat\text{-}3^{RIP}$ mycelial mats produced more spontaneous, oxygen-dependent, chemiluminescence than the Wt, as detected by a photon-counting system (Fig. 8).

During sexual development, ROS detected with dichlorofluorescein were higher in protoperithecia than in surrounding hyphae, indicating a localized oxidative stress. As expected, there were more protoperithecia formed by the *cat-3* mutant than in the Wt strain (Fig. 9).

4.2 CAT-2 Null Mutant Strains

CAT-2 is a heme-peroxidase from the microorganism/plant family of peroxidases. It is an enzyme completely different from monofunctional catalases, having catalase as well as peroxidase activity. Fungal catalase-peroxidases have a bacterial origin (Peraza and Hansberg, 2002; Klotz and Loewen, 2003) and are derived from gene duplication of a bacterial peroxidase gene and fusion of the two genes (Welinder, 1991). In fact, only prokaryotes and fungi have this enzyme.

CAT-2 is associated with cell lysis, and is induced during the late stationary phase of growth and during conidiation. In the course of conidiation, *cat-2* was

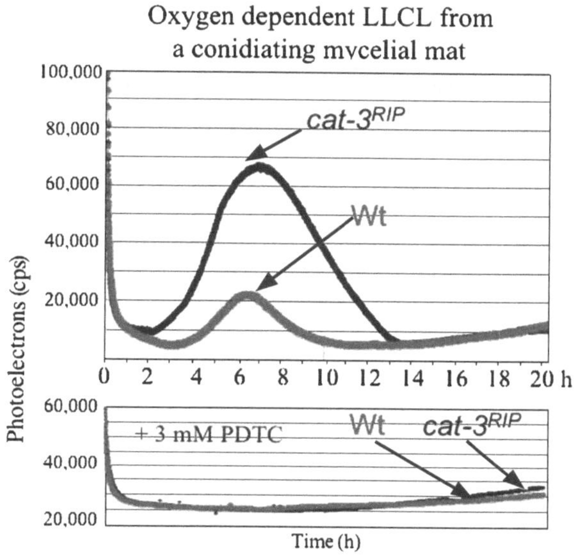

Figure 8 Oxygen-dependent LLCL during conidiation of Wt and *cat-3*^{RIP} mycelial mats. To enhance chemiluminescence, *bis-N*-Methylacridinium nitrate (lucigenine) (aldrich) was added to the cultures to a final concentration of 200 μM, 1 h before mycelial mats were made. Culture (100 ml) was filtered through whatman paper no. 1, and the harvested mycelium was washed twice with distilled water and then with 30 mM phosphate buffer, pH 6.1. Without compressing, the excess buffer was eliminated from the mycelial mats, which were then immediately transferred to small glass petri dishes and put upside down in front of the photomultiplier of the photon counting system. The antioxidant pyrrolidine dithicarbamate (PDTC) inhibits both LLCL and cell differentiation in both strains. **(See Colour Section)**

expressed during aerial hyphae formation and CAT-2 activity was present in adhered and aerial hyphae and was abundant in conidia (Peraza and Hansberg, 2002) (Peraza, Vega, Zamorano and Hansberg, unpublished).

Strains without CAT-2 were isolated by causing RIP to the *cat-2* gene. The *cat-2*^{RIP} strain showed increased formation of conidia under submerged conditions, as compared to Wt and mutants in other antioxidant enzymes (Peraza *et al.*, unpublished).

Conidiation of aerial and submerged hyphae follow similar steps. Both form cross-walls, delimitating cell compartments and pro-conidial chains by budding. Under submerged conditions, usually 4–6 conidia remained attached. Submerged conidiation did not take place in strains that avert formation of aerial conidia, such as *fluffy*, *aconidiate-2* or *aconidiate-3* strains (Peraza *et al.*, unpublished).

A transient increase in ROS production was indicated because dichlorofluorescein-dependent fluorescence was associated with conidiophores but not with conidia. There were numerous fluorescent conidiophores in the *cat-2*^{RIP} mutant strain but few in the Wt (Peraza *et al.*, unpublished).

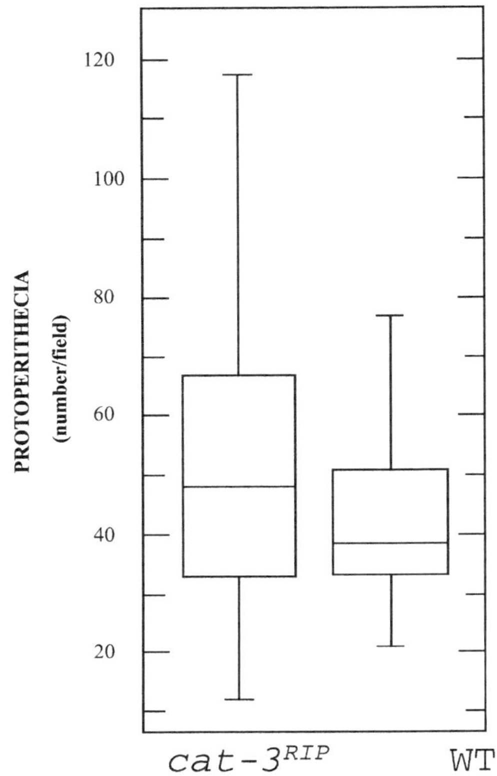

Figure 9 Number of protoperithecia formed by WT and *cat-3*^{RIP} strains. Spores from WT and the *cat-3*^{RIP} strain were inoculated on stacks of whatman no. 1 filter paper embedded in 10 ml liquid synthetic crossing medium. After 5 days incubation at 25°C, the paper with the colony was fixed with 4% formaldehyde in phosphate-buffered saline (PBS). The colony was rinsed with PBS and stained with 0.1% (w/v) coomassie-blue G250, 10% (v/v) acetic acid, and 40% (v/v) methanol; distaining was in 10% (v/v) acetic acid, 40% (v/v) methanol. Protoperithecia were counted in 100 randomly chosen fields per paper. Eight empty fields were discarded for each strain.

4.3 The Development of Mutant Strains Affected in *sod-1* and *ras-1* Genes

The "band" strain, *bd*, is characterized by cyclic conidiation; it alternates growth and conidiation within a period of 22 h. The *bd* strain has been used for more than 40 years to study this circadian rhythm (Dunlap, 1999), but until very recently, the gene affected in *bd* was unknown. We became interested in the *bd* strain because of three observations: (a) *sod-1* null mutant strains expressed a conidiation rhythm, similar to the *bd* strain, (b) germination in the *bd* strain was sensitive to Paraquat, a superoxide-generating agent, and (c) *bd* strain showed precocious and increased submerged conidiation, even higher than in the *cat-2* mutant strain (Ríos-Momberg and Hansberg, unpublished results).

ras-1 is a gene that maps in the same region as *bd*. RAS-1 is involved in ROS signalling and there are dominant mutants of *ras-1* that affect development in *N. crassa* (Sugisaki *et al.*, 1999). In fact, we found that *bd* is a semi-dominant mutant of *ras-1*. Simultaneously, the laboratories of Jennifer Loros and Jay Dunlap arrived to the same conclusion by following a single nucleotide polymorphism (SNP) associated with the *bd* mutant (Belden *et al.*, 2007).

4.4 Development in NADPH-Oxidase Null Mutant Strains

N. crassa has two *nox* genes (Aguirre *et al.*, 2005). Both genes were cancelled by substituting the gene with a hygromycin-resistance cassette. Although the Δ*nox-2* mutant did not show an evident phenotype, the ascospores from a homozygous Δ*nox-2* cross failed to germinate even in the presence of external oxidants (Cano-Domínguez *et al.*, unpublished).

In contrast, the Δ*nox-1* had a much richer phenotype; the strain formed scarce aerial mycelium and conidia and *nox-1* was female sterile because it failed to produce protoperithecia (Cano-Domínguez *et al.*, unpublished).

In summary, we have found that development in *N. crassa* is triggered by oxidative stress, which is characterized by an accentuated redox imbalance, protein oxidation and its degradation, rapid changes in enzyme activities, and increased production of ROS. Null mutants in antioxidant enzymes show increased levels of development, whereas development is reduced or completely abolished in mutants affected in pro-oxidant enzymes. Different developmental processes were affected; lack of *cat-3* increases the level of asexual and sexual development, inactivation of *cat-2* stimulated submerged conidiation, null *nox-1* strains showed inhibited asexual and abolished sexual development, *nox-2* mutants were impaired in ascospore germination. Increase in $O_2{\cdot}^-$ concentration in the *sod-1* phenocopied *bd* producing cyclic conidiation, and it was found that the *bd* strain is a semi-dominant *ras-1* mutant strain that promoted submerged conidiation.

REFERENCES

Accorsi, K., Giglione, C., Vanoni, M. and Parmeggiani, A. (2001). The Ras GDP/GTP cycle is regulated by oxidizing agents at the level of Ras regulators and effectors. *FEBS Letters*, **492**, 139–145.

Aguirre, J. and Hansberg, W. (1986). Oxidation of *Neurospora crassa* glutamine synthetase. *Journal of Bacteriology*, **166**, 1040–1045.

Aguirre, J., Ríos-Momberg, M., Hewitt, D. and Hansberg, W. (2005). Reactive oxygen species and development in microbial eukaryotes. *Trends in Microbiology*, **13**, 111–118.

Aguirre, J., Rodríguez, R. and Hansberg, W. (1989). Oxidation of *Neurospora crassa* NADP-specific glutamate dehydrogenase by activated oxygen species. *Journal of Bacteriology*, **171**, 6243–6250.

Belden, W.J., Larrondo, L.F., Froehlich, A.C., Shi, M., Chen, C.H., Loros, J.J. and Dunlap, J.C. (2007). The *band* mutation in *Neurospora crassa* is dominant allele of ras-1 implicating RAS signaling in circadian output. *Genes & Development*, **21**, 1494–1505.

Blokhina, O., Virolainen, E. and Fagerstedt, K.V. (2003). Antioxidants, oxidative damage and oxygen deprivation stress: A review. *Annals of Botany*, **91**, 179–194.

Burdon, R.H. (1995). Superoxide and hydrogen peroxide in relation to mammalian cell proliferation. *Free Radical Biology & Medicine*, **18**, 775–794.

Cambareri, E.B., Jensen, B.C., Schabtach, E. and Selker, E.U. (1989). Repeat-induced G-C to A-T mutations in *Neurospora*. *Science*, **244**, 1571–1575.

Canfield, D.E. (2005). The early history of atmospheric oxygen: Homage to Robert M. Garrels. *Annual Review of Earth and Planetary Sciences*, **33**, 1–36.

Cárdenas, M.E. and Hansberg, W. (1984a). Glutamine requirement for aerial mycelium growth in *Neurospora crassa*. *Journal of General and Applied Microbiology*, **130**, 1723–1732.

Cárdenas, M.E. and Hansberg, W. (1984b). Glutamine metabolism during aerial mycelium growth of *Neurospora crassa*. *Journal of General and Applied Microbiology*, **130**, 1733–1741.

Chauhan, N., Latge, J.-P. and Calderone, R. (2006). Signalling and oxidant adaptation in *Candida albicans* and *Aspergillus fumigatus*. *Nature Reviews Microbiology*, **4**, 435–444.

Chen, C. and Dickman, M.B. (2004). Dominant active Rac and dominant negative Rac revert the dominant active Ras phenotype in *Colletotrichum trifolii* by distinct signalling pathways. *Molecular Microbiology*, **51**, 1493–1507.

Daub, M.E. and Ehrenshaft, M. (2000). The photoactivated cercospora toxin cercosporin: Contributions to plant disease and fundamental biology. *Annual Review of Phytopathology*, **38**, 461–490.

Droge, W. (2002). Free radicals in the physiological control of cell function. *Physiological Review*, **82**, 47–95.

Dunlap, J.C. (1999). Molecular bases for circadian clocks. *Cell*, **96**, 271–290.

Esposito, F., Ammendola, R., Faraonio, R., Russo, T. and Cimino, F. (2004). Redox control of signal transduction, gene expression and cellular senescence. *Neurochemical Research*, **29**, 617–628.

Fang, F.C. (2004). Antimicrobial reactive oxygen and nitrogen species: concepts and controversies. *Nature Reviews Microbiology*, **2**, 820–832.

Gapper, C. and Dolan, L. (2006). Control of plant development by reactive oxygen species. *Plant Physiology*, **141**, 341–345.

Giaccia, A.J., Simon, M.C. and Johnson, R. (2004). The biology of hypoxia: the role of oxygen sensing in development, normal function, and disease. *Genes & Development*, **18**, 2183–2194.

Halliwell, B. and Gutteridge, J.M.C. (1999). *Free Radicals in Biology and Medicine*, 3rd edn. Oxford University Press.

Hansberg, W. (1996). A hyperoxidant state at the start of each developmental stage during *Neurospora crassa* conidiation. *Ciencia e Cultura*, **48**, 68–74.

Hansberg, W. and Aguirre, J. (1990). Hyperoxidant states cause microbial cell differentiation by cell isolation from dioxygen. *Journal of Theoretical Biology*, **142**, 201–221.

Hansberg, W., de Groot, H. and Sies, H. (1993). Reactive oxygen species associated with cell differentiation in *Neurospora crassa*. *Free Radical Biology & Medicine*, **14**, 287–293.

Heo, J. and Campbell, S.L. (2006). Ras regulation by reactive oxygen and nitrogen species. *Biochemistry*, **45**, 2200–2210.

Herve, C., Tonon, T., Collen, J., Corre, E. and Boyen, C. (2006). NADPH oxidases in eukaryotes: Red algae provide new hints! *Current Genetics*, **49**, 190–204.

Hordijk, P.L. (2006). Regulation of NADPH oxidases: The role of Rac proteins. *Circulation Research*, **98**, 453–462.

Irani, K., Xia, Y., Zweier, J.L., Sollott, S.J., Der, C.J., Fearon, E.R., Sundaresan, M., Finkel, T. and Goldschmidt-Clermont, P.J. (1997). Mitogenic signaling mediated by oxidants in Ras-transformed fibroblasts. *Science*, **275**, 1649–1652.

Klotz, M.G. and Loewen, P.C. (2003). The molecular evolution of catalatic hydroperoxidases: Evidence for multiple lateral transfer of genes between prokaryota and from bacteria into eukaryota. *Molecular Biology and Evolution*, **20**, 1098–1112.

Lambeth, J.D. (2004). NOX enzymes and the biology of reactive oxygen. *Nature Review Immunology*, **4**, 181–189.

Lander, H.M., Ogiste, J.S., Teng, K.K. and Novogrodsky, A. (1995). $p21^{ras}$ as a common signaling target of reactive free radicals and cellular redox stress. *The Journal of Biological Chemistry*, **270**, 21195–21198.

Lara-Ortiz, T., Riveros-Rosas, H. and Aguirre, J. (2003). Reactive oxygen species generated by microbial NADPH oxidase NoxA regulate sexual development in *Aspergillus nidulans*. *Molecular Microbiology*, **50**, 1241–1255.

Lardy, B., Bof, M., Aubry, L., Paclet, M.H., Morel, F., Satre, M. and Klein, G. (2005). NADPH oxidase homologs are required for normal cell differentiation and morphogenesis in *Dictyostelium discoideum*. *Biochimica et Biophysica Acta*, **1744**, 199–212.

Li, J., Stouffs, M., Serrander, L., Banfi, B., Bettiol, E., Charnay, Y., Steger, K., Krause, K.-H. and Jaconi, M.E. (2006). The NADPH oxidase NOX4 drives cardiac differentiation: Role in regulating cardiac transcription factors and MAP kinase activation. *Molecular Biology of the Cell*, **17**, 3978–3988.

Lledías, F., Rangel, P. and Hansberg, W. (1999). Singlet oxygen is part of a hyperoxidant state generated during spore germination. *Free Radical Biology & Medicine*, **26**, 1396–1404.

Malagnac, F., Lalucque, H., Lepere, G. and Silar, P. (2004). Two NADPH oxidase isoforms are required for sexual reproduction and ascospore germination in the filamentous fungus *Podospora anserina*. *Fungal Genetics and Biology*, **41**, 982–997.

Martindale, J.L. and Holbrook, N.J. (2002). Cellular response to oxidative stress: Signaling for suicide and survival. *Journal of Cellular Physiology*, **192**, 1.

Michán, S., Lledías, F., Baldwin, J.D., Natvig, D.O. and Hansberg, W. (2002). Regulation and oxidation of two large monofunctional catalases. *Free Radical Biology and Medicine*, **33**, 521–532.

Michán, S., Lledías, F. and Hansberg, W. (2003). Asexual development is increased in *Neurospora crassa cat-3*-null mutant strains. *Eukaryotic Cell*, **2**, 798–808.

Peraza, L. and Hansberg, W. (2002). *Neurospora crassa* catalases, singlet oxygen and cell differentiation. *Biological Chemistry*, **383**, 569–575.

Raymond, J. and Segre, D. (2006). The effect of oxygen on biochemical networks and the evolution of complex life. *Science*, **311**, 1764–1767.

Salmeen, A., Andersen, J.N., Myers, M.P., Meng, T.-C., Hinks, J.A., Tonks, N.K. and Barford, D. (2003). Redox regulation of protein tyrosine phosphatase 1B involves a sulphenyl-amide intermediate. *Nature*, **423**, 769–773.

Simon, H.U., Haj-Yehia, A. and Levi-Schaffer, F. (2000). Role of reactive oxygen species (ROS) in apoptosis induction. *Apoptosis*, **5**, 415–418.

Springer, M.L. and Yanofsky, C. (1989). A morphological and genetic analysis of conidiophore development in *Neurospora crassa*. *Genes & Development*, **3**, 559–571.

Sugisaki, M., Ito, S., Toh-e, A., Matsui, Y. and Inoue, H. (1999). Mutational analysis of the *ras* gene in *Neurospora crassa*. *Fungal Genetics Conference*, **20**, Abstract 368.

Sundaresan, M., Yu, Z.-X., Ferrans, V.J., Sulciner, D.J., Gutkind, J.S., Irani, K., Goldschmidt-Clermont, P.J. and Finkel, T. (1996). Regulation of reactive-oxygen-species generation in fibroblasts by Rac1. *The Biochemical Journal*, **318**, 379–382.

Tanaka, A., Christensen, M.J., Takemoto, D., Park, P. and Scott, B. (2006). Reactive oxygen species play a role in regulating a fungus-perennial ryegrass mutualistic interaction. *The Plant Cell*, **18**, 1052–1066.

Toledo, I., Aguirre, J. and Hansberg, W. (1986). Aerial growth in *Neurospora crassa*: Characterization of an experimental model system. *Experimental Mycology*, **10**, 114–125.

Toledo, I., Aguirre, J. and Hansberg, W. (1994). Enzyme inactivation related to a hyperoxidant state during conidiation of *Neurospora crassa*. *Microbiology*, **140**, 2391–2397.

Toledo, I. and Hansberg, W. (1990). Protein oxidation related to morphogenesis in *Neurospora crassa*. *Experimental Mycology*, **14**, 184–189.

Toledo, I., Noronha-Dutra, A.A. and Hansberg, W. (1991). Loss of NAD(P)-reducing power and glutathione disulfide excretion at the start of induction of aerial growth in *Neurospora crassa*. *Journal of Bacteriology*, **173**, 3243–3249.

Toledo, I., Rangel, P. and Hansberg, W. (1995). Redox imbalance at the start of each morphogenetic step of *Neurospora crassa* conidiation. *Archives of Biochemistry and Biophysics*, **319**, 519–524.

Torres, M. and Forman, H.J. (2003). Redox signaling and the MAP kinase pathways. *Biofactors*, **17**, 287–296.

Welinder, K.G. (1991). Bacterial catalase-peroxidases are gene duplicated members of the plant peroxidase superfamily. *Biochemica et Biophysica Acta*, **1080**, 215–220.

CHAPTER **16**

Signalling and Defences Against Oxidative Stress in *Candida albicans*

Rebeca Alonso-Monge, Elvira Román, David M. Arana, Susanne Gola and **Jesus Pla**

Contents

1. Introduction	260
2. Map Kinase Signal Transduction Pathways	261
2.1 The HOG Pathway	262
2.2 The Cell Integrity (*MKC1*-Mediated) Pathway	266
3. Transcription Factors	267
4. Defences Against Oxidative Stress	268
4.1 Non-Enzymatic Defences	268
4.2 Enzymatic Defences	268
5. Concluding Remarks	270
Acknowledgements	270
References	270

Abstract

Oxidative stress is one the major mechanisms by which immune cells are able to deal with microbial infections. Pathogens have therefore developed a set of antioxidant mechanisms, both non-enzymatic and enzymatic in nature. This process is adaptive and in the human pathogenic fungus *Candida albicans* different MAP kinase-mediated signal transduction pathways have been shown to play an essential role in transmitting the signal to downstream effectors, in addition to their roles in other processes such as morphogenesis and virulence. The cell integrity pathway, mediated by the Mkc1 MAP kinase, is activated in response to a wide range of conditions, including oxidative, saline and cell wall stress. The HOG pathway, mediated by the Hog1 MAP kinase, is also activated in response to different stress situations (osmotic and oxidative) and is able to trigger a transcriptional specific response, which is distinguishable from that due to other mediators such as the Yap1-homologue Cap1 or the Skn7 regulator. Manipulation of the oxidative stress response, either in the host or in the microbe, may therefore be an important therapeutic mechanism to control fungal infections.

British Mycological Society Symposia Series
Published by Elsevier Ltd.

1. INTRODUCTION

Candida albicans is one of the most frequent causes of fungal infection of humans. *C. albicans* lives as commensal in mucosal surfaces of healthy individuals, but is able to cause a wide variety of infections (called candidiasis) that range from mucosal or cutaneous to systemic. In developed countries this organism is still the fourth most common cause of nosocomial infections, which are generally linked to an impaired host immune system. The development of systemic candidiasis is thought to start at the intestinal mucosa; the microorganism is able to invade the epithelium penetrating as germ tubes and reach the blood stream, which enables dissemination of the yeast form to reach the target organs (kidney, brain and liver). Another common way to acquire candidiasis is via the use of catheters, allowing the direct entry of *C. albicans* cells into the blood stream. The ability of the fungus to adhere to different surfaces and to form biofilms is important mechanisms of dissemination and development of chronic disease (Douglas, 2003; Kuhn and Ghannoum, 2004).

C. albicans is rather resistant to adverse environmental conditions, at least when compared with other non-pathogenic yeasts such as *Saccharomyces cerevisiae* or *Schizosaccharomyces pombe*. Many and very diverse virulence factors have been reported (Calderone and Fonzi, 2001; Navarro-Garcia *et al.*, 2001). One of the most relevant was considered to be the ability of *C. albicans* to change cellular morphology from yeast to hypha, called the dimorphic transition, or other morphological programmes such as phenotypic switching (Soll *et al.*, 1993). The dimorphic transition is induced by environmental stimuli and, in addition to providing a mechanism for tissue penetration and invasion, results in the exposure or masking of different antigenic determinants on the outer surface, thereby influencing adhesion and the host's immune response. Other virulence factors have been described, such as secreted proteases and lipases which facilitate invasion, adhesins which mediate cell–cell interactions (Calderone *et al.*, 2000) and subsequently the adherence to tissues, central metabolic pathways (Lorenz and Fink, 2001; Lorenz *et al.*, 2004), and signal transduction pathways (Monge *et al.*, 2006).

As in other aerobic microorganisms, *C. albicans* must face oxidative stress from intra- and extracellular sources. Within the microorganism, reactive oxygen species (ROS) are generated during normal metabolic activity; these ROS may damage lipids, proteins and DNA if not properly eliminated. Oxidative stress arising from extracellular sources is the consequence of environmental perturbation such as the presence of oxidants, exposure to UV light or to phagocytic cells. The innate immune system is comprised of macrophages and neutrophils, which phagocytose microorganisms that invade the host tissues and which use reactive oxygen, nitrogen and chlorine species for eliminating them (reviewed by Missall *et al.*, 2004). Phagocytosis may result in an oxidative burst; the activation of the membrane-associated NADPH-dependent oxidase complex together with an increased rate of oxygen consumption lead to the reduction of oxygen to superoxide which is secreted to the phagosome. Superoxide readily dismutates to hydrogen peroxide or combines with nitric oxide (NO) to form the strong oxidant peroxynitrite, which has been shown to be fungicidal and more deleterious to

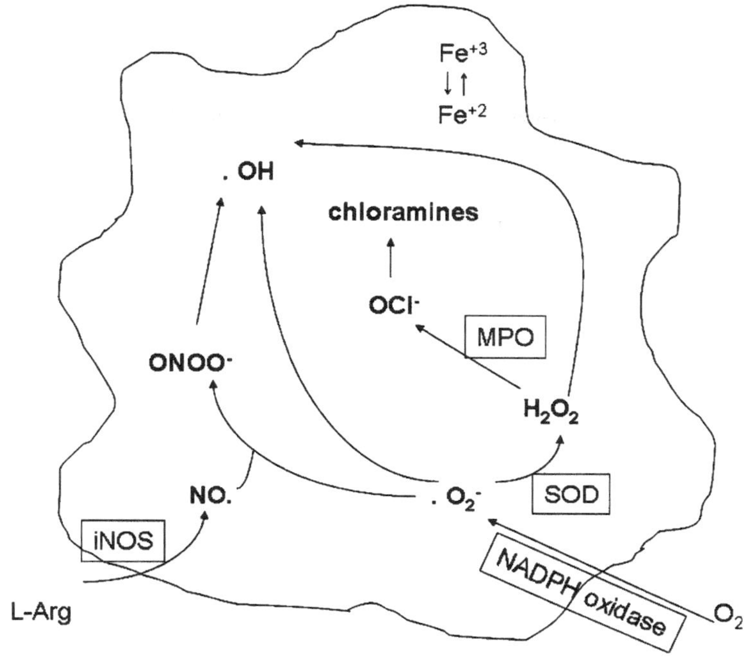

Figure 1 Main pathways that generate reactive oxygen, nitrogen and chlorine species by phagocytes. The enzymes that catalyse the reactions are boxed while the reactive species are written in bold. SOD, superoxide dismutase, iNOS, inducible nitric oxide synthase and MPO, myeloperoxidase.

microorganims than NO alone (Brunelli *et al.*, 1995). Many other reactions take place generating highly reactive compounds that are delivered to the phagosomes (review by Nathan and Shiloh, 2000) (Figure 1).

Fungal pathogens have developed diverse mechanisms to respond and minimize the effects of ROS and reactive nitrogen species (RNS). As in other aerobic organisms, this response in *C. albicans* is complex, comprising both non-enzymatic and enzymatic antioxidant defences, which probably reflects the intrinsic resistance of the fungus to oxidative stress and efficient adaptive mechanisms. This article summarizes the current knowledge on the role of two different MAPK pathways involved in sensing oxidative stress, transcription factors such as Cap1 and Skn7 that participate in the response, and the roles of some non-enzymatic and enzymatic antioxidants in this process.

2. MAP KINASE SIGNAL TRANSDUCTION PATHWAYS

Signal transduction pathways mediated by mitogen-activated kinases (MAPKs) are major mechanisms by which eukaryotic cells sense changes in their environment allowing adaptation. These pathways comprise a conserved module of three kinases: the MAP kinase kinase kinase (MAPKKK), the MAP kinase kinase (MAPKK) and the MAP kinase (MAPK). These MAPKs are activated by sequential phosphorylation; when the MAPKKK becomes activated by different mechanisms, it is able to phosphorylate the MAPKK, which in turn phosphorylates the MAPK. This MAPK transmits the signal to downstream targets (mainly transcription factors) which generate a specific response to the stimulus and trigger the adaptive response.

MAPKs have been classified into three major groups: (a) extracellular signal-regulated kinases (ERKs), (b) stress-activated kinases (SAPKs) and (c) other MAPKs (MAPK3). The MAPKs become phosphorylated in the conserved TXY dual-phosphorylation motif. Almost all members of the ERK subgroup and many of the MAPK3 have a TEY (Thr–Glu–Tyr) dual-phosphorylation motif characteristic of cell-growth kinases, such as mammalian p42/44 or Slt2/Mpk1, and Kss1 and Fus3 from *S. cerevisiae*. The majority of SAPK members have a TGY (Thr–Gly–Tyr) motif, such as the mammalian p38 and JNK families and the yeast Hog1 (Kultz and Burg, 1998). The SAPK subgroup has only been identified in animals and yeast, suggesting that this subgroup is phylogenetically the most recent MAPK group splitting from the ERK lineage after the protozoan and plant lineages separated from the metazoan/fungal lineages (Kultz and Burg, 1998). SAPK cascades are triggered by diverse stimuli as each biological species has evolved to adapt to the stresses they frequently encounter in their habitats. In *C. albicans* different signal transduction pathways have been found to play a role in resistance to oxidative stress.

2.1 The HOG Pathway

The pathway mediated by the MAP kinase Hog1 plays important roles in the physiology of *C. albicans*. Similar to the *S. cerevisiae* HOG pathway, the HOG pathway in *C. albicans* comprises two sensing branches (Figure 2), one involving Sho1, a transmembrane protein that probably plays a role as an adaptor or anchor of the MAPK complex rather than being a real sensor of the stimuli. In *S. cerevisiae* Sho1 transmits the signal to Ste11 and Pbs2. *STE11* has also been identified in the genome of *C. albicans* but the deletion mutants have not been obtained and therefore the function of this MAPKKK has not been characterized yet. The second branch comprises the three component system *SLN1-YPD1-SSK1*. This three component system transmits the signal through the MAPKKK Ssk2. Both branches converge at the MAPK kinase level, Pbs2 being able to phosphorylate the Hog1 MAP kinase (reviewed by Monge *et al.*, 2006).

As in *S. cerevisiae*, the *C. albicans* HOG pathway mediates the accumulation of glycerol in response to osmotic stress and is therefore involved in the adaptation to osmotic stress (San José *et al.*, 1996). In addition, Hog1 has a repressive effect on the yeast-to-hypha transition, controls the phosphorylation of other MAPKs and is crucial for the maintenance of infection in the mouse model of acute systemic infection (Alonso-Monge *et al.*, 1999). Hog1 as a SAPK becomes phosphorylated under a wide variety of stimuli such as osmotic stress, caffeine, Cd^{2+}, As^{3+} and oxidative agents (Alonso-Monge *et al.*, 2003; Smith *et al.*, 2004).

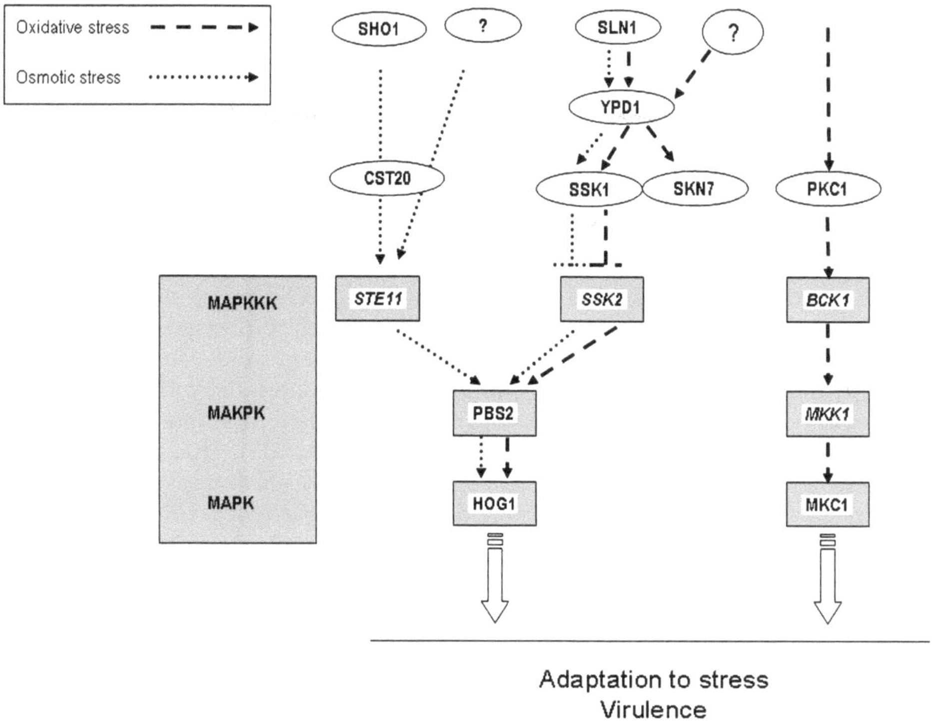

Figure 2 MAPK pathways activated by oxidative stress in *Candida albicans*. Genes are represented in normal characters where deletion and/or phenotypic characterisation of the corresponding strain has been carried out in *C. albicans* or, alternatively (in Italic) where its position and role is just presumed based on the *S. cerevisiae* model and/or inspection of the genome. The question mark indicates a putative additional input in the HOG pathway. The arrow from *HOG1* to *MKC1* (→) represents the activating effect exerted by Hog1 over the activation of Mkc1 although it is not known if this interaction is direct or indirect or what elements are implicated. The central module of MAPKs is boxed.

Differences have been observed between both branches regarding the stimuli sensed. The Sho1 branch is important in terms of adaptation to oxidative stress and *sho1* mutants are sensitive to oxidants; however, its relevance is minor regarding the transmission of the phospho signal to downstream effectors in response to an oxidative challenge (Román *et al.*, 2005). In contrast, the Ssk1 regulator protein is crucial in this process as Hog1 phosphorylation is clearly diminished under oxidative stress in a *ssk1* mutant (Chauhan *et al.*, 2003; Román *et al.*, 2005) and the slight phosphorylation signal in *sho1* cells is undetectable in the double mutant *ssk1 sho1*. Nevertheless, Hog1 phosphorylation in response to osmotic stress remains intact, indicating the existence of additional inputs to the pathway (Román *et al.*, 2005). Furthermore, the role of Sln1 is not clear since *sln1* mutants increase the basal level (i.e., under non-stress conditions) of Hog1 phosphorylation although they are still able to sense and increase the Hog1 phosphorylation in response to hydrogen peroxide (Román *et al.*, 2005). It is remarkable that deletion of *SLN1* is not lethal in *C. albicans*, in sharp contrast to *S. cerevisiae* where the lack of Sln1 leads to a fatal hyperactivation of Hog1 (Fassler *et al.*, 1997; Hohmann, 2002). Signalling converges at the MAPK kinase Pbs2 level, which is implicated in signalling of osmotic and oxidative stress, as well as in the phosphorylation of other MAPKs and the repression of the yeast-to-hypha transition and cell wall biogenesis (Arana *et al.*, 2005).

In *C. albicans* other elements have been described as putative members of a three-component system: *CHK1* and *NIK1* are both implicated in cell wall biosynthesis, morphogenesis and virulence (Yamada-Okabe *et al.*, 1999). *chk1* mutants are sensitive to oxidants, display a reduced virulence and are more easily killed by human neutrophils *in vitro* (Calera and Calderone, 1999; Torosantucci *et al.*, 2002; Román *et al.*, 2005). Nevertheless, Hog1 phosphorylation in response to oxidative stress was altered neither in single mutants (*hk1*, *nik1*) nor in double mutants (*chk1 nik1*) (Román *et al.*, 2005). This suggests either that these two proteins are not upstream sensors of the HOG pathway or that the presence of any component (Sln1, Chk1 or Nik1) is sufficient for the signalling to Ssk1 and Hog1 because the triple *chk1 nik1 sln1* mutant has not been obtained (Yamada-Okabe *et al.*, 1999).

Genome-wide transcription studies under oxidative stress, osmotic stress, and heavy metal stress show that oxidative stress influences the expression of more genes than the latter two conditions (429 induced and 448 repressed *versus* 179 induced and 190 repressed under osmotic stress and 101 induced and 85 repressed under heavy metal stress). As a SAPK, Hog1 controls the expression of many genes mainly under osmotic and heavy metal stresses, since gene transcription under oxidative stress was less affected in *hog1* mutants (Enjalbert *et al.*, 2006). This fact suggested that the HOG pathway acts together with other pathways to regulate the response to oxidative stress. Moreover, the exposure of *C. albicans* to a high concentration of H_2O_2 (5 mM) resulted in a gene expression profile clearly different from the one that was displayed as a consequence of treatment with a low dose of H_2O_2 (0.3 mM) (Enjalbert *et al.*, 2003, 2006). A similar observation has been reported for *S. cerevisiae* (O'Rourke and Herskowitz, 2004) and *S. pombe* (Quinn *et al.*, 2002). Intriguingly, 62 genes out of 246 genes induced by oxidative stress were induced to a greater extent in the *C. albicans hog1* mutant than in a wild-type strain. Some of these genes encode proteins with clear antioxidant function such as *GSH1*, *GPX1*, *TSA1* and *TRX1* (Table 1). This suggests a repressive role of Hog1 over these genes to maintain an appropriate redox balance (the sensitivity of *hog1* cells to oxidative damage indicates a defective redox balance in this mutant). In addition, the genome-wide transcription analysis in the absence of stress indicates a set of stress-related genes that are down-regulated in *HOG1* cells compared to the *hog1* mutant (*CTA1*, *HSP12*, *DDR48*, etc) (Table 2) (Enjalbert *et al.*, 2006). Hence, Hog1 may repress the expression of stress-related genes under standard growth conditions.

Transcription profiles obtained in the absence of stress also revealed significant down-regulation of hypha-specific genes such as *ECE1*, *HWP1* or *RBT2/4/5* in *HOG1* cells compared with *hog1* mutant cells (Table 2). This observation

Table 1 Genes up- or down-regulated during oxidative stress

Ca name[a]	Name	Function	wt	*hog1*	*Cap1*
CA0183	*CAP1*	Transcription factor	↑		↓
CA3059	*TRR1*	Thiorredoxin reductase	↑		↓
CA3086	*GLR1*	Glutathione reductase	↑		↓
CA4712	*GTT1*	Glutathione S-transferase	↑		↓
CA2719	*SOD2*	Manganese superoxide dismutase	↑		↓
CA0559	*GPX1*	glutathione peroxidase (by homology)	↑	↑	
CA6010	*TRX1*	thioredoxin (by homology)	↑	↑	
CA5714	*TSA1*	similar to *Saccharomyces cerevisiae* Tsa1p thiol-specific antioxidant-like protein (by homology)		↑	
CA0584	*GSH1*	Gamma-glutamylcysteine synthetase, 5prime end (by homology)		↑	
CA1782	*PHO84*	Phosphate transport protein	↓		
CA1975	*PLB1*	Phospholipase B	↓		
CA0752	*MNN22*	Mannosyltransferase	↓		

Source: Chauhan *et al.* (2003); Enjalbert *et al.* (2006); Wang *et al.* (2006).

[a] Activated (↑) or repressed (↓) genes in the corresponding mutant (***cap1***, ***hog1*** or wt) are shown in this Table together with the genome systematic name (first column), the normal name (second column) and the proposed function. ***Ca***, ***C. albicans***.

correlates with the hyperfilamentous phenotype described both in *hog1* and *pbs2* mutants (Alonso-Monge *et al.*, 1999; Arana *et al.*, 2005). Moreover, transcription profiles of cells defective for the Ssk1 regulator in the absence of stress compared to wild-type strains indicates two set of up-regulated genes: (1) genes implicated in the response/adaptation to oxidative stress (*HSP12, AHP1* and *PYC2*) and (2) genes that regulate cell wall biosynthesis (*FLO1, MNN4, CHK1*). The *ALS1* gene, which encodes a cell adhesin, was found to be down-regulated in a *ssk1* mutant, which correlates with the reduced adherence of *ssk1* mutants to human oesophageal tissues (Li *et al.*, 2002a; Chauhan *et al.*, 2003) (Table 2). In conclusion, genome-wide transcription analyses have provided evidence that the HOG pathway plays an important repressive role over morphological and stress-related genes even under basal growth conditions.

Presently, it remains unknown which is (are) the target transcription factor(s) activated by the Hog1 MAP kinase. The *C. albicans* homologue to the *S. cerevisiae* Yap1 transcription factor, Cap1, is a putative candidate, as it is activated by oxidative stress. However, the phenotypes of *hog1* and *cap1* mutants are clearly distinguishable (Alonso-Monge *et al.*, 2003), as are their genome-wide transcriptional responses (Enjalbert *et al.*, 2006) (see Table 1). Other candidates tested were Msn4 and Mnl1 (homologues to Msn2 and Msn4 from *S. cerevisiae*, respectively) but these transcription factors play no obvious role in the response to stress of *C. albicans* (Nicholls *et al.*, 2004). Hog1 could, finally, regulate additional non-transcriptional responses to oxidative stress in *C. albicans*, as has been described

Table 2 Genes up- or down-regulated in *ssk1* or *hog1* mutants under non-stress growth conditions

Ca name[a]	Name	Mutant	Regulation	Function
CA0183	*HSP12*	*ssk1*	↑	Heat shock protein (by homology)
CA3059	*AHP1*	*ssk1*	↑	Alkyl hydroxiperoxide reductase
CA2477	*MNN4*	*ssk1*	↑	Mannosyl transferase
CA0316	*ALS1*	*ssk1*	↓	Agglutinin-like protein
CA4676	*CHK1*	*ssk1*	↑	Histidine kinase
CA1678	*FLO1*	*ssk1*	↑	Putative cell wall protein
CA3011	*CTA1*	*hog1*	↑	Catalase A, peroxisomal (by homology)
CA1782	*PHO84*	*hog1*	↓	Phosphate transport protein
CA1975	*PLB1*	*hog1*	↓	Phospholipase B
CA0752	*MNN22*	hog1	↓	Mannosyltransferase
CA4336	*DDR48*	*hog1*	↑	Stress protein
CA1402	*ECE1*	*hog1*	↑	Cell elongation protein
CA2825	*HWP1*	*hog1*	↑	Hyphal growth protein
CA3957	*RBT2*	*hog1*	↑	Repressed by *TUP1* proteins
CA0104	*RBT4*	*hog1*	↑	Repressed by *TUP1* proteins
CA2558	*RBT5*	*hog1*	↑	Repressed by *TUP1* proteins

Source: Chauhan *et al.* (2003); Enjalbert *et al.* (2006).
Note : [a]Activated (↑) or repressed (↓) genes in the corresponding mutant are shown in this table together with the genome systematic name (first column), the normal name (second column) and the proposed function.

in *S. cerevisiae* and *S. pombe* (Bilsland-Marchesan *et al.*, 2000; Teige *et al.*, 2001), and could also be involved in cell cycle regulation (Escote *et al.*, 2004).

2.2 The Cell Integrity (*MKC1*-Mediated) Pathway

The MAPK pathway mediated by Mkc1 is homologous to the cell integrity pathway in *S. cerevisiae* (Cid *et al.*, 1995; Martin *et al.*, 2000). The *S. cerevisiae* PKC pathway has been shown to be involved in the construction and remodelling of the cell wall in response to many stimuli such as temperature, cell wall perturbing compounds, caffeine, and more recently, to oxidative stress (Vilella *et al.*, 2005). In *C. albicans* only two elements of the cell integrity pathway have been characterized to date: the protein kinase C, Pkc1, and the MAP kinase, Mkc1 (Figure 2). A *pkc1* mutant displays a constitutive lytic (osmotically remediable) phenotype that has complicated its study (Paravicini *et al.*, 1996) and it has only been recently established that the Pkc1 kinase influences the phosphorylation state of Mkc1 (Navarro-Garcia *et al.*, 2005).

Interestingly, in *C. albicans* Mkc1 is activated in response to several kinds of stress such as oxidative, nitrosative and osmotic. This MAPK is also activated in response to cell wall disturbing compounds such as Zymolyase 20-T and Congo Red, a decrease in growth temperature, the presence of divalent cations (Ca^{2+} and Mg^{2+}) (Navarro-Garcia *et al.*, 2005), and surface contact (Kumamoto, 2005).

Remarkably, *mkc1* mutants displayed attenuated virulence using a systemic model of infection in mice (Diez-Orejas *et al.*, 1997). The activation of the PKC pathway by oxidative stress in *S. cerevisiae* has been interpreted to result from cell wall alterations: both hydrogen peroxide and diamide provoked a marked depolarization of the actin cytoskeleton; diamide, in addition, induced the formation of disulfide bonds on newly secreted cell wall proteins with the subsequent activation of the PKC pathway to overcome these effects (Vilella *et al.*, 2005). Phosphorylation of Slt2 (the *S. cerevisiae* Mkc1 homologue) is observed after 40 min of incubation under oxidative stress conditions, while in *C. albicans* Mkc1 phosphorylation in response to oxidative/nitrosative stresses is detectable just 1 min after the challenge (Arana *et al.*, 2005). This suggests that the phosphorylation occurs directly rather than as a consequence of other phenomena. This fact, together with the reported phosphorylation of Mkc1 in the stationary phase of growth (Román *et al.*, 2005), may indicate that Mkc1 activation is required to arrest the cell cycle or at least, to slow down the speed of growth to allow cells to overcome the stress.

Moreover, the key pathway for responding to stress seems to be the HOG pathway since absence of the MAP kinase Hog1 leads to an altered phosphorylation pattern of Mkc1. Mkc1 is phosphorylated within a different range of stressor doses and its phosphorylation is lower in *hog1* and *pbs2* mutants (Arana *et al.*, 2005; Navarro-Garcia *et al.*, 2005).

3. TRANSCRIPTION FACTORS

A key regulator of the oxidative stress response in *C. albicans* is the basic region-leucine zipper (bZip) transcription factor Cap1. This protein, homologous to Yap1 in *S. cerevisiae*, belongs to the AP-1 family and is required for adaptation to oxidative, drug and metal stresses (Alarco and Raymond, 1999). Cap1 localizes in the cytoplasm and is translocated to the nucleus under oxidative stress (Zhang *et al.*, 2000). The relationship between Cap1 and Hog1 has been analysed. The absence of *HOG1* did not affect the translocation of Cap1 to the nucleus while in the *cap1* mutant the phosphorylation pattern of Hog1 was not altered (Alonso-Monge *et al.*, 2003; Enjalbert *et al.*, 2006). Genome-wide transcription studies of a wild-type strain and *cap1* mutant under oxidative stress (0.5 mM H_2O_2) indicate that 89 genes change their expression in the wild-type strain, 76 of them being Cap1 dependent (Wang *et al.*, 2006). This suggests a major role of Cap1 in the response to oxidative stress. Cap1 controls the gene transcription both at low- and high-hydrogen peroxide concentrations while Hog1 may be involved in the response to high doses of oxidant (Alonso-Monge *et al.*, 2003; Enjalbert *et al.*, 2006). Transcription of Cap1 is induced upon oxidative stress (Wang *et al.*, 2006) but also in the presence of human neutrophils (Fradin *et al.*, 2005).

Skn7 is another transcription factor potentially involved in the response to oxidative stress. In addition to its function as transcription factor, Skn7 acts as the response-regulator protein of a three-component system. In *S. cerevisiae*, Skn7 is activated by the intermediate protein Ypd1, which shuttles between the cytoplasm and the nucleus and is involved in the response of Skn7 to osmotic and oxidative stresses (Lu *et al.*, 2003). Skn7 also mediates the transcription of cell wall-related genes such as *OCH1* (Li *et al.*, 2002b). In *C. albicans*, *skn7* mutants are sensitive to hydrogen peroxide, but not to menadione and other oxidative agents, and their virulence is only slightly attenuated (Singh *et al.*, 2004). The function of this gene can be, however, wider that previously reported.

4. DEFENCES AGAINST OXIDATIVE STRESS

4.1 Non-Enzymatic Defences

Polyols and polyhydroxyalcohols are present in all organisms from bacteria to animals. These compounds serve as reserve carbon source and play protective roles. Glycerol, as mentioned above, is a compatible osmolyte that protects *C. albicans* and other fungi such as *S. cerevisiae* from osmotic stress (Albertyn *et al.*, 1994; San José *et al.*, 1996). *C. albicans* also synthesizes D-arabinitol in response to oxidative stress and high temperatures (Kayingo and Wong, 2005). Remarkably, *C. albicans* produces large amounts of D-arabinitol during infection, a feature that has been used to diagnose systemic candidiasis (Christensson *et al.*, 1999). Nevertheless, the role of D-arabinitol is unclear and no direct evidence for a scavenger function or protection against oxidative stress has been reported so far. Non-enzymatic defences against oxidative and nitrosative stresses are metabolites that act as scavengers of reactive species. *C. albicans* accumulates trehalose in response to heat and oxidative stress (Arguelles, 1997; Zaragoza *et al.*, 2003). Trehalose (α-D-glucopyranosyl-α-D-glucopyranoside) is a non-reducing disaccharide that acts by stabilizing membranes and native proteins as well as by suppressing the aggregation of denatured proteins (Singer and Lindquist, 1998). Trehalose plays a role during direct oxidative stress but not in the acquisition of tolerance, since the adaptive response induced in cells pretreated with a low (sublethal) dose of hydrogen peroxide or incubated at 37°C is not concomitant with an accumulation of trehalose (Alvarez-Peral *et al.*, 2002).

4.2 Enzymatic Defences

All aerobic organisms have developed enzymatic defences against metabolic ROS; in addition, pathogenic fungi must cope with the ROS produced by phagocytes. Superoxide dismutases (SODs) are enzymes that convert superoxide (O_2^-) to hydrogen peroxide (H_2O_2), although this reaction can also take place spontaneously. Protection by SODs is explained by the formation of ROS other than lipid hydroperoxides and the reactive nitric intermediate (RNI) peroxynitrite, which are more detrimental to cells due to the absence of defences against theses types of compounds. Depending on the cofactor used, four classes of SODs have been described: Mn, Fe, Ni and Cu, Zn. In *C. albicans* a total of six SODs have been identified. Sod1 is a Cu, Zn enzyme that localizes in the cytoplasm and is expressed in the exponential phase of growth (Rhie *et al.*, 1999). *sod1* mutants were sensitive to menadione, were less virulent in mice and showed reduced survival in macrophages (Hwang *et al.*, 2002). *SOD2* encodes a mitochondrial Mn SOD and is thought to scavenge superoxide generated by the mitochondrial respiratory chain (Hwang *et al.*, 2003). *SOD3* encodes a Mn cytoplasmic isoenzyme and is expressed mainly during the stationary phase of growth rather than exponential phase, contrary to *SOD1* (Lamarre *et al.*, 2001). The recently identified *SOD4, SOD5* and *SOD6* genes encode putative Cu,Zn proteins which are presumably cytoplasmic (Martchenko *et al.*, 2004). Sod5 is expressed during the yeast-to-hypha transition, under oxidative and osmotic stress, and when growing on non-fermentable carbon sources. *sod5* mutants are also less virulent in a systemic infection in mice.

Since hydrogen peroxide (formed by SOD activity) is still toxic for the cell, it must be eliminated. Catalases, ubiquitous in all aerobic organisms, can fulfil this role and convert hydrogen peroxide to water and molecular oxygen. *C. albicans* only possesses one catalase gene, named *CTA1, CCT1* or *CAT1*. Cta1 has been shown to protect cells from peroxide stress, killing by neutrophils and is required for virulence (Wysong *et al.*, 1998; Nakagawa *et al.*, 2003).

Other antioxidant enzymes are thiol peroxidase, glutaredoxins, glutathione (GSH) peroxidases and GSH S-transferases. Thiol peroxidases and glutaredoxins are important for many normal metabolic reactions involving disulfide bond formation as part of the catalytic cycle, and protein folding or prosthetic group insertion during protein assembly (Grant, 2001). In addition, many of these enzymes are involved in responses to oxidative or nitrosative stresses. The GSH peroxidases and GSH S-transferases are involved in the breakdown of organic hydroperoxides, with GSH acting as reducing agent or as a conjugant to toxic lipophilic compounds, respectively (Grant, 2001). Thiol peroxidases remove peroxides and provide defence against oxidative damage; some of these enzymes are known as thiol-specific antioxidants (TSAs). In *C. albicans*, the *TSA1* gene is highly expressed during hyphal growth and, interestingly, the localization of the protein depends on the morphology of the cell. While Tsa1 is localized in the cytoplasm in yeast cells it was found in the cell wall of hyphae. Tsa1 confers resistance towards oxidative stress and also affects hyphal cell wall composition (Choi *et al.*, 2003; Urban *et al.*, 2005). As mentioned above, the expression of *TSA1* is induced in response to hydrogen peroxide, and to a higher extent in *hog1* mutants (Enjalbert *et al.*, 2006).

Among the enzymes responsible for resistance to nitrosative stress, three flavohemoglobin denitrosylase homologues (CaYhb1, CaYhb4 and CaYhb5) have been identified in the *C. albicans* genome. Only one of them, Yhb1 is responsible for NO consumption and detoxification (Ullmann *et al.*, 2004). *YHB1* expression is highly induced in response to NO and this is dependent on amphotericin B but not peroxide or superoxide (Ullmann *et al.*, 2004; Liu *et al.*, 2005). *yhb1* mutants are hypersensitive to NO and are hyperfilamentous. The virulence of *yhb1* mutants in the mice systemic infection model is moderately attenuated, but this defect was not restored when the *NOS2* gene was deleted in the host, suggesting that ·NO is not a prime determinant of virulence in systemic candidiasis in mice (Hromatka *et al.*, 2005).

5. CONCLUDING REMARKS

Signalling oxidative and nitrosative stress is crucial in pathogenic microorganisms. Pathogens have developed multiple and flexible mechanisms of signalling and resistance against oxidative stress, which are used by immune cells to fight against disease but are also generated as a consequence of normal metabolism. Some signal transduction pathways and downstream antioxidants have been identified in *C. albicans* as mediators of the oxidative and nitrosative stress response. Although many components and mechanisms are still not known, the current data suggest that these elements are important during infection and phagocytosis. Mutants deficient in antioxidant enzymes (*cta1*, *sod5* and *sod2*) or signalling (*ssk1*, *cap1* and *chk1*) elements are most susceptible to neutrophils and many of them are less virulent in the acute mouse infection model (*hog1*, *mkc1*, *ssk1*, *cta1*, among others). In addition, some of them (*hog1*) have been shown to generate a protective immune response (Fernandez-Arenas *et al.*, 2004) suggesting that other stress response-deficient mutants may be useful in the generation of a candidiasis vaccine. Future studies may help to complete the picture of the strategies and mechanisms for resistance to oxidative and nitrosative stress, and should lead also to a deeper understanding of the host–pathogen interaction.

ACKNOWLEDGEMENTS

Work in our group is supported by grants BIO2003-0992 and NIH grant RX4215-030-UM. R. Alonso-Monge (R.A.M.) is recipient of a Ramón y Cajal postdoctoral position from the Ministerio de Educación y Ciencia (M.E.C).

REFERENCES

Alarco, A.M. and Raymond, M. (1999). The bZip transcription factor Cap1p is involved in multidrug resistance and oxidative stress response in *Candida albicans*. *Journal of Bacteriology*, **181**(3), 700–708.

Albertyn, J., Hohmann, S., Thevelein, J.M. and Prior, B.A. (1994). *GPD1*, which encodes glycerol-3-phosphate dehydrogenase, is essential for growth under osmotic stress in *Saccharomyces cerevisiae*, and its expression is regulated by the high-osmolarity glycerol response pathway. *Molecular and Cellular Biology*, **14**(6), 4135–4144.

Alonso-Monge, R., Navarro-García, F., Molero, G., Diez-Orejas, R., Gustin, M., Pla, J., Sánchez, M. and Nombela, C. (1999). Role of the mitogen-activated protein kinase Hog1p in morphogenesis and virulence of *Candida albicans*. *Journal of Bacteriology*, **181**(10), 3058–3068.

Alonso-Monge, R., Navarro-García, F., Román, E., Negredo, A.I., Eisman, B., Nombela, C. and Pla, J. (2003). The Hog1 mitogen-activated protein kinase is essential in the oxidative stress response and chlamydospore formation in *Candida albicans*. *Eukaryotic Cell*, **2**(2), 351–361.

Alvarez-Peral, F.J., Zaragoza, O., Pedreno, Y. and Arguelles, J.C. (2002). Protective role of trehalose during severe oxidative stress caused by hydrogen peroxide and the adaptive oxidative stress response in *Candida albicans*. *Microbiology*, **148**(Pt 8), 2599–2606.

Arana, D.M., Nombela, C., Alonso-Monge, R. and Pla, J. (2005). The Pbs2 MAP kinase kinase is essential for the oxidative-stress response in the fungal pathogen *Candida albicans*. *Microbiology*, **151**(Pt 4), 1033–1049.

Arguelles, J.C. (1997). Thermotolerance and trehalose accumulation induced by heat shock in yeast cells of *Candida albicans*. *FEMS Microbiological Letters*, **146**(1), 65–71

Bilsland-Marchesan, E., Arino, J., Saito, H., Sunnerhagen, P. and Posas, F. (2000). Rck2 kinase is a substrate for the osmotic stress-activated mitogen-activated protein kinase *HOG1*. *Molecular and Cellular Biology*, **20**(11), 3887–3895.

Brunelli, L., Crow, J.P. and Beckman, J.S. (1995). The comparative toxicity of nitric oxide and peroxynitrite to *Escherichia coli*. *Archives of Biochemistry and Biophysics*, **316**(1), 327–334.

Calderone, R.A. and Fonzi, W.A. (2001). Virulence factors of *Candida albicans*. *Trends in Microbiology*, **9**(7), 327–335.

Calderone, R., Suzuki, S., Cannon, R., Cho, T., Boyd, D., Calera, J., Chibana, H., Herman, D., Holmes, A., Jeng, H. W., Kaminishi, H., Matsumoto, T., Mikami, T., O'Sullivan, J.M., Sudoh, M., Suzuki, M., Nakashima, Y., Tanaka, T., Tompkins, G.R. and Watanabe, T. (2000). *Candida albicans*: Adherence, signaling and virulence. *Medical Mycology*, **38**(Suppl 1), 125–137.

Calera, J.A. and Calderone, R. (1999). Histidine kinase, two-component signal transduction proteins of *Candida albicans* and the pathogenesis of candidosis. *Mycoses*, **42**(Suppl 2), 49–53.

Chauhan, N., Inglis, D., Román, E., Pla, J., Li, D., Calera, J.A. and Calderone, R. (2003). *Candida albicans* response regulator gene *SSK1* regulates a subset of genes whose functions are associated with cell wall biosynthesis and adaptation to oxidative stress. *Eukaryotic Cell*, **2**(5), 1018–1024.

Choi, W., Yoo, Y.J., Kim, M., Shin, D., Jeon, H.B. and Choi, W. (2003). Identification of proteins highly expressed in the hyphae of *Candida albicans* by two-dimensional electrophoresis. *Yeast*, **20**(12), 1053–1060.

Christensson, B., Sigmundsdottir, G. and Larsson, L. (1999). D-arabinitol—A marker for invasive candidiasis. *Medical Mycology*, **37**(6), 391–396.

Cid, V.J., Durán, A., del Rey, F., Snyder, M.P., Nombela, C. and Sánchez, M. (1995). Molecular basis of cell integrity and morphogenesis in *Saccharomyces cerevisiae*. *Microbiological Reviews*, **59**, 345–386.

Diez-Orejas, R., Molero, G., Navarro-García, F., Pla, J., Nombela, C. and Sánchez-Pérez, M. (1997). Reduced virulence of *Candida albicans MKC1* mutants: A role for a mitogen-activated protein kinase in pathogenesis. *Infection and Immunity*, **65**,(2), 833–837.

Douglas, L.J. (2003). *Candida* biofilms and their role in infection. *Trends in Microbiology*, **11**(1), 30–36.

Enjalbert, B., Nantel, A. and Whiteway, M. (2003). Stress-induced gene expression in *Candida albicans*: Absence of a general stress response. *Molecular Biology of the Cell*, **14**(4), 1460–1467.

Enjalbert, B., Smith, D.A., Cornell, M.J., Alam, I., Nicholls, S., Brown, A.J. and Quinn, J. (2006). Role of the Hog1 stress-activated protein kinase in the global transcriptional response to stress in the fungal pathogen *Candida albicans*. *Molecular Biology of the Cell*, **17**(2), 1018–1032.

Escote, X., Zapater, M., Clotet, J. and Posas, F. (2004). Hog1 mediates cell-cycle arrest in G1 phase by the dual targeting of Sic1. *Nature Cell Biology*, **6**(10), 997–1002.

Fassler, J.S., Gray, W.M., Malone, C.L., Tao, W., Lin, H. and Deschenes, R.J. (1997). Activated alleles of yeast *SLN1* increase Mcm1-dependent reporter gene expression and diminish signaling through the hog1 osmosensing pathway. *Journal of Biological Chemistry*, **272**(20), 13365–13371.

Fernandez-Arenas, E., Molero, G., Nombela, C., Diez-Orejas, R. and Gil, C. (2004). Contribution of the antibodies response induced by a low virulent *Candida albicans* strain in protection against systemic candidiasis. *Proteomics*, **4**(4), 1204–1215.

Fradin, C., de Groot, P., Maccallum, D., Schaller, M., Klis, F., Odds, F.C. and Hube, B. (2005). Granulocytes govern the transcriptional response, morphology and proliferation of *Candida albicans* in human blood. *Molecular Microbiology*, **56**(2), 397–415.

Grant, C.M. (2001). Role of the glutathione/glutaredoxin and thioredoxin systems in yeast growth and response to stress conditions. *Molecular Microbiology*, **39**,(3), 533–541.

Hohmann, S. (2002). Osmotic stress signaling and osmoadaptation in yeasts. *Microbiology and Molecular Biology Reviews*, **66**(2), 300–372.

Hromatka, B.S., Noble, S.M. and Johnson, A.D. (2005). Transcriptional response of *Candida albicans* to nitric oxide and the role of the *YHB1* gene in nitrosative stress and virulence. *Molecular Biology of the Cell*, **16**(10), 4814–4826.

Hwang, C.S., Baek, Y.U., Yim, H.S. and Kang, S.O. (2003). Protective roles of mitochondrial manganese-containing superoxide dismutase against various stresses in *Candida albicans*. *Yeast*, **20**(11), 929–941.

Hwang, C.S., Rhie, G.E., Oh, J.H., Huh, W.K., Yim, H.S. and Kang, S.O. (2002). Copper- and zinc-containing superoxide dismutase (Cu/Zn SOD) is required for the protection of *Candida albicans* against oxidative stresses and the expression of its full virulence. *Microbiology,* **148**(Pt 11), 3705–3713.

Kayingo, G. and Wong, B. (2005). The MAP kinase Hog1p differentially regulates stress-induced production and accumulation of glycerol and D-arabitol in *Candida albicans. Microbiology,* **151**(Pt 9), 2987–2999.

Kuhn, D.M. and Ghannoum, M.A. (2004). *Candida* biofilms: Antifungal resistance and emerging therapeutic options. *Current Opinion Investigational Drugs,* **5**(2), 186–197.

Kultz, D. and Burg, M. (1998). Evolution of osmotic stress signaling via MAP kinase cascades. *Journal of Experimental Biology,* **201**(Pt 22), 3015–3021.

Kumamoto, C.A. (2005). A contact-activated kinase signals *Candida albicans* invasive growth and biofilm development. *Proceeding of the National Academy of Sciences of the United States of America,* **102**(15), 5576–5581.

Lamarre, C., LeMay, J.D., Deslauriers, N. and Bourbonnais, Y. (2001). *Candida albicans* expresses an unusual cytoplasmic manganese-containing superoxide dismutase (*SOD3* gene product) upon the entry and during the stationary phase. *Journal of Biological Chemistry,* **276**(47), 43784–43791.

Li, D., Bernhardt, J. and Calderone, R. (2002a). Temporal expression of the *Candida albicans* genes *CHK1* and *CSSK1*, adherence, and morphogenesis in a model of reconstituted human esophageal epithelial candidiasis. *Infection and Immunity,* **70**(3), 1558–1565.

Li, S., Dean, S., Li, Z., Horecka, J., Deschenes, R.J. and Fassler, J.S. (2002b). The eukaryotic two-component histidine kinase Sln1p regulates OCH1 via the transcription factor, Skn7p. *Molecular Biology of the Cell,* **13**(2), 412–424.

Liu, T.T., Lee, R.E., Barker, K.S., Lee, R.E., Wei, L., Homayouni, R. and Rogers, P.D. (2005). Genome-wide expression profiling of the response to azole, polyene, echinocandin, and pyrimidine antifungal agents in *Candida albicans. Antimicrobial Agents and Chemotherapy,* **49**(6), 2226–2236.

Lorenz, M.C., Bender, J.A. and Fink, G.R. (2004). Transcriptional response of *Candida albicans* upon internalization by macrophages. *Eukaryotic Cell,* **3**(5), 1076–1087.

Lorenz, M.C. and Fink, G.R. (2001). The glyoxylate cycle is required for fungal virulence. *Nature,* **412**(6842), 83–86.

Lu, J.M., Deschenes, R.J. and Fassler, J.S. (2003). *Saccharomyces cerevisiae* histidine phosphotransferase Ypd1p shuttles between the nucleus and cytoplasm for *SLN1*-dependent phosphorylation of Ssk1p and Skn7p. *Eukaryotic Cell,* **2**(6), 1304–1314.

Martchenko, M., Alarco, A.M., Harcus, D. and Whiteway, M. (2004). Superoxide dismutases in *Candida albicans*: Transcriptional regulation and functional characterization of the hyphal-induced *SOD5* gene. *Molecular Biology of the Cell,* **15**(2), 456–467.

Martin, H., Rodriguez-Pachon, J.M., Ruiz, C., Nombela, C. and Molina, M. (2000). Regulatory mechanisms for modulation of signaling through the cell integrity Slt2-mediated pathway in *Saccharomyces cerevisiae. Journal of Biological Chemistry,* **275**(2), 1511–1519.

Missall, T.A., Lodge, J.K. and McEwen, J.E. (2004). Mechanisms of resistance to oxidative and nitrosative stress: Implications for fungal survival in mammalian hosts. *Eukaryotic Cell,* 3(4), 835–846.

Monge, R.A., Román, E., Nombela, C. and Pla, J. (2006). The MAP kinase signal transduction network in *Candida albicans. Microbiology,* **152**(Pt 4), 905–912.

Nakagawa, Y., Kanbe, T. and Mizuguchi, I. (2003). Disruption of the human pathogenic yeast *Candida albicans* catalase gene decreases survival in mouse-model infection and elevates susceptibility to higher temperature and to detergents. *Microbiology Immunology,* **47**(6), 395–403.

Nathan, C. and Shiloh, M.U. (2000). Reactive oxygen and nitrogen intermediates in the relationship between mammalian hosts and microbial pathogens. *Proceedings of the National Academy of Sciences of the United States of America,* **97**(16), 8841–8848.

Navarro-Garcia, F., Eisman, B., Fiuza, S.M., Nombela, C. and Pla, J. (2005). The MAP kinase Mkc1p is activated under different stress conditions in *Candida albicans. Microbiology,* **151**(Pt 8), 2737–2749.

Navarro-García, F., Sánchez, M., Nombela, C. and Pla, J. (2001). Virulence genes in the pathogenic yeast *Candida albicans. FEMS Microbiological Reviews,* **25**(2), 245–268.

Nicholls, S., Straffon, M., Enjalbert, B., Nantel, A., Macaskill, S., Whiteway, M. and Brown, A.J. (2004). Msn2- and Msn4-like transcription factors play no obvious roles in the stress responses of the fungal pathogen *Candida albicans*. *Eukaryotic Cell*, **3**(5), 1111–1123.

O'Rourke, S.M. and Herskowitz, I. (2004). Unique and redundant roles for HOG MAPK pathway components as revealed by whole-genome expression analysis. *Molecular Biology of the Cell*, **15**(2), 532–542.

Paravicini, G., Mendoza, A., Antonsson, B., Cooper, M., Losberger, C. and Payton, M. (1996). The *Candida albicans PKC1* gene encodes a protein kinase C homolog necessary for cellular integrity but not dimorphism. *Yeast*, **12**, 741–756.

Quinn, J., Findlay, V.J., Dawson, K., Millar, J.B., Jones, N., Morgan, B.A. and Toone, W.M. (2002). Distinct Regulatory Proteins Control the Graded Transcriptional Response to Increasing H_2O_2 Levels in Fission Yeast *Schizosaccharomyces pombe*. *Molecular Biology of the Cell*, **13**(3), 805–816.

Rhie, G.E., Hwang, C.S., Brady, M.J., Kim, S.T., Kim, Y.R., Huh, W.K., Baek, Y.U., Lee, B.H., Lee, J.S. and Kang, S.O. (1999). Manganese-containing superoxide dismutase and its gene from *Candida albicans*. *Biochemistry and Biophysics Acta*, **1426**(3), 409–419.

Román, E., Nombela, C. and Pla, J. (2005). The Sho1 adaptor protein links oxidative stress to morphogenesis and cell wall biosynthesis in the fungal pathogen *Candida albicans*. *Molecularand Cellular Biology*, **25**(23), 10611–10627.

San José, C., Alonso-Monge, R., Pérez-Díaz, R.M., Pla, J. and Nombela, C. (1996). The mitogen-activated protein kinase homolog *HOG1* gene controls glycerol accumulation in the pathogenic fungus *Candida albicans*. *Journal of Bacteriology*, **178**(19), 5850–5852.

Singer, M.A. and Lindquist, S. (1998). Thermotolerance in *Saccharomyces cerevisiae*: The Yin and Yang of trehalose. *Trends in Biotechnology*, **16**(11), 460–468.

Singh, P., Chauhan, N., Ghosh, A., Dixon, F. and Calderone, R. (2004). *SKN7* of *Candida albicans*: Mutant construction and phenotype analysis. *Infection and Immunity*, **72**(4), 2390–2394.

Smith, D.A., Nicholls, S., Morgan, B.A., Brown, A.J. and Quinn, J. (2004). A conserved stress-activated protein kinase regulates a core stress response in the human pathogen *Candida albicans*. *Molecular Biology of the Cell*, **15**(9), 4179–4190.

Soll, D.R., Morrow, B. and Srikantha, T. (1993). High-frequency phenotypic switching in *Candida albicans*. *Trends in Genetics*, **9**(2), 61–65.

Teige, M., Scheikl, E., Reiser, V., Ruis, H. and Ammerer, G. (2001). Rck2, a member of the calmodulin-protein kinase family, links protein synthesis to high osmolarity MAP kinase signaling in budding yeast. *Proceedings of the National Academy of Sciences of the United States of America*, **98**(10), 5625–5630.

Torosantucci, A., Chiani, P., De Bernardis, F., Cassone, A., Calera, J.A. and Calderone, R. (2002). Deletion of the two-component histidine kinase gene (*CHK1*) of *Candida albicans* contributes to enhanced growth inhibition and killing by human neutrophils *in vitro*. *Infection and Immunity*, **70**(2), 985–987.

Ullmann, B.D., Myers, H., Chiranand, W., Lazzell, A.L., Zhao, Q., Vega, L.A., Lopez-Ribot, J.L., Gardner, P.R. and Gustin, M.C. (2004). Inducible defense mechanism against nitric oxide in *Candida albicans*. *Eukaryotic Cell*, **3**(3), 715–723.

Urban, C., Xiong, X., Sohn, K., Schroppel, K., Brunner, H. and Rupp, S. (2005). The moonlighting protein Tsa1p is implicated in oxidative stress response and in cell wall biogenesis in *Candida albicans*. *Molecular Microbiology*, **57**(5), 1318–1341.

Vilella, F., Herrero, E., Torres, J. and De La Torre-Ruiz, M.A. (2005). Pkc1 and the upstream elements of the cell integrity pathway in *Saccharomyces cerevisiae*, Rom2 and Mtl1, are required for cellular responses to oxidative stress. *Journal of Biological Chemistry*, **280**(10), 9149–9159.

Wang, Y., Cao, Y.Y., Jia, X.M., Cao, Y.B., Gao, P.H., Fu, X. P., Ying, K., Chen, W.S. and Jiang, Y.Y. (2006). Cap1p is involved in multiple pathways of oxidative stress response in *Candida albicans*. *Free Radical Biology & Medicine*, **40**(7), 1201–1209.

Wysong, D.R., Christin, L., Sugar, A.M., Robbins, P.W. and Diamond, R.D. (1998). Cloning and sequencing of a *Candida albicans* catalase gene and effects of disruption of this gene. *Infection and Immunity*, **66**(5), 1953–1961.

Yamada-Okabe, T., Mio, T., Ono, N., Kashima, Y., Matsui, M., Arisawa, M. and Yamada-Okabe, H. (1999). Roles of three histidine kinase genes in hyphal development and virulence of the pathogenic fungus *Candida albicans*. *Journal of Bacteriology*, **181**(23), 7243–7247.

Zaragoza, O., Gonzalez-Parraga, P., Pedreno, Y., Alvarez-Peral, F.J. and Arguelles, J.C. (2003). Trehalose accumulation induced during the oxidative stress response is independent of TPS1 mRNA levels in *Candida albicans*. *Internal Microbiology*, **6**(2), 121–125.

Zhang, X., De Micheli, M., Coleman, S.T., Sanglard, D. and Moye-Rowley, W.S. (2000). Analysis of the oxidative stress regulation of the *Candida albicans* transcription factor, Cap1p. *Molecular Microbiology*, **36**(3), 618–629.

CHAPTER 17

Oxidant-Specific Protein Folding During Fungal Oxidative Stress: Activation and Function of the Yap1p Transcription Factor in *Saccharomyces cerevisiae*

W. Scott Moye-Rowley

Contents

1. Introduction 276
 1.1 Yap1p Role in the Response to Stress 277
 1.2 Molecular Details of Yap1p Activation by Oxidants 277
 1.3 Oxidant-Specific Downstream Function of Yap1p 279
 1.4 Bifunctional Role of the H_2O_2-Oxidized Form of Yap1p 282
2. Discussion 285
Acknowledgments 287
References 288

Abstract

Fungal cells are eliminated from animal hosts through oxidative stress-mediated mechanisms. The ability to tolerate oxidative stress has been correlated with virulence in pathogenic fungi like *Candida* and *Aspergillus* species. The yeast *Saccharomyces cerevisiae* has served as an excellent model system for both fungal oxidative stress tolerance in particular and eukaryotic oxidant detoxification in general. A central determinant of oxidative-stress resistance in *S. cerevisiae* is the transcriptional activator protein Yap1p. Yap1p is normally found in the cytoplasm in the absence of oxidative stress but rapidly recruits to the nucleus upon challenge by oxidants such as H_2O_2 or diamide. Detailed mutagenesis and functional studies have demonstrated that two different cysteine-rich domains (CRDs) present in Yap1p are involved in detecting and responding to changes in the redox potential of the cell. The carboxy-terminal CRD (c-CRD) is sufficient to provide diamide regulation of nuclear localization to a chimeric green fluorescent protein (GFP) reporter fusion and contains an important nuclear export signal of Yap1p. Biochemical analyses have provided evidence that different folding states of Yap1p are found during diamide or H_2O_2 exposure. In the context of wild-type Yap1p, a

British Mycological Society Symposia Series
Published by Elsevier Ltd.

second amino-terminal CRD (n-CRD) is required for normal regulation and function of Yap1p in response to H_2O_2-induced oxidative stress. Unlike the situation during diamide stress, correct folding of Yap1p in the presence of H_2O_2 requires the presence of ancillary factors like the glutathione peroxidase Gpx3p and the Yap1p-interacting protein Ybp1p. While regulation of the nuclear export of Yap1p is a key modulatory step during both diamide and H_2O_2-induced oxidative stress, recent experiments have shown that a second feature of the H_2O_2-folded state of Yap1p is required to assure normal function in the presence of this oxidant. This review will summarize the state of understanding of oxidant-specific folding of Yap1p and the requirement for the folded state of Yap1p that is generated in the presence of H_2O_2 during the response to this oxidant.

1. INTRODUCTION

Microorganisms face a range of different environmental challenges but one of the most fundamental issues for aerobic microbes is the detoxification of reactive oxygen species (ROS). ROS are generated as a course of normal aerobic metabolism, often through inefficient reduction of molecular oxygen to H_2O during oxidative phosphorylation (recently reviewed in Adam-Vizi, 2005). Failure to adequately handle accumulated ROS can lead to apoptosis and cell death. The toxic effect of ROS is exploited by animals during the innate immune response as a means eliminating microbes from a potential host (reviewed by Kobayashi and Flavell, 2004). Monocytes and macrophages internalize microbial cells and subject the would-be invaders to a large bolus of ROS produced by the NADPH-oxidase system of these immune cells (Park, 2003). This represents the first line of animal defense against the onslaught of pathogenic organisms.

Fungi represent important clinical pathogens and are a special problem in terms of chemotherapy as the spectrum of antifungal agents is relatively limited compared to antibacterial drugs (Odds *et al.*, 2003). An important goal of current research is to understand the pathways of cellular defense used by fungi to survive the hostile mammalian host environment. Since a major route of fungal killing in the host comes from subjecting these microbial pathogens to toxic-oxidative stress, changes in the ability of these organisms to resist this environmental challenge would be expected to have an influence on the ultimate pathogenicity of fungal cells (see for example Wysong *et al.*, 1998). This prediction has been tested in the major nosocomial fungal pathogen *Candida albicans*. Loss of a variety of genes involved in oxidative-stress tolerance reduced the virulence of the mutant *C. albicans* versus wild type in mouse models of candidiasis (reviewed in Chauhan *et al.*, 2006). Furthermore, patients with chronic granulomatosis disease show a greatly enhanced susceptibility to morbidity from fungal infection relative to normal individuals (Dennis *et al.*, 2006). Chronic granulomatosis disease is most commonly caused by a defect in the NADPH-oxidase complex (Gorlach *et al.*, 1997) that renders neutrophils less effective at generating the oxidative burst used to kill invading microorganisms (Fang, 2004).

Together, these observations argue for the importance of fungal oxidative-stress tolerance in the pathogenicity of these organisms. We are using the yeast

Saccharomyces cerevisiae as a model fungus in order to employ the sophisticated genetics, molecular biology and biochemistry available in this yeast as tools to dissect oxidative-stress tolerance. Our main focus has been on the transcription factor called Yap1p that is a key determinant in oxidative-stress tolerance. The goal of this review is to discuss recent findings that shed new light on the differential molecular mechanisms required for Yap1p to mediate resistance to oxidative stresses elicited by different oxidative agents.

1.1 Yap1p Role in the Response to Stress

Yap1p of *S. cerevisiae* was first identified through its biochemical activity as a DNA-binding protein that was capable of specifically recognizing the AP-1 recognition element from the SV-40 early enhancer (Harshman *et al.*, 1988). Analysis of the primary sequence of Yap1p indicated that the DNA-binding domain of this factor was located in the amino-terminus and that two separable transcriptional activation domains could be identified (Moye-Rowley *et al.*, 1989). Given the basis of the initial identification of Yap1p, an early challenge was to place this regulator into a physiological context in which it could act. This was accomplished by several laboratories in which it was shown that amplification of the *YAP1* gene strongly increased resistance to diverse compounds like cycloheximide, 1,10-phenanthroline and 4-nitroquinoline-*N*-oxide (Leppert *et al.*, 1990; Hertle *et al.*, 1991; Schnell and Entian, 1991). Strikingly, *yap1Δ* mutants were not typically found to be hypersensitive to these compounds suggesting that other factors might provide sufficient transcriptional activation to suppress the phenotype caused by loss of Yap1p. Analysis of the *S. cerevisiae* genome and functional experiments established that a family of proteins related to Yap1p is found in this organism (Fernandes *et al.*, 1997). Many of these proteins are able to bind to Yap1p response elements (YREs) and function analogously to Yap1p.

The first phenotype identified for cells lacking Yap1p was oxidant sensitivity (Schnell *et al.*, 1992). While wild-type cells were able to tolerate certain dosages of oxidants like H_2O_2 and diamide, isogenic *yap1Δ* strains failed to grow at these same concentrations. Later work demonstrated that *yap1Δ* cells were also hypersensitive to the redox active metal cadmium (Bossier *et al.*, 1993; Wu *et al.*, 1993) although not to most other metals that have an oxidative component in their toxicity. Two well-known examples of this Yap1p independent metal tolerance include copper and iron that require different systems to ensure normal resistance to these essential nutrients (see Van Ho *et al.*, 2002; De Freitas *et al.*, 2003 for reviews).

From these results, a pattern of the role of Yap1p can be seen. In the case of drug resistance, loss of Yap1p is well tolerated by cells and overproduction of Yap1p leads to a dramatic increase in resistance. Conversely, Yap1p is required for normal oxidative-stress resistance and its loss generates a growth phenotype.

1.2 Molecular Details of Yap1p Activation by Oxidants

The finding that Yap1p is a key participant in tolerance of oxidative-stress focused investigation of the role of this transcription factor on this phenotype. The first information describing the molecular details of the Yap1p response to oxidants came from studies on Yap1p regulation of the *TRX2* gene. *TRX2* encodes one of the three thioredoxin proteins in *S. cerevisiae* (reviewed in Wheeler and Grant, 2004). Thioredoxins provide reducing potential for reactions ranging from DNA replication (Muller, 1991) to vacuole fusion (Xu and Wickner, 1996). A high-copy-number plasmid containing *TRX2* conferred high-level resistance to oxidants in a Yap1p-dependent fashion (Kuge and Jones, 1994). Analysis of the expression of the *TRX2* mRNA demonstrated that transcription of this locus was oxidant inducible provided that the *YAP1* gene was intact. Similarly, expression of the *GSH1* gene was also found to be regulated by Yap1p (Wu and Moye-Rowley, 1994). *GSH1* encodes the γ-glutamylcysteine synthetase protein of *S. cerevisiae*, which is the rate-limiting enzyme controlling glutathione biosynthesis (Ohtake *et al.*, 1990). *GSH1* transcription is also induced by oxidants in a Yap1p-dependent manner (Stephen *et al.*, 1995) consistent with the important role Yap1p has in redox-responsive gene expression. DNA-microarray experiments have confirmed these early observations implicating Yap1p as a major determinant of oxidant inducible gene expression (Gasch *et al.*, 2000; Carmel-Harel *et al.*, 2001).

Since many Yap1p-regulated genes were induced in the presence of oxidants, experiments aimed at the mechanism of regulation of Yap1p activity by these oxidants were carried out. Use of a green fluorescent protein (GFP)-Yap1p fusion protein provided important insight into Yap1p regulation during oxidative stress (Kuge *et al.*, 1997). GFP-Yap1p was found primarily in the cytoplasm during normal growth conditions but rapidly recruited to the nucleus during oxidative stress by diamide challenge. Western blot analyses of Yap1p levels during oxidative stress indicated that this protein was controlled at a post-translational step (Wemmie *et al.*, 1997). Mutagenesis of a c-CRD caused the resulting mutant GFP-Yap1p to be localized in the nucleus in an oxidant independent fashion. (Kuge *et al.*, 1997) These same constitutively nuclear mutant proteins were able to elevate diamide resistance to a level higher than that seen in the wild-type strain. Transfer of the Yap1p c-CRD region to a heterologous GFP protein led to this chimeric factor exhibiting oxidant-regulated nuclear localization.

This analysis provided a strong argument that Yap1p function was regulated by differential localization in response to diamide-induced oxidative stress. Study of the *S. cerevisiae* nuclear exportin Crm1p determined that the localization of Yap1p was altered by changing the rate at which this transcription factor exited the nucleus (Kuge *et al.*, 1998; Yan *et al.*, 1998). Temperature-sensitive mutants of Crm1p rapidly accumulated Yap1p in the nucleus when shifted to the restrictive temperature in the absence of added oxidants. Interaction studies demonstrated that Crm1p associated with Yap1p in an oxidant-sensitive manner. In the absence of oxidative stress, Yap1p bound to Crm1p and was rapidly transported out of the nucleus. When oxidants were present, Yap1p-Crm1p association was inhibited and the transcription factor accumulated in the nucleus. This allowed target-gene transcription to be induced. More recent experiments studying the biochemical properties of Yap1p after diamide stress have shown that disulfide bonds form between the cysteine residues present in the c-CRD (Kuge *et al.*, 2001). These disulfide bonds are believed to interfere with

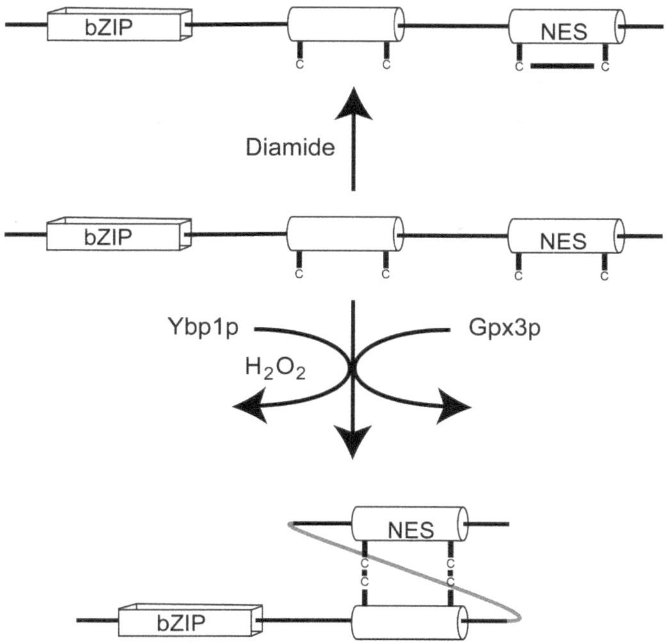

Figure 1 Oxidant-specific protein folding of Yap1p during diamide and H_2O_2 exposure. The differential folding of Yap1p upon oxidative stress onset is diagrammed here. Several domains of the factor are indicated including the basic region-leucine zipper (bZIP) DNA-binding domain as a box and the two cysteine rich domains (CRDs) as cylinders. The CRD closest to the bZIP domain is the n-CRD while the nuclear export signal-containing CRD is the C-CRD. The parallel arrangement of the H_2O_2-oxidized Yap1p is taken from the structure determined by Wood *et al.* (2004).

Crm1p access to the nuclear export signal (NES) present in the Yap1p c-CRD and permit the buildup of this transcriptional activator in the nucleus during diamide stress (Figure 1).

1.3 Oxidant-Specific Downstream Function of Yap1p

The experiments described above paint a relatively simple picture detailing regulation of Yap1p by oxidative stress. Essentially all of these studies focused on oxidative stress elicited by diamide and results for Yap1p localization and diamide resistance were very consistent. Manipulations that led to increased Yap1p levels in the nucleus were invariably linked with elevated levels of diamide resistance. In minimalistic terms, nuclear localization appeared to be the only requirement for normal Yap1p-mediated transactivation of downstream target genes required for tolerance to this oxidant. However, when oxidants other than diamide were included in the analysis of Yap1p response to oxidative stress, this simple model was no longer adequate to explain Yap1p regulation.

The first experiments demonstrating the oxidant-specific response of Yap1p came from analyses of the behavior of various mutant derivatives of this factor in terms of their ability to complement the phenotype of a *yap1Δ* mutant strain. As mentioned above, strains lacking *YAP1* are hypersensitive to both diamide and H_2O_2 stress. Low-copy-number plasmids expressing various deletion derivatives of *YAP1* were introduced and compared to the wild-type gene for their ability to complement the H_2O_2 and diamide sensitivity of the *yap1Δ* mutant (Wemmie *et al.*, 1997). Several different internal deletion derivatives lacking a second n-CRD were found to exhibit a similar altered phenotype: H_2O_2 sensitivity but diamide hyper-resistance. Additionally, site-directed mutations replacing cysteine residues in the c-CRD with alanine led to the same altered phenotype. An artificial Yap1p-dependent reporter gene (Wemmie *et al.*, 1994) consisting of concatemerized SV-40 AP-1 recognition elements used to replace the normal upstream elements of a *TRP5-lacZ* reporter gene was used to analyze Yap1p-dependent transactivation. This reporter gene was constitutively derepressed in these c-CRD mutants irrespective of the presence of oxidative stress. This analysis suggested that negative regulatory elements existed within the Yap1p polypeptide chain that acted to inhibit diamide resistance but were required for wild-type H_2O_2 tolerance. This led us to suggest that authentic *S. cerevisiae* target genes would be found that placed differential demands on Yap1p depending on the type of oxidative stress faced by the cell. We found that the *TRX2* gene represented a locus of this type.

Earlier analyses of *TRX2* indicated that this gene was required for wild-type H_2O_2 resistance (Kuge and Jones, 1994) but dispensable and possibly even inhibitory (Muller, 1996) for diamide resistance. We constructed a *TRX2-lacZ* reporter gene and used this to assess the oxidant inducibility of *TRX2* in the presence of various Yap1p mutants (Coleman *et al.*, 1999). Briefly, any mutant that was found to be H_2O_2 hypersensitive also failed to normally induce *TRX2* in the presence of this oxidant. Importantly, cysteines residues present in both the n- and c-CRD regions were required for normal H_2O_2 inducibility and resistance. These data confirmed our hypothesis that Yap1p did not function identically in all genes that lie downstream from this transcriptional regulatory protein.

Later work of others provided a molecular basis for these findings. Using a rapid trichloroacetic acid-based protein extraction method and non-reducing gel electrophoresis, it was shown that a disulfide bond formed between cysteine residues present in the n- and c-CRD when cells were challenged with H_2O_2 but not with diamide (Delaunay *et al.*, 2000). Failure to form this interdomain disulfide bond prevented normal H_2O_2 resistance and *TRX2* induction. Further analysis by this group discovered that presumed glutathione peroxidase protein encoded by the *GPX3* gene was required to facilitate formation of this interdomain disulfide (Delaunay *et al.*, 2002). Using a n-CRD mutant, the authors were able to trap a covalent intermediate with the Gpx3p protein covalently linked to a cysteine residue in the Yap1p c-CRD. Gpx3p may serve as a leaving group to accelerate the rate of interdomain disulfide bond formation, a common feature of reactions in organic chemistry. Gpx3p has been argued to serve exclusively as a sensor for oxidant since cysteine residues important in the usual Gpx catalytic mechanism may be removed from this protein with little effect on its ability to confer H_2O_2 tolerance (Delaunay *et al.*, 2002).

These studies clearly implicated both the n- and c-CRD regions as required participants in the oxidant-selective protein folding of Yap1p but only implicated single cysteine residues in each CRD as being required for normal H_2O_2 tolerance. This model did not fit the genetic data as two different cysteines in the c-CRD and two cysteines in the n-CRD are required for normal H_2O_2 resistance (Coleman *et al.*, 1999; Delaunay *et al.*, 2000). There are three cysteine residues in the c-CRD present as cysteine-serine-glutamate (CSE) repeats. Alanine scanning mutagenesis of these three CSE repeats in the c-CRD demonstrated that loss of these repeats had distinct phenotypic effects (Wemmie *et al.*, 1997). Loss of the CSE repeat starting with the cysteine at position 598 resulted in a strain that was very sensitive to H_2O_2 and normally tolerant of diamide. Replacement of the repeat starting with the cysteine at position 620 had no significant phenotypic effect on function although this replacement is thought to overlap the NES in Yap1p (Yan *et al.*, 1998). Finally, loss of the CSE at position 629 dramatically elevated diamide resistance and reduced H_2O_2 tolerance.

The mutational analysis of the c-CRD strongly implicated two cysteines as being required for normal H_2O_2 resistance with one (cysteine 629) behaving as a negative regulator of diamide resistance. Dissection of the three-cysteine residues present in the n-CRD (only 6 cysteines are present in the entire *S. cerevisiae* Yap1p sequence) determined that two of these amino acids were required for normal H_2O_2 tolerance (Coleman *et al.*, 1999; Delaunay *et al.*, 2000). Interestingly, loss of the entire n-CRD by deletion mutagenesis produced a Yap1p derivative that conferred diamide hyper-resistance and H_2O_2 hypersensitivity while removal of a single cysteine in the n-CRD produced the same H_2O_2 hypersensitivity while failing to influence diamide tolerance (Wemmie *et al.*, 1997; Coleman *et al.*, 1999; Delaunay *et al.*, 2000). This indicates a functional requirement for additional sequences in the n-CRD beyond the cysteine residues.

Structural studies of the oxidized form of Yap1p clarified these data (Wood *et al.*, 2004). NMR structural determination of a H_2O_2-oxidized Yap1p derivative indicated the presence of two disulfide bonds between the n- and c-CRD. This folded conformation is thought to obscure access of Crm1p to the Yap1p NES and allow nuclear accumulation of this factor. However, the simple masking of the NES is inadequate to explain the role of the H_2O_2-folded form of Yap1p. Mutant forms of Yap1p that are constitutively nuclear, like CSE629AAA Yap1p (Wemmie *et al.*, 1997; Coleman *et al.*, 1999), are hypersensitive to H_2O_2. This indicates that control of nuclear localization by H_2O_2-induced oxidative folding is only one function of this particular conformation of Yap1p (Figure 1). A second requirement for this oxidant-specific folded form of Yap1p is discussed below.

The identification of Gpx3p as an essential participant in H_2O_2-induced folding of Yap1p clearly demonstrated that auxiliary factors were required for this oxidant-specific process. Strains lacking *GPX3* are hypersensitive to H_2O_2 but normally resistant to diamide (Inoue *et al.*, 1999; Avery and Avery, 2001; Delaunay *et al.*, 2002). A second protein factor required for induction of Yap1p by H_2O_2 exposure was isolated in two different groups as either a high-copy-number mediator of increased H_2O_2 resistance or in a screen of the haploid disruption library for genes required for normal peroxide tolerance (Veal *et al.*,

2003; Gulshan *et al.*, 2004). The *YBP1* (Yap1p-binding protein) gene was shown to be required for normal H_2O_2 tolerance but dispensable for diamide resistance. Yap1p nuclear localization was induced in *ybp1Δ* strains by exposure to diamide but not H_2O_2. Ybp1p clearly interacts with Yap1p in the absence of stress but opinions vary as to the level of Ybp1p–Yap1p interaction after H_2O_2 treatment (Veal *et al.*, 2003; Gulshan *et al.*, 2004).

Analysis of a range of fungal genomes indicates the presence of Ybp1p homologues in many other species (Figure 2). Use of a yeast two-hybrid assay demonstrated that the Yap1p-Ybp1p interaction could be recapitulated in this system and that the carboxy-terminal 258 amino acids of Ybp1p were required for this interaction (Gulshan *et al.*, 2004). As can be seen in Figure 2, amino acid identity is conserved throughout the lengths in this region of the various fungal Ybp1p molecules. The region immediately downstream of residue 416 is well conserved and is a likely candidate for the Yap1p interaction region, an idea we are currently testing.

Along with the conservation of Ybp1p, Yap1p polypeptides from these same fungi exhibit several regions of high sequence similarity (Figure 3). These include the basic region-leucine zipper (bZIP) DNA binding domain, the n- and c-CRD regions as well as other less well documented stretches of sequence conservation. Mapping of the minimal Ybp1p interacting domain in Yap1p indicated that the C-terminal 271 amino acids were required for full two-hybrid interaction (Veal *et al.*, 2003). It is likely that these additional conserved regions are required for the Ybp1p catalyzed protein folding of Yap1p during H_2O_2 stress in these other fungi. Further experiments are required to confirm this suggestion.

1.4 Bifunctional Role of the H_2O_2-Oxidized Form of Yap1p

As described above, Ybp1p- and Gpx3p-catalyzed folding of Yap1p is required to modulate the nuclear export of Yap1p during H_2O_2 stress. However, regulation of the function of the Yap1p NES during H_2O_2 treatment is inadequate to explain defects in the formation of the specific folded form that occurs during this type of oxidative stress. This is not the case for diamide-induced nuclear recruitment. Every mutant form of Yap1p that is constitutively trapped in the nucleus confers diamide hyper-resistance (Kuge *et al.*, 1997; Coleman *et al.*, 1999). This argues that the control of nuclear localization of Yap1p is the sole parameter that must be regulated in order that Yap1p mount an appropriate response to diamide stress. Conversely, Yap1p mutants that are trapped in the nucleus typically elicit H_2O_2 hypersensitivity. This observation is difficult to explain solely by invoking regulation of nuclear localization and strongly suggests the existence of a second role for the H_2O_2-folded form of Yap1p during oxidative stress invoked by this agent.

Insight into this secondary function of the H_2O_2-folded form of Yap1p came from analysis of the expression of the *TRX2* gene in the presence of various mutant forms of Yap1p (Coleman *et al.*, 1999). Yap1p derivatives that are constitutively nuclear fail to induce *TRX2* transcription normally when challenged with H_2O_2. This is also true for *ybp1Δ* or *gpx3Δ* mutant strains (Delaunay *et al.*, 2002; Veal *et al.*, 2003; Gulshan *et al.*, 2004). Either cis- or trans-defects in H_2O_2

```
417 I T D A I Y L Y L R F T T P S L Y S K V Y Y N V A V E S V S R Y W L W Y A I - - Sc
400 I H D S L V M T L R L I I P Q M V N P K F I N A G N H D V V V F W V W F A L Y Q Ca
407 I A D A L Y L Y L R F A S E S L F S P T C H N V T I E G V A R Y W I W A A L - - Cg
400 F K D A L V L T L R S V I P S L I H P S F R S R G V A D V C L F W S W L A I - H Dh
366 L D D A I Y M F L R F N T P E I Y S K S F G N I A V T D A S L Y F V L L A L - - Eg
450 L Q E A I Y M Y L R F V T P S M F S A L F E N R S S H D L A R T W I L F A L - - Kl

455 - T T E P L E D V K K E L K N L S V F V T K T L L H V L L - Q K N C I Q V N Q Q Sc
440 Q Q I I N S K N L Q L E I S Y I P K V L L T T F F Q C L L F I V I K S E G K P N Ca
445 - T T T D N N I L K E K L A E L S P L V L H S V L N L L L - V K N C H Q V N E E Cg
439 K L T I N N K K I E L E I T K I P S T L L S I Y Y Q A L L F I C T T L P S A S N Dh
404 - E G N S H E E N H K L V R E F P Q T V F A V F I Q V L L - L N M C R L P S R E Eg
488 - T N N S T H D L M D S L K D L P S Y I I T V Y L Q T E L - I R A C L Q I N D N Kl

493 L R M I T F T L L T R L L C L I P E K V A F E F I L D V L K T S P L P L A K T S Sc
480 F K Y M L L T L L T K L L T L S P - D T G Y E F I K D S L N N C P Y E S V Y P S Ca
483 I R M I T F T L I T R I L C L L P E N C S Y E F L M D E L D N C A V V F G K S C Cg
479 F R Y G T L T L L T K I L T L S P E D I S F S F I K D S L M N C P Y D N V K A V Dh
442 I G R M T H S I L A R V L F L A S E T T A F D F V I D T L L Q C P Y E T A K Q I Eg
526 L R R T Q F S I L T R I L C L L P E D F A F N F I R D T L L S C P Y E Q A K C C Kl

533 V L C V F K D L S R R - R I S T K D N D S E T D L I V E K L S K L K V N D S N K Sc
519 L I G V Y K Q L L L N E K W D V N S I E L E K L N I S S S S S N T P P K L P P R Ca
523 V L G I L R D L V I K V D H S V S S N N T D T E D L S E S M A Q L K I N N E K R Cg
519 L I G V L K E L L T K V K P T S L T E N M M N L N I K D T S - K S A P I L P S R Dh
482 I L E I A K I M M V S D Q Y T I C T R - - D V P S H T R T D T K - - - - - - A P Eg
566 A L A I L K D M M Q H E R K V P Q K S - - D E D D L A K D M E K L K I K N S P P Kl

572 A Q Q S N I R H Y I Q L D S S K M K A V H D C C L Q T I Q D S F T A D A K K S D Sc
559 N G I K R - - K H F S L T N E S L N D L V D L I N N S S K N A F V E D - - - - - Ca
563 A K - - - - K T F I T L D P K R A G E I E D L A I K T L K E T - K K S M K K D Y Cg
558 D V S K N S N K S L T L N K E K A D D I F E L I D E A I E L T F V E D P E S G D Dh
514 P L P P - - R P F I K L N D D R M A A F H T L A M M S A K F C T E T E D P K Q - Eg
604 P L P S - - R A Y M L L N D D R I A T L H S I T L L A I D S C A A D P E S K K - Kl

612 - - - - - - - - - - - I L L L L T Y L N I F I V L K K - - - - T W D E D L L K I V Sc
592 - - N S K I D P S K L S T I A A Y L N L L V A I K K D P V I V E N K E K L T T L Ca
598 - - - - - - - - - - - I L L V L N Y I K F F S T F A H - - - - K W N K S K L N E F Cg
598 K V S V E L N Q S K F S T L S A Y L N L L V V L K D D P A I K T D P S R I D K I Dh
551 - - - - - - - - - - - L T T L L S Y L N L I T V L R K - - - - K W D Q V L L R E L Eg
641 - - - - - - - - - - - V K T L L T Y L N F L N A F L T - - - - K W D S V F L K E I Kl

638 C S K I D S N L - - - K S V E P D K L P K Y K E I V D K N E S L N D Y F T G - I K 674 Sc
630 I S S I E N K I K S V K K S S Q - - - - - N Q F E L N A A G M L E I T I E R F N E 665 Ca
624 T T L V A T N F - - - S - - D S K M L P E I N A I I D A N E K L R S L - - - - T E 655 Cg
638 L E L M D E N I S Q I K Q N R K P D S D A D K A E L N A V D M L N I T L D R I R A 678 Dh
577 T V A V S N C I E P R K E - - - - D I P E I G F I K I A N D G L K E F L G G N S T 613 Eg
667 C D A V N D K L I K N E K V G D K D E P H Y S L L V S T V A S I S S K L         702 Kl
```

Figure 2 Alignment of the Yap1p interacting segments of fungal Ybp1p proteins. The one letter code is used throughout and only the region of alignment is shown that corresponds to the known Yap1p interacting domain in the *S. cerevisiae* protein. When at least four sequences share amino acid identity with the *S. cerevisiae* protein, the residue is highlighted. The numbers on the left denote the position along the polypeptide chain. The different species of fungi are indicated as: Sc, *Saccharomyces cerevisiae*; Ca, *Candida albicans*; Cg, *Candida glabrata*; Dh, *Debaryomyces hansenii*; Eg, *Eremothecium gossypii*; Kl, *Kluveromyces lactis*.

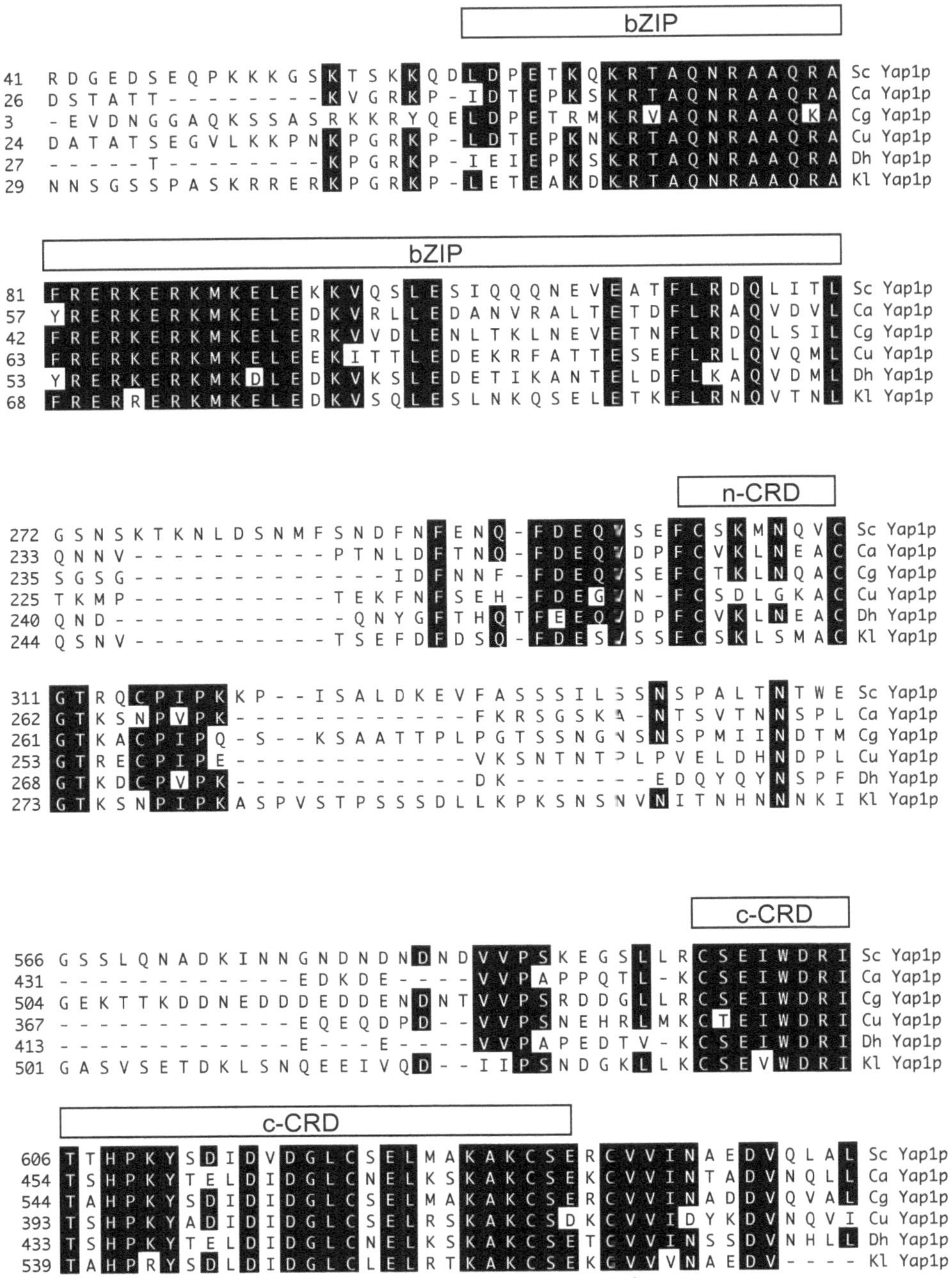

Figure 3 Sequence similarity of fungal Yap1p homologues. The three regions of fungal Yap1p proteins sharing the highest degree of sequence identity are shown. The approximate positions of important functional domains in each segment are indicated. Abbreviations and highlighting are as in Figure 2 with the addition of Cu, *Candida utilus*.

folding of Yap1p preclude normal H_2O_2 induction of *TRX2* expression. It is important to note that other Yap1p target genes do not place the same requirements on either the same domains in Yap1p or its trans-regulators. *GSH1* gene expression is constitutively high in nuclear export-defective forms of Yap1p and is normally regulated in *ybp1*Δ cells (Gulshan *et al.*, 2004). These observations strongly suggest that some specific transcriptional defect occurs at *TRX2* during H_2O_2 stress when normal folding of Yap1p is prevented.

The nature of this transcriptional defect was uncovered during an analysis of mediator components required for Yap1p transactivation (Gulshan *et al.*, 2005). The mediator complex is a group of general transcription factors that act to permit communication between DNA-bound upstream regulatory proteins and the core RNA polymerase II machinery (see Myers and Kornberg, 2000; Lewis and Reinberg, 2003; Bjorklund and Gustafsson, 2005 for representative reviews). Several different subcomplexes are present in the mediator and we tested removal of various proteins that are contained within these subcomplexes for the resulting effect on oxidative-stress tolerance. Loss of the Rox3p mediator component caused a dramatic increase in sensitivity to H_2O_2 stress with an associated defect in *TRX2* activation. Cycloheximide resistance was not altered in *rox3*Δ cells indicating that loss of this mediator component did not yield a general decrease in fitness of the resulting strain. Chromatin immunoprecipitation experiments demonstrated that increased levels of Rox3p were found associated with the *TRX2* promoter after H_2O_2 challenge. However, this increment was only seen so long as Yap1p was capable of being normally folded during H_2O_2 exposure. Mutant derivatives of Yap1p that lacked cysteines in either the n- or c-CRD failed to exhibit elevated Rox3p recruitment during the response to H_2O_2 stress.

Since Rox3p is part of a the multiprotein mediator complex (Gustafsson *et al.*, 1997), we tested the ability of this protein to associate with Yap1p *in vivo*. Using a strain (Ghaemmaghami *et al.*, 2003) expressing a functional Rox3p-TAP (tandem affinity purification (Puig *et al.*, 2001)) fusion protein, we recovered Rox3p-containing protein complexes from the cell. Yap1p was found in this complex in a H_2O_2-dependent fashion. This association was blocked if a n-CRD mutant form of Yap1p was used. Interestingly, a c-CRD Yap1p derivative was found to coprecipitate with Rox3p under non-stress conditions but this degree of association was unaffected by H_2O_2 treatment. Together these data are most consistent with a model in which correct H_2O_2-mediated folding of Yap1p recruits Rox3p to the *TRX2* promoter. This recruitment is required for normal H_2O_2 induction of *TRX2* expression and provides an explanation for the second function of the H_2O_2-induced folding of Yap1p beyond simply masking the NES of this transcriptional regulator.

2. DISCUSSION

Early experiments aimed at understanding Yap1p regulation by oxidative-stress presented an attractively simple model in which the key feature consisted of modulation of subcellular localization of this transcriptional regulatory protein

(Kuge *et al.*, 1997; Yan *et al.*, 1998). As further efforts tested this model, fractures appeared in the simple view owing to the discrepant behavior of identical mutant forms of Yap1p on diamide and H_2O_2. Initially, genetic data (Wemmie *et al.*, 1997; Coleman *et al.*, 1999) argued against this reductionist view of Yap1p regulation and later, biochemical analyses (Delaunay *et al.*, 2000) confirmed that the response of this transcription factor to different oxidants was quite different.

Our current picture of Yap1p regulation in response to oxidative stress is more nuanced and reflects the abundance of experimental analysis of this protein over the years. Diamide induction of Yap1p still is most simply explained by alteration in the subcellular distribution of this gene activator protein. Biochemical analysis of diamide-stressed cells argues that changes in the redox status of cysteine residues in the c-CRD seems to be sufficient for decreased Crm1p interaction, increased nuclear localization and target gene activation (Kuge *et al.*, 2001). C-terminal truncation mutant forms of Yap1p are constitutively located in the nucleus and able to up-regulate strongly both diamide resistance and reporter gene expression (Kuge *et al.*, 1997), suggestive that the sole function of this region of Yap1p is to allow cytoplasmic retention in the absence of diamide stress. However, these same mutant proteins are defective in the ability to tolerate H_2O_2 and induce expression of *TRX2*, clearly indicating that nuclear localization alone is inadequate for the normal response to this oxidant.

During the response to H_2O_2 exposure, Ybp1p and Gpx3p act to catalyze formation of a twin disulfide-bonded form of Yap1p. This unique conformer of Yap1p modified the behavior of the factor in two essential ways. First, the structure of H_2O_2-oxidized Yap1p masks the NES of this factor (Wood *et al.*, 2004). This feature occludes Crm1p recognition and allows nuclear accumulation of Yap1p. Second, the H_2O_2-induced structural conformation of Yap1p allows the recruitment of appropriate mediator components to key target promoters. This recruitment is required for normal gene induction during H_2O_2 stress but is dispensable during diamide exposure.

The additional machinery required for triggering the H_2O_2 response may be explained by the likely physiological nature of this type of stress. During the innate immune response, fungi will be exposed to H_2O_2 *in vivo*, which acts to generate oxidative damage directly. Diamide is a chemical that has the useful property of causing depletion of reduced glutathione, thereby sensitizing cells to oxidative stress (Kosower and Kosower, 1987). Perhaps cells have developed this additional machinery to allow increased sensitivity to H_2O_2 challenge and a more rapid response to this oxidant that will be encountered during their normal existence.

The roles of the n- and c-CRD are still not fully understood. Biochemical examination of H_2O_2-induced Yap1p folding indicated that replacement of a cysteine at 598 with alanine completely eliminated inter-CRD disulfide bond formation while similar removal of a cysteine at 629 still allowed some disulfide bond formation to be detected (Delaunay *et al.*, 2000). These two mutants also have very different diamide phenotypes as loss of cysteine 598 exhibits normal diamide tolerance while removal of cysteine 629 causes dramatic elevation of resistance to this oxidant (Wemmie *et al.*, 1997). Interestingly, deletion of the n-CRD renders cells extremely sensitive to H_2O_2 but hyper-resistant to diamide.

A double mutant lacking both the n-CRD deletion and cysteine 629 is more resistant to diamide than either single mutant alone although no increased H_2O_2 sensitivity is seen (Coleman *et al.*, 1999). This suggests that these mutations affect different regulatory features of Yap1p and are unlikely to be explained by alteration of a single parameter such as subcellular location.

Since nuclear localization of Yap1p is required for both diamide and H_2O_2 resistance, it was surprising that most constitutively nuclear Yap1p mutants exhibited a diamide hyper-resistant but H_2O_2 hypersensitive phenotype. One possible explanation for this H_2O_2 hypersensitivity may be provided by the observation that properly H_2O_2-folded Yap1p is required to recruit Rox3p to enable *TRX2* promoter activation (Gulshan *et al.*, 2005). This folding requires the participation of both Ybp1p and Gpx3p. The available data indicate that Ybp1p is located in the cytoplasm (Veal *et al.*, 2003; Gulshan *et al.*, 2004) where Yap1p is normally found prior to oxidant exposure. Constitutive nuclear location of Yap1p in mutants that lack normal NES function may compartmentalize this transcription factor away from the folding machinery that is required to correctly modify Yap1p during H_2O_2 stress. Another possibility is that the C-terminus of Yap1p is required for both Crm1p and Ybp1p binding and that mutants that disrupt binding to Crm1p, also prevent Ybp1p recognition. There is no detectable sequence similarity between these two Yap1p-interacting proteins (unpublished data) but this notion can be directly tested using already existing interaction assays.

While much has been learned of the mechanisms underlying the Yap1p-mediated response to oxidative stress, important questions remain. The ability of Yap1p to activate *TRX2* requires normal folding as discussed above but even in the presence of intact folding machinery, loss of another transcription factor called Skn7p eliminates the ability of Yap1p to induce *TRX2* in the presence of H_2O_2 (Morgan *et al.*, 1997). This indicates that folding of Yap1p is necessary but not sufficient for the normal H_2O_2 response. While Yap1p induces the expression of many target genes during oxidative stress, other Yap1p-regulated genes are not oxidant inducible. An example of such an oxidant-insensitive but Yap1p-responsive gene is the membrane protein-encoding locus *ATR1* (Coleman *et al.*, 1997). The first-known function of Yap1p was provided by the observation that overproduction of this factor caused cells to become multidrug resistant (Leppert *et al.*, 1990) but the rationale underlying Yap1p co-regulation of oxidative stress genes (i.e. *TRX2*) and drug resistance loci (*ATR1*) remains a mystery. Understanding the interaction and molecular logic behind the control of multidrug resistance and oxidative stress by Yap1p will shed light on the general mechanisms through which a eukaryotic cell responds to stress and modulates its intracellular environment.

ACKNOWLEDGMENTS

I thank Kailash Gulshan for helpful discussions and hard work. Work on Yap1p in my laboratory has been supported by NIH GM49825 and GM57007 and grants-in-aid from the American Heart Association.

REFERENCES

Adam-Vizi, V. (2005). Production of reactive oxygen species in brain mitochondria: Contribution by electron transport chain and non-electron transport chain sources. *Antioxidants & Redox Signal*, **7**, 1140–1149.

Avery, A.M. and Avery, S.V. (2001). *Saccharomyces cerevisiae* expresses three phospholipid hydroperoxide glutathione peroxidases. *Journal of Biological Chemistry*, **276**, 33730–33735.

Bjorklund, S. and Gustafsson, C.M. (2005). The yeast Mediator complex and its regulation. *Trends in Biochemical Science*, **30**, 240–244.

Bossier, P., Fernandes, L., Rocha, D. and Rodrigues-Pousada, C. (1993). Overexpression of YAP2, coding for a new yAP protein, and YAP1 in *S. cerevisiae* alleviates growth inhibition caused by 1,10-phenanthroline. *Journal of Biological Chemistry*, **268**, 23640–23645.

Carmel-Harel, O., Stearman, R., Gasch, A.P., Botstein, D., Brown, P.O. and Storz, G. (2001). Role of thioredoxin reductase in the Yap1p-dependent response to oxidative stress in *Saccharomyces cerevisiae*. *Molecular Microbiology*, **39**, 595–605.

Chauhan, N., Latge, J.P. and Calderone, R. (2006). Signalling and oxidant adaptation in *Candida albicans* and *Aspergillus fumigatus*. *Nature Reviews Microbiology*, **4**, 435–444.

Coleman, S.T., Epping, E.A., Steggerda, S.M. and Moye-Rowley, W.S. (1999). Yap1p activates gene transcription in an oxidant-specific fashion. *Molecular and Cellular Biology*, **19**, 8302–8313.

Coleman, S.T., Tseng, E. and Moye-Rowley, W.S. (1997). *Saccharomyces cerevisiae* basic region-leucine zipper protein regulatory networks converge at the *ATR1* structural gene. *The Journal of Biological Chemistry*, **272**, 23224–23230.

De Freitas, J., Wintz, H., Kim, J.H., Poynton, H., Fox, T. and Vulpe, C. (2003). Yeast, a model organism for iron and copper metabolism studies. *Biometals*, **16**, 185–197.

Delaunay, A., Isnard, A.D. and Toledano, M.B. (2000). H_2O_2 sensing through oxidation of the Yap1 transcription factor. *The EMBO Journal*, **19**, 5157–5166.

Delaunay, A., Pflieger, D., Barrault, M.B., Vinh, J. and Toledano, M.B. (2002). A thiol peroxidase is an H_2O_2 receptor and redox-transducer in gene activation. *Cell*, **111**, 471–481.

Dennis, C.G., Greco, W.R., Brun, Y., Youn, R., Slocum, H.K., Bernacki, R.J., Lewis, R., Wiederhold, N., Holland, S.M., Petraitiene, R. *et al.* (2006). Effect of amphotericin B and micafungin combination on survival, histopathology, and fungal burden in experimental aspergillosis in the p47phox-/- mouse model of chronic granulomatous disease. *Antimicrobial Agents and Chemotherapy*, **50**, 422–427.

Fang, F.C. (2004). Antimicrobial reactive oxygen and nitrogen species: Concepts and controversies. *Nature Reviews Microbiology*, **2**, 820–832.

Fernandes, L., Rorigues-Pousada, C. and Struhl, K. (1997). Yap, a novel family of eight bZIP proteins in *Saccharomyces cerevisiae* with distinct biological functions. *Molecular and Cellular Biology*, **17**, 6982–6993.

Gasch, A.P., Spellman, P.T., Kao, C.M., Carmel-Harel, O., Eisen, M.B., Storz, G., Botstein, D. and Brown, P.O. (2000). Genomic expression programs in the response of yeast cells to environmental changes. *Molecular Biology of the Cell*, **11**, 4241–4257.

Ghaemmaghami, S., Huh, W.K., Bower, K., Howson, R.W., Belle, A., Dephoure, N., O'Shea, E.K. and Weissman, J.S. (2003). Global analysis of protein expression in yeast. *Nature*, **425**, 737–741.

Gorlach, A., Lee, P.L., Roesler, J., Hopkins, P.J., Christensen, B., Green, E.D., Chanock, S.J. and Curnutte, J.T. (1997). A p47-phox pseudogene carries the most common mutation causing p47-phox- deficient chronic granulomatous disease. *The Journal of Clinical Investigation*, **100**, 1907–1918.

Gulshan, K., Rovinsky, S.A., Coleman, S.T. and Moye-Rowley, W.S. (2005). Oxidant-specific folding of Yap1p regulates both transcriptional activation and nuclear localization. *The Journal of Biological Chemistry*, **280**, 40524–40533.

Gulshan, K., Rovinsky, S.A. and Moye-Rowley, W.S. (2004). YBP1 and its homologue YBP2/YBH1 influence oxidative-stress tolerance by nonidentical mechanisms in *Saccharomyces cerevisiae*. *Eukaryotic Cell*, **3**, 318–330.

Gustafsson, C.M., Myers, L.C., Li, Y., Redd, M.J., Lui, M., Erdjument-Bromage, H., Tempst, P. and Kornberg, R.D. (1997). Identification of Rox3 as a component of mediator and RNA polymerase II holoenzyme. *The Journal of Biological Chemistry*, **272**, 48–50.

Harshman, K.D., Moye-Rowley, W.S. and Parker, C.S. (1988). Transcriptional activation by the SV40 AP-1 recognition element in yeast is mediated by a factor similar to AP-1 that is distinct from GCN4. *Cell*, **53**, 321–330.

Hertle, K., Haase, E. and Brendel, M. (1991). The SNQ3 gene of *Saccharomyces cerevisiae* confers hyper-resistance to several functionally unrelated chemicals. *Current Genetics*, **19**, 429–433.

Inoue, Y., Matsuda, T., Sugiyama, K., Iszawa, S. and Kimura, A. (1999). Genetic analysis of glutathione peroxidase in oxidative stress response of *Saccharomyces cerevisiae. The Journal of Biological Chemistry*, **274**, 27002–27009.

Kobayashi, K.S. and Flavell, R.A. (2004). Shielding the double-edged sword: Negative regulation of the innate immune system. *Journal of Leukocyte Biology*, **75**, 428–433.

Kosower, N.S. and Kosower, E.M. (1987). Formation of disulfides with diamide. *Methods Enyzmol*, **143**, 264–270.

Kuge, S., Arita, M., Murayama, A., Maeta, K., Izawa, S., Inoue, Y. and Nomoto, A. (2001). Regulation of the yeast Yap1p nuclear export signal is mediated by redox signal-induced reversible disulfide bond formation. *Molecular and Cellular Biology*, **21**, 6139–6150.

Kuge, S. and Jones, N. (1994). *YAP1*-dependent activation of *TRX2* is essential for the response of *S. cerevisiae* to oxidative stress by hydroperoxides. *The EMBO Journal*, **13**, 655–664.

Kuge, S., Jones, N. and Nomoto, A. (1997). Regulation of yAP-1 nuclear localization in response to oxidative stress. *The EMBO Journal*, **16**, 1710–1720.

Kuge, S., Toda, T., Iizuka, N. and Nomoto, A. (1998). Crm1 (Xpo1) dependent nuclear export of the budding yeast transcription factor yAP-1 is sensitive to oxidative stress. *Genes to Cells*, **3**, 521–532.

Leppert, G., McDevitt, R., Falco, S.C., Van Dyk, T.K., Ficke, M.B. and Golin, J. (1990). Cloning by gene amplification of two loci conferring multiple drug resistance in *Saccharomyces. Genetics*, **125**, 13–20.

Lewis, B.A. and Reinberg, D. (2003). The mediator coactivator complex: Functional and physical roles in transcriptional regulation. *Journal of Cell Science*, **116**, 3667–3675.

Morgan, B.A., Banks, G.R., Toone, W.M., Raitt, D., Kuge, S. and Johnston, L.H. (1997). The Skn7 response regulator controls gene expression in the oxidative stress response of the budding yeast *Saccharomyces cerevisiae. The EMBO Journal*, **16**, 1035–1044.

Moye-Rowley, W.S., Harshman, K.D. and Parker, C.S. (1989). Yeast YAP1 encodes a novel form of the jun family of transcriptional activator proteins. *Genes & Development*, **3**, 283–292.

Muller, E.G.D. (1991). Thioredoxin deficiency in yeast prolongs S phase and shortens the G1 interval of the cell cycle. *The Journal of Biological Chemistry*, **266**, 9194–9202.

Muller, E.G.D. (1996). A glutathione reductase mutant of yeast accumulates high levels of oxidized glutathione and requires thioredoxin for growth. *Molecular Biology of the Cell*, **7**, 1805–1813.

Myers, L.C. and Kornberg, R.D. (2000). Mediator of transcriptional regulation. *Annual Review of Biochemistry*, **69**, 729–749.

Odds, F.C., Brown, A.J. and Gow, N.A. (2003). Antifungal agents: Mechanisms of action. *Trends in Microbiology*, **11**, 272–279.

Ohtake, Y., Satou, A. and Yabuuchi, S. (1990). Isolation and characterization of glutathione biosynthesis-deficient mutants in *Saccharomyces cerevisiae. Agricultural and Biological Chemistry*, **54**, 3145–3150.

Park, J.B. (2003). Phagocytosis induces superoxide formation and apoptosis in macrophages. *Experimental & Molecular Medicine*, **35**, 325–335.

Puig, O., Caspary, F., Rigaut, G., Rutz, B., Bouveret, E., Bragado-Nilsson, E., Wilm, M. and Seraphin, B. (2001). The tandem affinity purification (TAP) method: A general procedure of protein complex purification. *Methods*, **24**, 218–229.

Schnell, N. and Entian, K.-D. (1991). Identification and characterization of a *Saccharomyces cerevisiae* gene (PAR1) conferrring resistance to iron chelators. *European Journal of Biochemistry*, **200**, 487–493.

Schnell, N., Krems, B. and Entian, K.-D. (1992). The PAR1 (YAP1/SNQ3) gene of *Saccharomyces cerevisiae*, a c-jun homolog, is involved in oxygen metabolism. *Current Genetics*, **21**, 269–273.

Stephen, D.W.S., Rivers, S.L. and Jamieson, D.J. (1995). The role of the *YAP1* and *YAP2* genes in the regulation of the adaptive oxidative stress responses of *Saccharomyces cerevisiae. Molecular Microbiology*, **16**, 415–423.

Van Ho, A., Ward, D.M. and Kaplan, J. (2002). Transition metal transport in yeast. *Annual Review of Microbiology*, **56**, 237–261.

Veal, E.A., Ross, S.J., Malakasi, P., Peacock, E. and Morgan, B.A. (2003). Ybp1 is required for the hydrogen peroxide-induced oxidation of the Yap1 transcription factor. *The Journal of Biological Chemistry*, **278**, 30896–30904.

Wemmie, J.A., Steggerda, S.M. and Moye-Rowley, W.S. (1997). The *Saccharomyces cerevisiae* AP-1 protein discriminates between oxidative stress elicited by the oxidants H_2O_2 and diamide. *The Jounal of Biological Chemistry*, **272**, 7908–7914.

Wemmie, J.A., Wu, A.-L., Harshman, K.D., Parker, C.S. and Moye-Rowley, W.S. (1994). Transcriptional activation mediated by the yeast AP-1 protein is required for normal cadmium tolerance. *The Journal of Biological Chemistry*, **269**, 14690–14697.

Wheeler, G.L. and Grant, C.M. (2004). Regulation of redox homeostasis in the yeast *Saccharomyces cerevisiae. Physiologia Plantarum*, **120**, 12–20.

Wood, M.J., Storz, G. and Tjandra, N. (2004). Structural basis for redox regulation of Yap1 transcription factor localization. *Nature*, **430**, 917–921.

Wu, A. and Moye-Rowley, W.S. (1994). *GSH1*, which encodes g-glutamylcysteine synthetase, is a target gene for yAP-1 transcriptional regulation. *Molecular and Cellular Biology*, **14**, 5832–5839.

Wu, A., Wemmie, J.A., Edgington, N.P., Goebl, M., Guevara, J.L. and Moye-Rowley, W.S. (1993). Yeast bZIP proteins mediate pleiotropic drug and metal resistance. *The Journal of Biological Chemistry*, **268**, 18850–18858.

Wysong, D.R., Christin, L., Sugar, A.M., Robbins, P.W. and Diamond, R.D. (1998). Cloning and sequencing of a *Candida albicans* catalase gene and effects of the disruption of this gene. *Infection and Immunity*, **66**, 1953–1961.

Xu, Z. and Wickner, W. (1996). Thioredoxin is required for vacuole inheritance in *Saccharomyces cerevisiae. The Journal of Cellular Biology*, **132**, 787–794.

Yan, C., Lee, L.H. and Davis, L.I. (1998). Crm1p mediates regulated nuclear export of a yeast AP-1-like transcription factor. *The EMBO Journal*, **17**, 7416–7429.

Colour Section

Please refer to Fig. 1 p. 39 in text for figure legend.

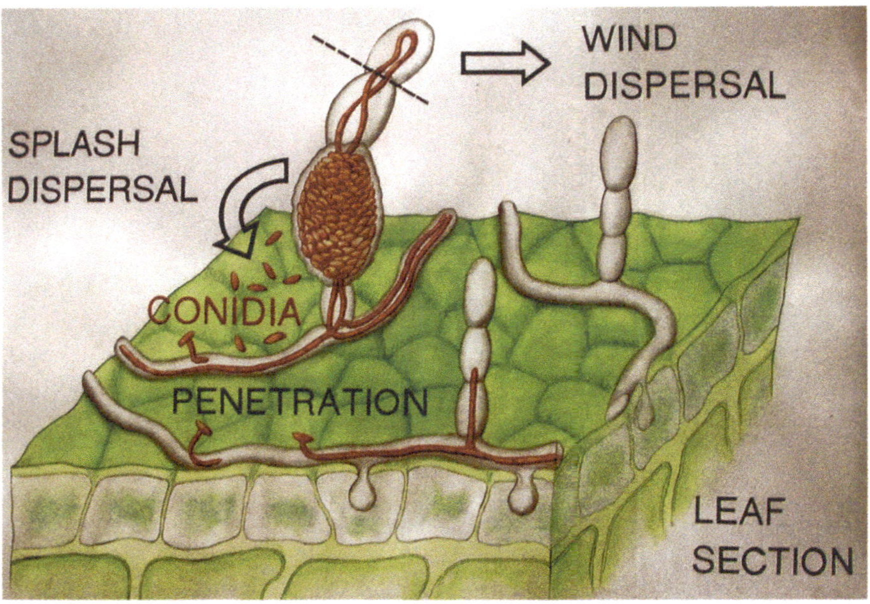

Please refer to Fig. 2 p. 40 in text for figure legend.

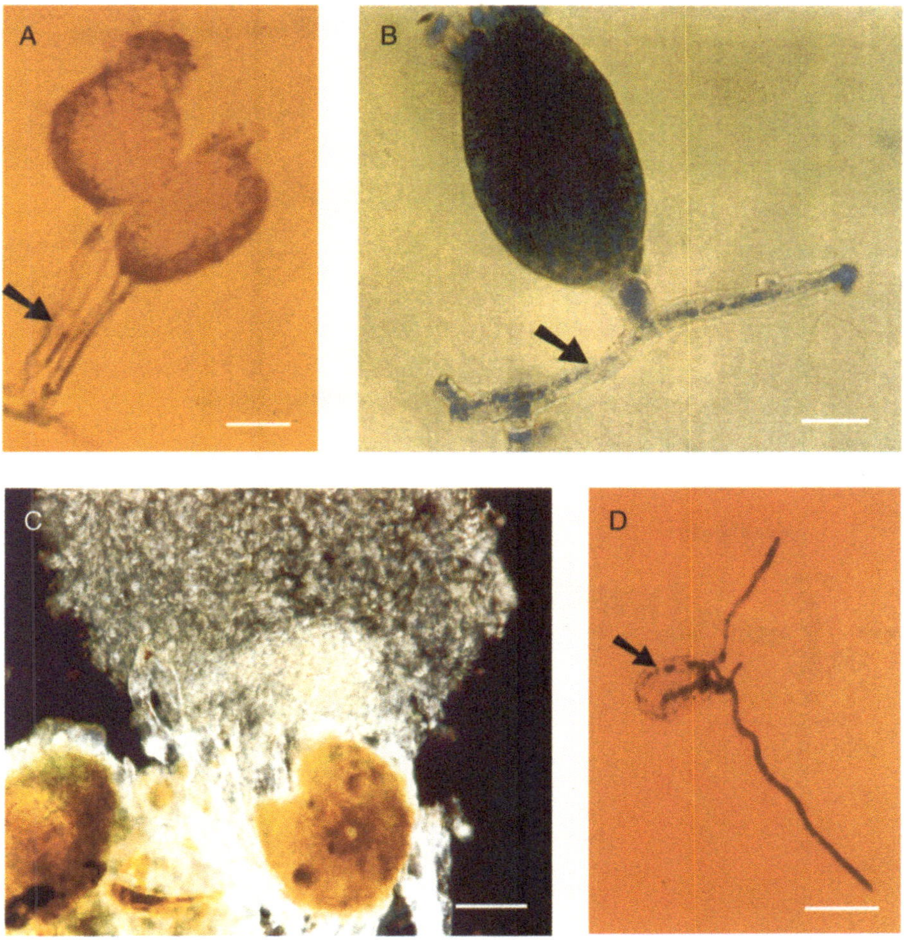

Please refer to Fig. 3 p. 41 in text for figure legend.

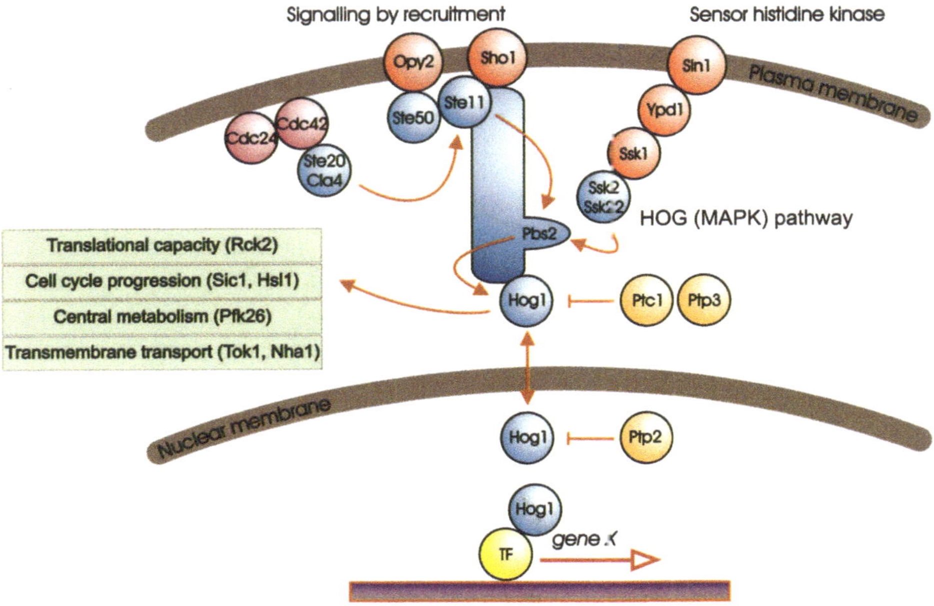

Please refer to Fig. 1 p. 111 in text for figure legend.

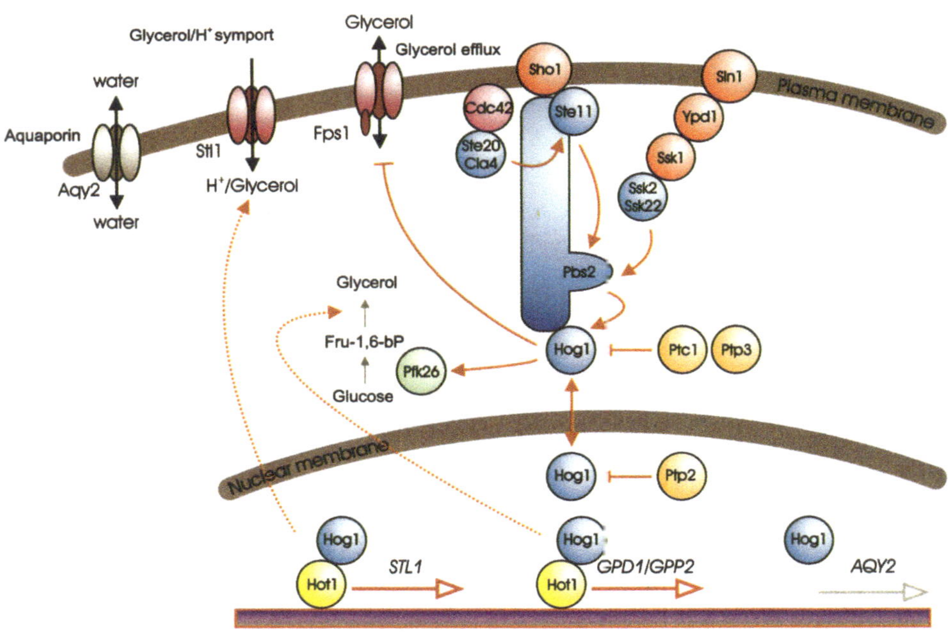

Please refer to Fig. 2 p. 115 in text for figure legend.

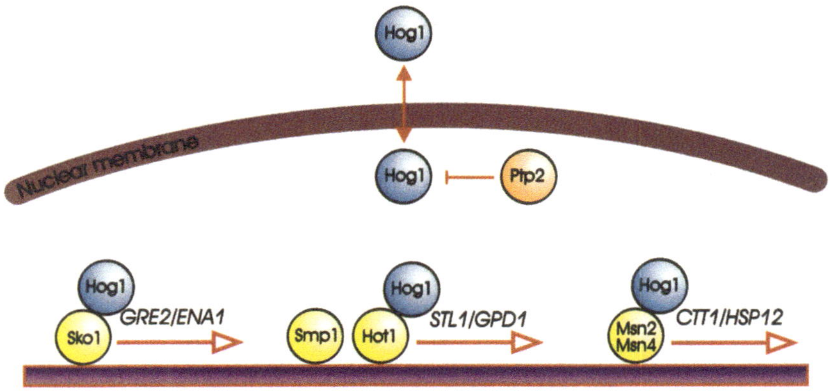

Please refer to Fig. 3 p. 118 in text for figure legend.

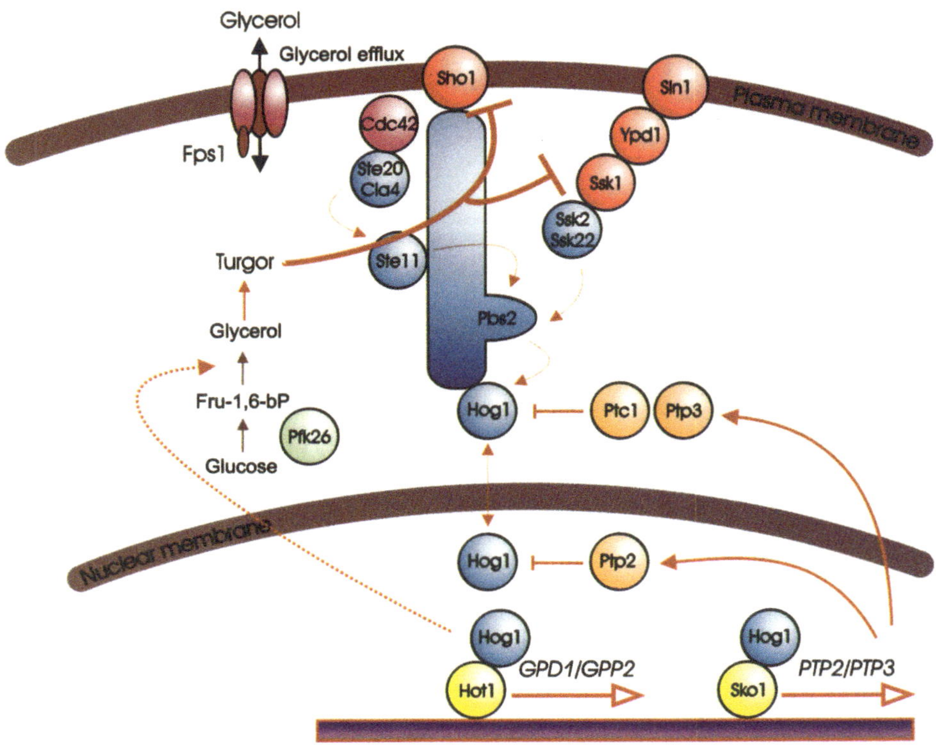

Please refer to Fig. 5 p. 121 in text for figure legend.

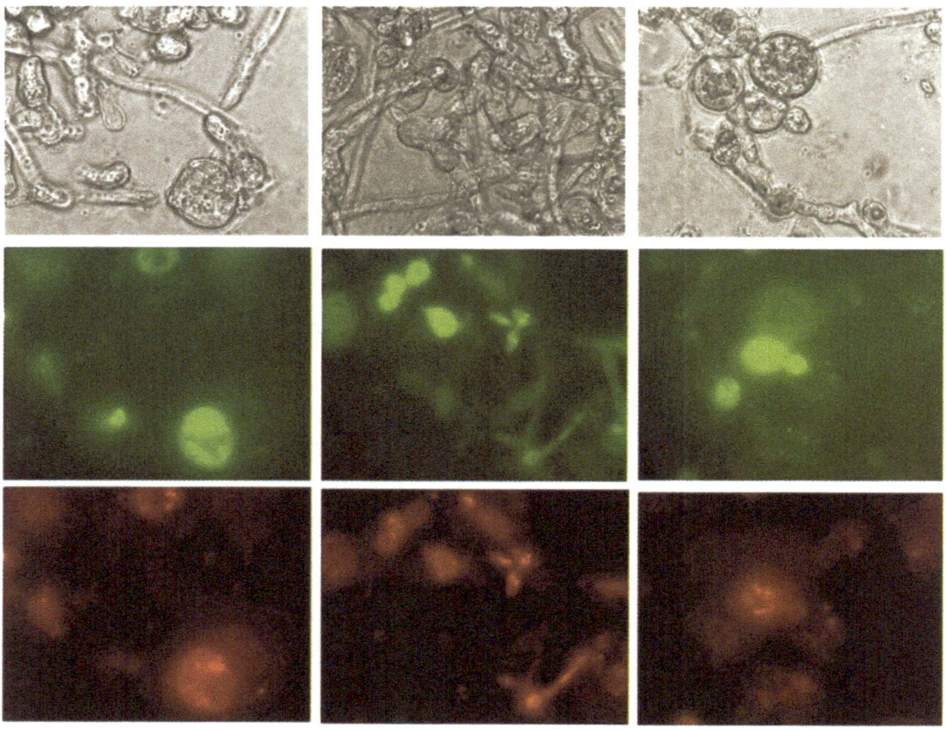

Please refer to Fig. 2 p. 133 in text for figure legend.

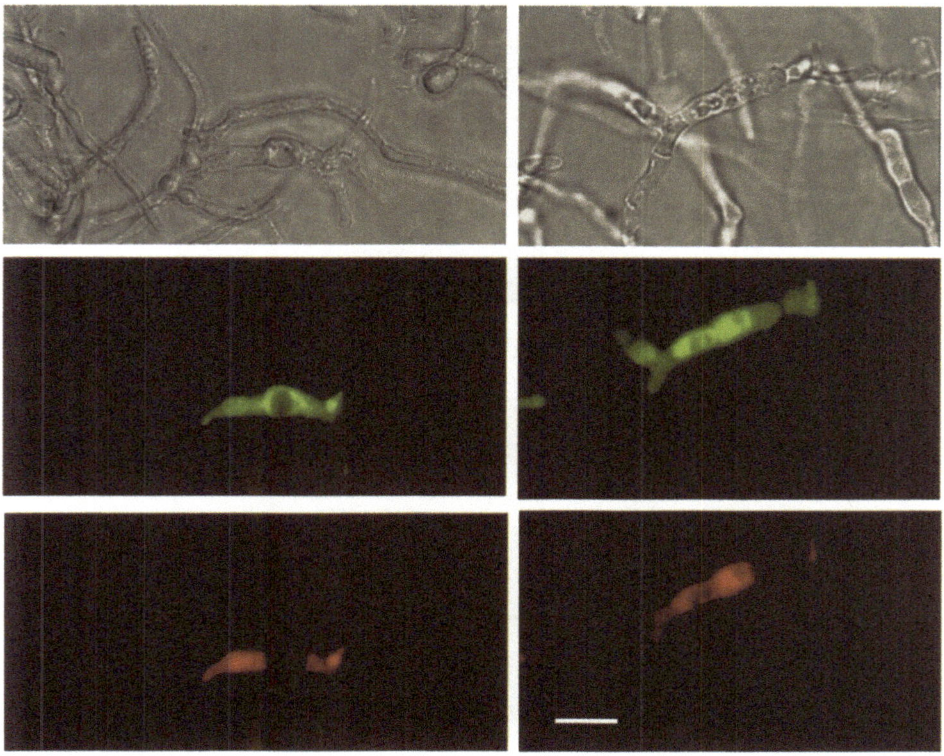

Please refer to Fig. 3 p. 134 in text for figure legend.

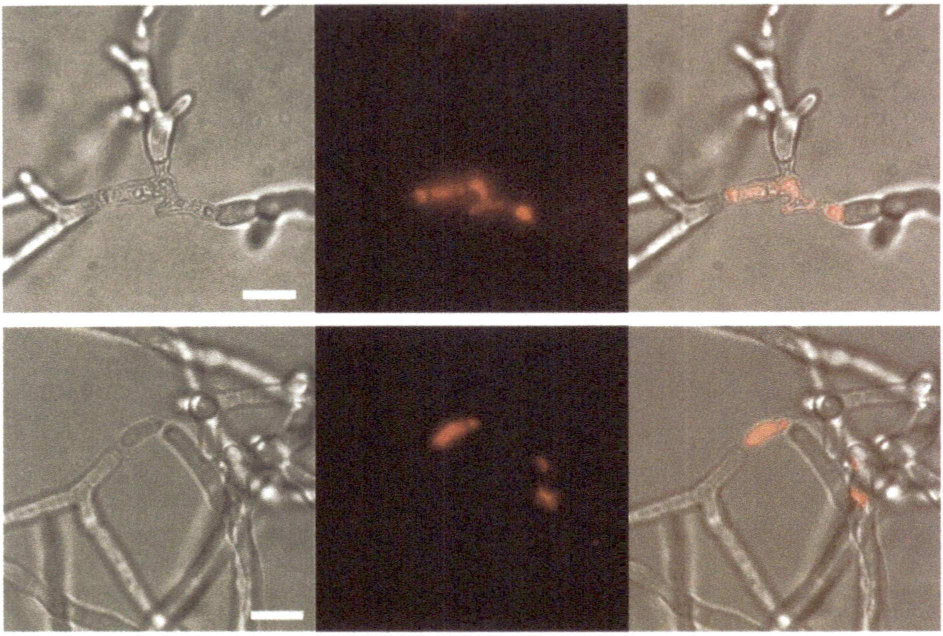

Please refer to Fig. 4 p. 135 in text for figure legend.

Please refer to Fig. 1 p. 161 in text for figure legend.

Please refer to Fig. 2 p. 181 in text for figure legend.

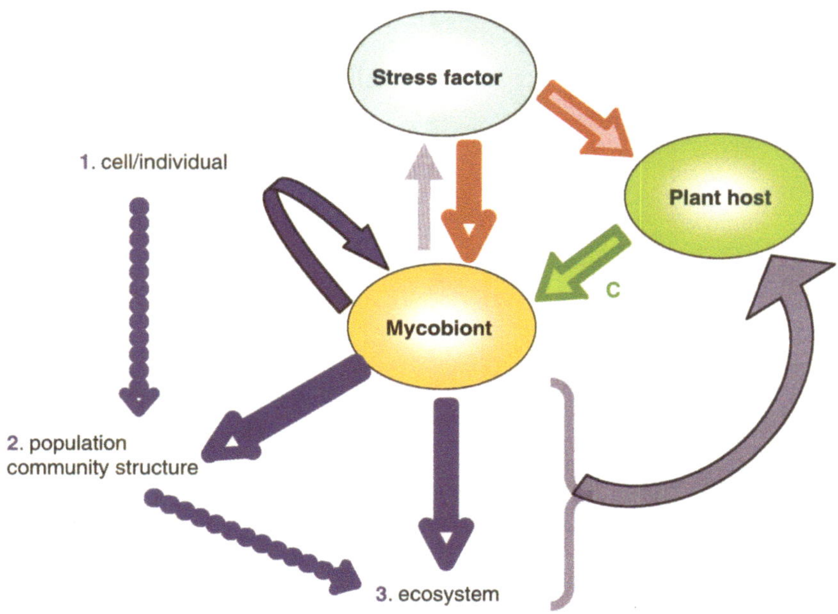

Please refer to Fig. 1 p. 203 in text for figure legend.

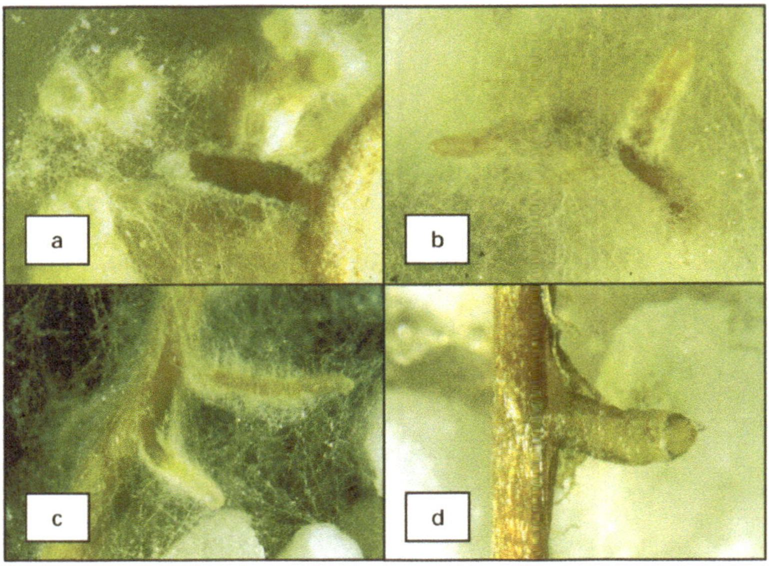

Please refer to Fig. 2 p. 205 in text for figure legend.

Please refer to Fig. 3 p. 208 in text for figure legend.

Please refer to Fig. 4 p. 213 in text for figure legend.

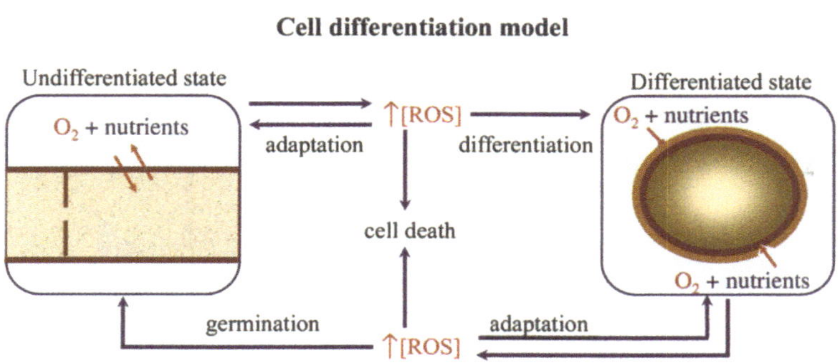

Please refer to Fig. 1 p. 238 in text for figure legend.

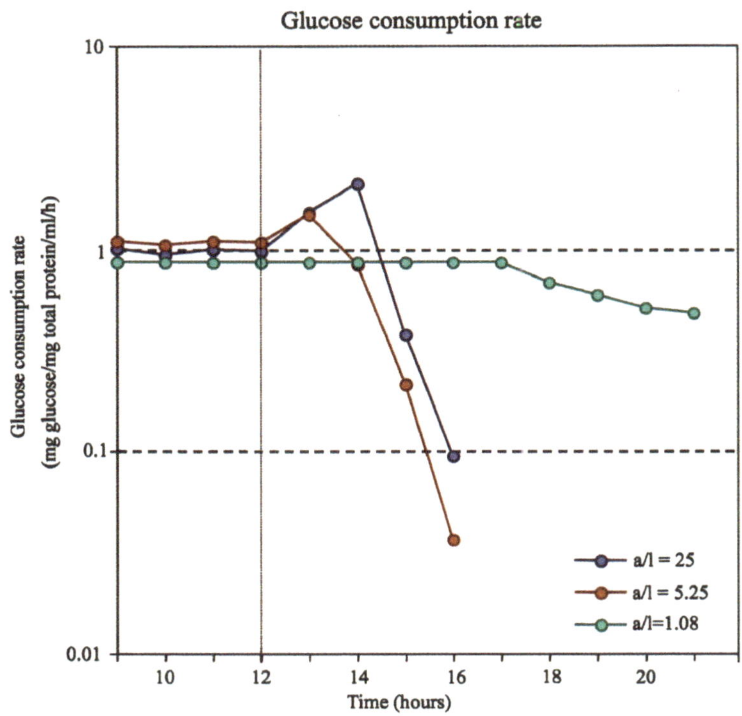

Please refer to Fig. 4 p. 246 in text for figure legend.

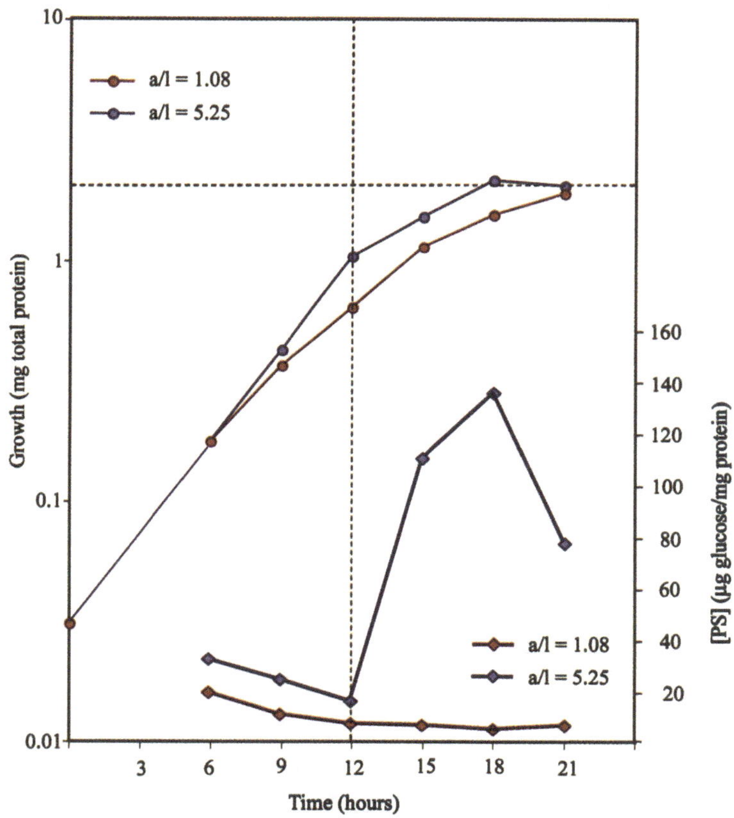

Please refer to Fig. 5 p. 247 in text for figure legend.

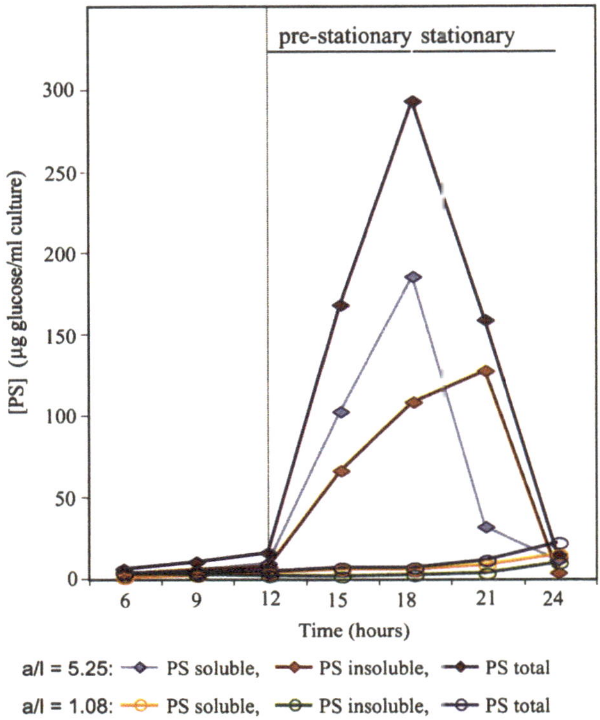

Please refer to Fig. 6 p. 248 in text for figure legend.

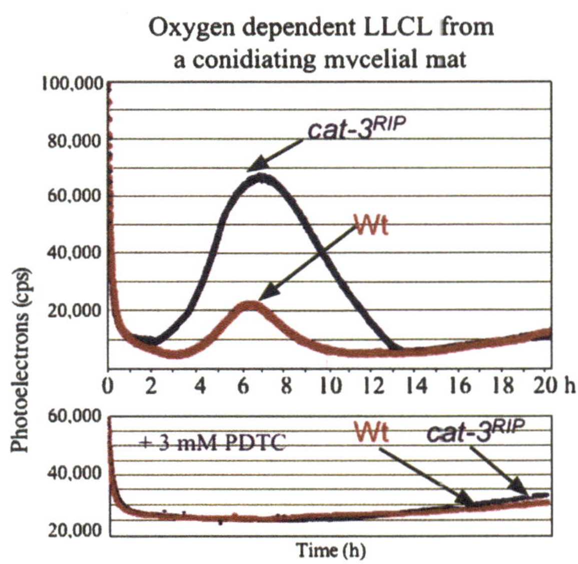

Please refer to Fig. 8 p. 253 in text for figure legend.

Subject Index

ABC-transporter, 143, 148
Acarospora, 179, 181, 186
Ace, 2, 97–103
acetic acid, 143–152, 254, 280
1-4, β-N-acetylglucosaminidase, 7
acid stress, 143, 146, 148–152
adaptation, 7, 54, 62–63, 68, 71, 75, 82, 110, 113, 119–123,143, 147–148, 157–159, 163–164, 166–167, 169–170, 187, 191, 221, 224, 228–229,231, 236, 238, 261–263, 265, 267
adaptive response, 221, 229–230, 262, 268
aflatoxin, 29, 31
Agaricus bisporus, 1–3, 5, 7, 9, 11, 13
Agrobacterium tumefaciens, 9
aluminium (Al), 163
Amanita, 162, 168
Ampelomyces, 37–48, 59
antagonism, 22–24
antibiosis, 46
anti-fungal, 68, 80
antimicrobial, 144
antioxidant, 189–190, 222, 229, 235–238, 251, 253, 255, 259,261, 264–265, 269–270
apoptosis, 239, 276
appressorium, 45
aquaglyceroporin, 114, 116, 122, 143, 147, 152
aquaporin, 114, 210
arabidopsis, 87–90, 92–93, 158
arms race, 87, 93
arsenic (As), 75, 175
arsenite stress, 114
ascomycete, 4, 12, 56, 160, 167
ascospores, 59–60, 255
Aspergillus, 20–22, 26–27, 32, 99, 129–130, 239, 275
  nidulans, 99, 239
ATR1, 87, 91–93, 287
ATR13, 87, 90–93
Aureobasidium, 187
autolysis, 129, 131–132, 139
avirulence gene, 88

basal immune response, 88–89
basidiomycete, 3, 12, 161–162
benzoic acid, 143–147, 150–152
biocontrol agent, 3, 22, 24, 38, 47–48
biogeochemical cycling, 175
biomonitor, 175, 177, 179, 188
bioreactor, 129, 131, 133, 135, 137
biosorption, 166, 187
biotic stress, 21, 32, 37–38, 45, 49, 53–54, 201, 215
biotrophic, 37–38, 45, 56, 160
Botrytis, 22, 28, 54, 58, 251
Buellia, 186

cadmium (Cd), 73, 75, 224, 277
Candida, 22, 67–69, 81–82, 97, 99, 101–102, 259–261,263, 265, 267, 269, 275–276, 283–284
  albicans, 67–68, 97, 99, 259–261, 263, 265,267, 269, 276, 283
  glabrata, 67–68, 97, 101, 283
candidiasis, 68–69, 102, 260, 268–270, 276
cap spotting, 2, 4, 7–8
CAP1, 73–74, 76, 78–79, 259, 261, 265, 267, 270
carboxylic acid, 143–146, 148, 150, 187, 244
catalase, 68–69, 78, 228, 236, 238, 251–252, 266, 269
cell,
  cycle, 97–100, 103, 109–110, 114, 119, 266–267
  cycle arrest, 114
  cycle progression, 97, 100
  death, 89, 129, 131–132, 136, 139, 149, 215, 222,236–238, 276
  fragmentation, 139
  homeostasis, 109
  integrity, 100, 111, 259, 266
  polarity, 99
  separation, 97–103
  wall, 2–3, 6–8, 10, 12, 38, 45, 58, 101, 111, 113, 115,166, 180, 185, 187–188, 244–246, 259, 264–269
  wall stress, 259
CER, 71
CESR, 71, 74
chaperone, 71
chitin, 7, 10, 13–14, 88, 98, 103, 167
chitinolytic fungi, 5
Cladonia, 181, 189
co-evolved, 89
co-immunoprecipitation, 100

cold stress, 74
commensal, 68, 71, 260
common environmental response, 71
competition, 24, 28–29, 32–33, 53–54, 57, 62, 201–202,212–214
competitiveness, 19, 21, 23–24, 32
conidiation (asexual reproduction), 236
Coniothyrium, 3, 45
copper (Cu), 175–177, 179, 181, 184, 186–187, 189, 277
'copper lichen', 179
core environmental stress response, 71
Cortinarius, 162, 204
co-tolerance, 164

daughter cell, 98, 100–103
density dependence, 53, 55–56, 58, 60
Dermocybe, 162
diamide, 267, 275–282, 286–287
differentiation, 2–3, 62, 163, 215, 235–241, 243, 245, 247,249, 251, 253
dimorphic transition, 260
Diploschistes, 186–187, 190
diversifying selection, 88, 90–91, 93
DNA damage repair, 69
DON, 29–31
downy mildew, 87, 89
dry bubble, 2–3, 7–8
dsRNA, 13

ectomycorrhizal, 7, 157, 161, 204, 208
effector, 87, 89–94
eIF2, 222–225
endochitinase, 7, 97
endoglucanase, 98, 100
entomopathogenic fungi, 7
environmental,
fluxes, 19, 21, 23, 25, 27, 29, 31
stress, 19–20, 30, 71, 82, 129, 201–203, 212,214, 228
stress response (ESR), 228
Eurotium, 20, 26
evolution, 62, 158–159, 162–163, 165–166, 176, 201, 211,214, 236
EXAFS, 187
exochitinase, 7

fed-batch cultivation, 129, 132
fermentation, 129–136, 138–139
phase, 129, 134
feedback control, 109, 119–120, 122–123
fermentation, 129–136, 138–139, 208
filamentous fungi, 9, 21, 116, 129–131, 133, 135, 137, 139,170, 201, 203–204, 215
fitness, 38, 61, 157, 159–162, 168, 285
fragmentation, 129–132, 134, 139, 201–202, 211
fumonisin, 28, 31
fungal,
community, 20, 25, 206, 214
transformation, 1, 9
virulence, 97, 99, 101
fungicidal, 25, 260
fungicides, 3, 165
fungistatic, 25

GCN2, 223–227
GCN4, 224
gene,
disruption, 12–13
knockout, 2, 9, 13
silencing, 10
genetic diversity, 163
genomics, 69, 87, 94, 120
GFP-reporter, 9, 275
Glomus, 165
1,3-β-glucanase, 6, 45
glutaredoxin, 69
glutathione (GSH), 189, 222, 242, 269
peroxidase, 222, 265, 276, 280
reductase, 242, 265
glycerol, 71, 75, 77, 109–111, 114–116, 118, 120, 122–124,147–149, 262, 268
Gompertz model, 57
GPX3, 280–281
grazing, 201, 211–212
growth curve, 57
GSH1, 264–265, 278, 285

heat shock, 71, 224, 228–229, 266
heatshock protein homothallic, 71, 266
heavy metal stress, 72–74, 76–77, 81, 159, 164, 190, 264
heterothallic, 53, 58–59
HOG pathway, 109, 111–112, 115, 117, 119–123, 151, 259,262–265, 267
HOG signalling network, 109
Hog1 SAPK pathway, 67, 74, 76
HOG1, 67–68, 71–72, 74–81, 109–123, 149, 227, 259,262–267, 269–270
host-pathogen interaction, 14, 89–90, 94, 270
Hot1, 118–119
Hyaloperonospora parasitica, 87, 89, 91, 93
Hydnum, 162

hydrogen peroxide (H_2O_2), 237, 268
hydrogen peroxide, 131, 151, 225, 230, 237, 260, 264,267–269
hydrolases, 129–130
hydrolytic enzymes, 3, 5, 7, 28, 49, 103
hydroxyl radical, 131, 151, 222, 237, 243
hyper osmotic response, 73
hyperosmotic shock, 110–111, 113, 117, 119
hyperparasite, 61
hypervirulence, 102
hyphal adhesion, 235, 240–241, 244
Hypogymnia, 177, 184, 186, 188
hypo-osmotic shock, 110–111, 115

inhibitor, 3, 30, 40, 222–223
innate immune response, 101, 276, 286
insertional mutagenesis, 12–13
iron (Fe), 163, 181, 187, 188, 277

Laccaria, 162
Lactarius, 162, 205
lead (Pb), 175, 245
Lecanicillium lecanii, 48
Lecanora, 176, 181, 186–187, 189
Lecidea, 179, 181, 186
leucine-rich repeats, 88
lichen, 175–191
  community, 184
  diversity, 177, 182, 184
life cycle, 37–38, 40, 43–44, 46, 53–54, 56–57, 94, 162
liming, 201, 206–208
lipase, 6
lipophilic acid, 143, 148, 151–152
LRR, 88, 90, 92

M/G1 transition, 97
manganese (Mn), 163
MAP kinase, 10, 143, 148–149, 227, 240, 259, 261–262,265–267
MAPK pathway, 240, 261, 263, 266
mating, 58–61, 98
melanin, 175, 183, 187
MEN, 99
metal bioavailability, 175
  chelation, 179
  complexation, 180, 190
  detoxification, 166, 186, 189–190
  dissolution, 209
  homeostasis, 165, 167, 170
  hyperaccumulation, 186
  resistance, 158, 182, 188–189
  smelter, 158, 176
  solubilisation, 209
  solubility, 167
  tolerance, 77, 157–160, 162–165, 168, 170, 209, 277
  toxicity, 157–160, 163, 167–168, 180, 188
metallophyte, 160, 176, 179
metallothionein, 168
microarray analysis, 224, 227–228, 230
mildew, 37–49, 59, 87, 89
mitotic exit network, 99
morphogenesis, 97, 99, 240, 259, 264
mother cell, 101
Msn2, 76, 79, 81, 118–119, 265
Msn4, 76, 79, 81, 118–119, 265
Mucor, 20
multidrug resistance, 287
multi-element determination, 191
multi-trophic interactions, 53, 59
mycoparasite, 3, 7, 40, 44
mycoparasitism, 5, 7–8, 37–38, 45–47
mycorrhiza, 167, 182, 214
Mycosphaerella, 54
mycotoxigenic, 21, 25, 28–31
mycotoxin, 25, 28–31
mycovirus, 59,61,62

NADPH oxidase, 235, 238
necrotrophic, 37, 45, 58
Neurospora, 235–236, 240
niche overlap, 19, 24–25, 27
nitric oxide, 68, 131, 237, 260–261
nitrosative stress, 267–270
NIV, 29–30
nivalenol, 29–30
nucleus, 75–77, 79, 97–98, 103, 110, 113–114, 118, 120,267–268, 275, 278–279, 282, 286
nutrient competition, 54
  limitation, 74, 224
  stress, 58, 169
nutritional stress, 201, 204, 212

obligate, 38, 87, 159, 176
ochratoxin A, 28, 31
oomycete, 87, 89, 93–94
opportunistic pathogen, 68
organic acids, 143–148, 150, 166–167, 179–180, 187, 189, 209
osmolyte, 110, 114, 116, 148, 268
osmo-protectant, 71, 75
osmoregulation, 109–111, 113, 115–119, 121–123, 148
osmosensitivity, 117

osmotic stress, 70, 73–77, 80–81, 112, 119, 122, 148–149,224, 227–228, 262–264, 268–269
osmotolerance, 114, 117
OTA, 28–32, 120–121
oxalic acid, 58, 209
oxidative burst, 69, 73–74, 260, 276
oxidative stress, 69, 71, 73–74, 76–77, 79–81, 129, 131–132,146, 151, 164, 188–190, 201, 209–210, 221–231, 235–237, 239, 241, 243, 245–249,251–253, 255, 259–270, 275–283, 285–287

PAMPS, 88
Paraquat, 252, 254
Parmelia, 185
pathogen associated molecular patterns, 88
Paxillus, 162, 165, 167–168
PDR12, 143, 148, 150–151
Penicillium, 20–23, 25–27, 29–32, 130, 132, 138
peroxide stress, 73, 76, 80–81, 269
Pertusaria, 186
pH stress, 70
Phaeosphaeria, 55
phagocytosis, 69–70, 239, 260, 270
phenotypic plasticity, 159, 201
pheromone response pathway, 112, 120
Phoma, 48
phyochelatin, 167, 168, 175, 180, 189
physico-chemical factors, 158
Phytophthora, 93–94
Piloderma, 207–209
Pisolithus, 162, 164, 167
Plasmodium, 93–94
Podospora, 239–240
pollution, 157, 159–161, 163–167, 169, 177–178, 182, 184
polyphosphate granules, 188
population stress, 53, 62
potato blight, 87
powdery mildew, 37–49
preservative, 143–144, 146–148, 150
programmed cell death pathway, 139
protein,
  folding, 73, 269, 275, 277, 279, 281–283, 285
  oxidation, 244, 249–250, 255
  synthesis, 10, 12, 71, 73, 221–230, 250
proteome, 101, 222, 230
proton efflux, 146, 148
pseudohyphal, 101, 112, 121
Pseudozyma, 46, 48
Psilolechia, 178, 181
Puccinia, 54–55, 61
purifying selection, 90
pycnidia, 37, 39–41, 43–44, 48, 55
quorum sensing, 29, 58, 75

R gene, 2, 4, 9–11, 13, 32, 59, 69, 88–90, 92–93, 120, 146,162, 229, 280–281, 286
race-specific immune response, 87
RAM network, 99–100
Ramalina, 184, 189
Ras, 235–236, 240, 254–255
RCK2, 114, 227
reactive oxygen, 76, 129, 131, 210, 222, 235, 260–261, 276
reactive oxygen species (ROS), 129, 210, 222, 235, 260, 276
regulator, 68, 76, 78, 97–99, 102–103, 113, 259, 263, 265,267, 277, 281, 285
repressor, 118
resistance gene, 87–90, 92
response regulator protein, 113
Rhizopus, 20, 32
rhizosphere, 46, 160
RNAi, 2, 13
root colonisation, 162, 165, 210
ROS, 129, 131, 133–134, 136, 138–139, 210, 222, 224–226,229, 231, 235–240, 242, 249–253, 255, 260–261, 268, 276
ROX3,285, 287
RPP1, 87, 89, 92–93
RPP13, 87, 89–92
rubidium (Rb), 185,187
rust, 54, 56–57, 59, 179, 187–188
RXLR motif, 87, 93

Saccharomyces, 20, 67, 70, 97, 109–110, 143–144, 222,260, 265, 275, 277, 283
Saccharomyces cerevisiae, 67, 70, 97, 109–110, 143–144, 222, 260, 265, 275, 277, 283
salt stress, 210
Schizosaccharomyces, 67, 70, 224, 260
Schizosaccharomyces pombe, 67, 70, 224, 260
Scleroderma, 162
secondary metabolites, 19, 21, 24–26, 28–30, 32–33, 179–180,190
septum, 98, 100, 103, 132
sexual reproduction, 37, 58, 63, 163, 236
siderophore, 204
SKN7, 259, 261, 267–268
Sko1, 118–119
SLN1, 80–81, 111–113, 120, 123, 262–264
sorbic acid, 145–148, 150–153
species diversity, 158, 160, 214
spindle pole body, 99
spoilage fungi, 20, 26–27, 29

spoilage yeasts, 143–145, 147, 149, 151
Sporidesmium, 3
sporocarp, 161–162, 205–206
SSK1, 68, 111, 113, 121, 123, 262–266, 270
starvation, 69, 223–228
starvation response, 69–70, 226
stationary phase, 28, 131–132, 134, 136, 252, 267, 269
stipe blowout, 2–3
stramenopiles, 94
stress,
  activated protein kinase (SAPK), 68, 224
  cross protection, 71, 75, 77
  granules, 227
  metabolites, 176
  protective enzymes, 68
Sty1 pathway, 75
submerged culture, 129–131, 139
Suillus, 161, 165–166, 169, 205, 213
superoxide anion, 131, 222, 229
superoxide dismutase (SOD), 190
suppressive subtractive hybridization, 10
suppressor, 93, 112
Swi5, 97–98, 102
symbiosis, 159, 164, 168, 176, 203, 210
synergism, 53, 58

T-DNA, 9
thiols, 165, 189
thioredoxin, 69, 228, 265, 278
Tilletia, 58
Tilletiopsis, 46, 48
toxin, 37, 45, 101
transcriptome, 77, 222, 230
translational elongation, 119
translational initiation, 221–227
Trapelia, 181, 183, 187, 191
trehalose, 68, 71, 74, 152, 268
Trichoderma harzianum, 7
Trichoderma, 3, 7, 20, 22, 31, 45, 129–130, 136, 251
trichothecene deoxynivalenol, 29
tri-trophic relationship, 37, 39, 47
TRX2, 278, 280, 282, 285–287
turgor, 110, 112–113, 120–121, 146, 149

unfolded protein response (UPR), 223
uranium (U), 175
Usnea, 184, 189

Verticillium fungicola, 1–3, 5, 7, 9, 11, 13
volatiles, 19, 21

Warlp, 143, 148, 150, 152
water stress, 201, 210
weak acid adaption, 143,147,150
weak organic acid resistance, 143, 145, 147, 149, 151

Xanthoria, 187–189
xenobiotic, 33, 57

YAP1, 79, 259, 265, 267, 277–278, 280
YBP1, 282
yeast, 70–74, 76, 78, 89, 101–102, 109–117, 119–123, 131,138, 143, 145–146, 149, 151–152, 164, 221–227, 229–231, 260, 262, 264, 269, 275–277, 282
yeast two hybrid, 89
YHB1, 68, 269

zinc (Zn), 168–169
zygosaccharomyces, 143–145, 152

FSC
www.fsc.org
MIX
Papier | Fördert gute Waldnutzung
FSC® C083411

Kolibri 360 GmbH
Ferdinand-Jühlke-Straße 7
99095 Erfurt, Deutschland
produktsicherheit@kolibri360.de